Sustainable Materials and Manufacturing Technologies

Materials and manufacturing techniques are a few of the vital processes in production industries. Most of the materials processing and manufacturing techniques currently used in industries are a major cause of environmental pollution and are hence unsustainable. This book provides in-depth knowledge about challenges faced during the processing of advanced materials and discusses possible ways to achieve sustainability in manufacturing.

This book:

- Covers advances in cryogenic machining, optimization, and economical and energy assessment of machining;
- Provides case studies and numerical design with analysis using computational fluid dynamics of minimum quantity lubrication mist droplets;
- Reviews metalworking fluids, laser micro-texturing, materials and manufacturing in sustainability, biofuels additives, nano-materials, and additive manufacturing of waste plastic;
- Explores the use of artificial intelligence and machine learning-based manufacturing techniques; and
- Covers the latest challenges and future trends in sustainable manufacturing.

Sustainable Materials and Manufacturing Technologies is primarily written for senior undergraduate and graduate students, as well as researchers in mechanical, manufacturing, industrial, and production engineering, and material science.

Dr. Navneet Khanna is an Assistant Professor at the IITRAM Ahmedabad, India.

Dr. Kishor Kumar Gajrani is an Assistant Professor at the Indian Institute of Information Technology, Design, and Manufacturing (IIITDM), Kancheepuram, India.

Dr. Khaled Giasin is a lecturer at the University of Portsmouth, England.

Prof. J. Paulo Davim is a Full Professor at the University of Aveiro, Portugal.

Science, Technology, and Management Series
Series Editor: J. Paulo Davim
Professor, Department of Mechanical Engineering, University of Aveiro, Portugal

This book series focuses on special volumes from conferences, workshops, and symposiums, as well as volumes on topics of current interested in all aspects of science, technology, and management. The series will discuss topics such as, mathematics, chemistry, physics, materials science, nanosciences, sustainability science, computational sciences, mechanical engineering, industrial engineering, manufacturing engineering, mechatronics engineering, electrical engineering, systems engineering, biomedical engineering, management sciences, economical science, human resource management, social sciences, engineering education, etc. The books will present principles, models techniques, methodologies, and applications of science, technology and management.

Multi-Criteria Decision Modelling
Applicational Techniques and Case Studies
Edited by Rahul Sindhwani, Punj Lata Singh, Bhawna Kumar, Varinder Kumar Mittal, and J. Paulo Davim

High-k Materials in Multi-Gate FET Devices
Edited by Shubham Tayal, Parveen Singla, and J. Paulo Davim

Advanced Materials and Manufacturing Processes
Edited by Amar Patnaik, Malay Kumar, Ernst Kozeschnik, Albano Cavaleiro, J. Paulo Davim, and Vikas Kukshal

Computational Technologies in Materials Science
Edited by Shubham Tayal, Parveen Singla, Ashutosh Nandi, and J. Paulo Davim

Industry 4.0 and Climate Change
Edited by Rajeev Agrawal, J. Paulo Davim, Maria L.R. Varela and Monica Sharma

Sustainable Materials and Manufacturing Technologies
Edited by Navneet Khanna, Kishor Kumar Gajrani, Khaled Giasin and J. Paulo Davim

For more information about this series, please visit: www.routledge.com/Science-Technology-and-Management/book-series/CRCSCITECMAN

Sustainable Materials and Manufacturing Technologies

Edited by
Navneet Khanna, Kishor Kumar Gajrani,
Khaled Giasin and J. Paulo Davim

CRC Press
Taylor & Francis Group
Boca Raton London New York Leiden

CRC Press is an imprint of the
Taylor & Francis Group, an **informa** business
A BALKEMA BOOK

Designed cover image: © Shutterstock

First published 2023
by CRC Press/Balkema
4 Park Square, Milton Park, Abingdon, Oxon, OX14 4RN
e-mail: enquiries@taylorandfrancis.com
www.routledge.com—www.taylorandfrancis.com

CRC Press/Balkema is an imprint of the Taylor & Francis Group, an informa business

Library of Congress Cataloging-in-Publication Data
Names: Khanna, Navneet, editor. | Gajrani, Kishor Kumar, editor. | Giasin, Khaled, editor. | Davim, J. Paulo, editor.
Title: Sustainable materials and manufacturing technologies / edited by Navneet Khanna, Kishor Kumar Gajrani, Khaled Giasin, J. Paulo Davim.
Description: Boca Raton : CRC Press, 2023. | Series: Science, technology, and management | Includes bibliographical references and index.
Subjects: LCSH: Machining. | Manufacturing processes. | Materials—Technological innovations. | Green products. | Green chemistry.
Classification: LCC TJ1185 .S844 2023 (print) | LCC TJ1185 (ebook) | DDC 670.28/6—dc23/eng/20221116
LC record available at https://lccn.loc.gov/2022041625
LC ebook record available at https://lccn.loc.gov/2022041626

ISBN: 978-1-032-27243-6 (hbk)
ISBN: 978-1-032-27249-8 (pbk)
ISBN: 978-1-003-29196-1 (ebk)

DOI: 10.1201/9781003291961

Typeset in Times New Roman
by Apex CoVantage, LLC

Contents

About the editors

Dr. Navneet Khanna is Associate Dean (Career Development and Promotion of Innovation and Industrial Research). He also acted as a Department coordinator of Mechanical Engineering (July 2017–November 2020) and Coordinator of the Siemens Centre of Excellence at IITRAM Ahmedabad. He worked at Birla Institute of Technology and Science, Pilani (BITS Pilani), from January 2007 to July 2014 and recently taught at Mechanical Engineering Department, Indian Institute of Technology (IIT) Gandhinagar, as an adjunct faculty. He is a specialist in building a strong pool of trained engineering graduates and is involved in solving manufacturing-related problems of Indian industries and government organizations. He bagged the Early Career Research Award of SERB (Government of India). He has successfully led 70+ projects defined by the Indian industries and published 40+ research papers in peer-reviewed journals. In addition, he has created several avenues of collaborative work with ISRO, DRDO, IPR, IITs, and renowned institutions in the USA, Australia, the UK, Spain, Slovenia, Russia, Brazil, Poland, etc. He has expertise in experimental, analytical, and computational analysis of machining processes.

Dr. Kishor Kumar Gajrani is Assistant Professor in the Department of Mechanical Engineering at the Indian Institute of Information Technology, Design, and Manufacturing (IIITDM), Kancheepuram, Chennai, India. He obtained his M.Tech. and Ph.D. degrees from the Department of Mechanical Engineering at the Indian Institute of Technology Guwahati. Thereafter, he worked as a post-doctoral researcher at the Indian Institute of Technology Bombay. He has published 32 international journals and book chapters of repute and has attended numerous conferences. He is co-editor of two books titled *Advances in Sustainable Machining and Manufacturing Processes* and *Biodegradable Composites for Packaging Applications*. His research interests are broadly related to the advancement of sustainable manufacturing, additive manufacturing, advanced materials, micro-manufacturing, coatings, green lubricants, and biodegradable packaging.

Dr. Khaled Giasin is a lecturer in the School of Mechanical and Design Engineering (SMDE) at the faculty of Technology. He joined the school back in May 2019 with teaching and research duties. He is also one of the school outreach coordinators. His main research background is in machining aerospace materials using experimental and numerical techniques, machining coolants, and characterization of composite and fiber metal laminate materials. Dr. Khaled obtained his Ph.D. degree from the University of Sheffield, M.Sc. degree from the University of Liverpool, and B.Sc. degree from Jordan University of

Science and Technology. He is a chartered engineer with IMechE and an associate fellow with Advanced Higher Education-UK.

Before joining the University of Portsmouth, he worked at Cardiff University as a postdoctoral research associate (project officer) in ASTUTE2020 (A European-funded project). He worked with leading academic experts in close collaboration with industry, developing impactful, state-of-the-art research-based solutions to the real-world problems facing advanced manufacturing businesses and their supply chains throughout West Wales and the Welsh Valleys. Some of the role activities included the provision of substantive, quantifiable, practical assistance to manufacturing enterprises through high-quality applied research and development projects in collaboration with industry partners, which combines industrial and academic experience.

Prof. J. Paulo Davim is Full Professor at the University of Aveiro, Portugal. He is also distinguished as an honorary professor in several universities/colleges/institutes in China, India, and Spain. He received his Ph.D. degree in Mechanical Engineering in 1997, M.Sc. degree in Mechanical Engineering (materials and manufacturing processes) in 1991, Mechanical Engineering degree (5 years) in 1986 from the University of Porto (FEUP), the Aggregate title (Full Habilitation) from the University of Coimbra in 2005 and the D.Sc. (Higher Doctorate) from London Metropolitan University in 2013. He is a Senior Chartered Engineer in the Portuguese Institution of Engineers with an MBA and Specialist titles in Engineering and Industrial Management as well as in Metrology. He is also Eur Ing at FEANI-Brussels and a Fellow (FIET) of IET-London. He has more than 30 years of teaching and research experience in Manufacturing, Materials, Mechanical, and Industrial Engineering, with a special emphasis on Machining & Tribology. He also has an interest in Management, Engineering Education, and Higher Education for Sustainability. He has guided many postdoc, Ph.D., and master's students and coordinated and participated in several financed research projects. He has received several scientific awards and honors. He has worked as an evaluator of projects for the European Research Council (ERC) and other international research agencies, as well as an examiner of Ph.D. theses for many universities in different countries. He is the Editor in Chief of several international journals, Guest Editor of journals, book Editor, book Series Editor, and Scientific Advisory for many international journals and conferences. Presently, he is an Editorial Board member of 30 international journals and acts as a reviewer for more than 100 prestigious Web of Science journals. In addition, he has also published as editor (and co-editor) of more than 200 books and as the author (and co-author) of more than 15 books, 100 book chapters, and 500 articles in various journals and conferences (more than 280 articles in journals indexed in Web of Science core collection/h-index 59+/11500+ citations, SCOPUS/h-index 64+/14000+ citations, Google Scholar/h-index 82+/24000+ citations). He has been listed in World's Top 2% Scientists by a Stanford University study.

Contributors

Darshit Shah Advanced Manufacturing Laboratory, Institute of Infrastructure Technology Research and Management (IITRAM), Ahmedabad, Gujarat, India

R. A. Rahman Rashid School of Engineering, Swinburne University of Technology, Hawthorn, Victoria, Australia

Muhammad Jamil College of Mechanical and Electrical Engineering, Nanjing University of Aeronautics and Astronautics, Nanjing, China

Salman Pervaiz Department of Mechanical and Industrial Engineering, Rochester Institute of Technology—Dubai Campus, Dubai, United Arab Emirates

Kishor Kumar Gajrani Department of Mechanical Engineering, Indian Institute of Information Technology, Design and Manufacturing, Kancheepuram, Chennai, India And Centre for Smart Manufacturing, Indian Institute of Information Technology, Design and Manufacturing, Kancheepuram, Chennai, India

M. Azizur Rahman Department of Mechanical and Production Engineering, Ahsanullah University of Science and Technology, Dhaka, Bangladesh

Navneet Khanna Advanced Manufacturing Laboratory, Institute of Infrastructure Technology Research and Management (IITRAM), Ahmedabad, Gujarat, India

Nagella Lahari Department of Mechanical Engineering, Indian Institute of Information Technology, Design and Manufacturing, Kancheepuram, Chennai, India

Jogendra Kumar NIMS School of Mechanical and Aerospace Engineering, NIMS University, Rajasthan, Jaipur, India

Rajneesh Kumar Singh Department of Mechanical Engineering, School of Engineering, Harcourt Butler Technical University (HBTU) Kanpur, Kanpur, India

Jinyang Xu State Key Laboratory of Mechanical System and Vibration, School of Mechanical Engineering, Shanghai Jiao Tong University, Shanghai, China

Karuturi Manohar Sai Krishna Department of Mechanical Engineering, Indian Institute of Information Technology, Design and Manufacturing, Kancheepuram, Chennai, India

Satyam Dwivedi Department of Mechanical Engineering, Indian Institute of Information Technology, Design and Manufacturing, Kancheepuram, Chennai, India

Rajendra Soni Department of Mechanical Engineering, Visvesvaraya National Institute of Technology Nagpur, Nagpur, Maharashtra, India

Pandi Jyothish Department of Mechanical Engineering, National Institute of Technology Warangal, India

Upendra Maurya Department of Mechanical Engineering, National Institute of Technology Warangal, India

V. Vasu Department of Mechanical Engineering, National Institute of Technology Warangal, India

S. Purushothaman Department of Mechanical Engineering, Indian Institute of Technology Tirupati, Andhra Pradesh, India

M. S. Srinivas Department of Mechanical Engineering, Indian Institute of Technology Tirupati, Andhra Pradesh, India

R. R. Behera School of Mechanical Engineering, KIIT (Deemed to be University), Bhubaneswar, India

N. Venkaiah Department of Mechanical Engineering, Indian Institute of Technology Tirupati, Andhra Pradesh, India

M. R. Sankar Department of Mechanical Engineering, Indian Institute of Technology Tirupati, Andhra Pradesh, India

Anuj Kumar Department of Mechanical Engineering, National Institute of Technology Warangal, Warangal, Telangana, India

Arun Kumar Bambam Department of Mechanical Engineering, Indian Institute of Information Technology, Design and Manufacturing, Kancheepuram, Chennai, India

Şenol Şirin Department of Machine and Metal Technologies, Duzce University, Duzce, Turkey

M. B. Kiran Department of Mechanical Engineering, School of Technology, Pandit Deendayal Energy University, Gandhinagar, Gujarat

Anil Dhanola Department of Mechanical Engineering, Chandigarh University, Mohali, Punjab, India

Vijay Kumar Department of Mechanical Engineering, Guru Jambheshwar University of Science and Technology, Hisar, Haryana, India

Vikrant Singh Sant Longowal Institute of Engineering and Technology, Longowal, Sangrur, Punjab, India

Anuj Bansal Sant Longowal Institute of Engineering and Technology, Longowal, Sangrur, Punjab, India

Anil Kumar Singla Sant Longowal Institute of Engineering and Technology, Longowal, Sangrur, Punjab, India

Deepak Kumar Goyal IK Gujral Punjab Technical University, Main Campus, Kapurthala, Punjab, India

Rampal Sant Longowal Institute of Engineering and Technology, Longowal, Sangrur, Punjab, India

Jonny Singla Sant Longowal Institute of Engineering and Technology, Longowal, Sangrur, Punjab, India

Rohit Prajapati Department of Chemistry, School of Technology, Pandit Deendayal Energy University, Raisan, Gandhinagar, Gujarat, India

Divya Jadav Institute of Infrastructure, Technology, Research and Management, IITRAM, Maninagar, Ahmedabad, Gujarat, India

Parikshit Paredi Institute of Infrastructure, Technology, Research and Management, IITRAM, Maninagar, Ahmedabad, Gujarat, India

Rajib Bandyopadhyay Department of Chemistry, School of Technology, Pandit Deendayal Energy University, Raisan, Gandhinagar, Gujarat, India

Mahuya Bandyopadhyay Institute of Infrastructure, Technology, Research and Management, IITRAM, Maninagar, Ahmedabad, Gujarat, India

Balram Jaiswal Materials & Morphology Laboratory, Department of Mechanical Engineering, Madan Mohan Malaviya University of Technology, Gorakhpur, India

Kuldeep Kumar Materials & Morphology Laboratory, Department of Mechanical Engineering, Madan Mohan Malaviya University of Technology, Gorakhpur, India

Kaushlendra Kumar Materials & Morphology Laboratory, Department of Mechanical Engineering, Madan Mohan Malaviya University of Technology, Gorakhpur, India

Rajesh Kumar Verma Department of Mechanical Engineering, School of Engineering, Harcourt Butler Technical University (HBTU) Kanpur, Kanpur, India

Pankaj Sawdatkar Academy of Scientific and Innovation Research, Ghaziabad, India

Dilpreet Singh Academy of Scientific and Innovation Research, Ghaziabad, India CSIR-CMERI-Centre of Excellence for Farm Machinery, Ludhiana, India

Preface

Day by day, the burden of humanity is increasing on the earth. As the earth is a finite source, we are reaching the limits of growth due to planetary boundaries. In the name of growth and economic output, humanity is exploiting the earth's resources. Our natural equilibria have a limit to which they can store carbon-based emissions. Also, the consumption of earth's resources at an alarming rate is causing difficulty in timely replenishments. These are a few unresolved challenges of these times. Even though the idea of sustainable development is quite popular and has received attention from the government and industrialists, the challenges and costs to achieve sustainability are enormous. Therefore, this book presents a group of efforts exemplifying the adaptation of sustainable materials and manufacturing technologies to reduce detrimental environmental impact.

Sustainable materials and manufacturing technologies are the need of today's world. The chapter in this book is categorized into two broad parts: (I) Sustainable manufacturing technologies and (II) Sustainable materials and processing. Part I includes the work of numerous researchers as a review, analytical modeling, and experimental work. Chapter 1 discusses an industry-supported study on cryogenic assisted drilling of difficult-to-cut material, specifically titanium alloy. Chapter 2 deals with the analytical modeling of machining economics and energy usage under a cryo-MQL environment during the machining of titanium alloy. Chapter 3 discusses an optimization-based case study of drilling laminated nanocomposites for sustainability. Chapter 4 analyzes a computation fluid dynamic model of MQL mist droplet characteristics for selecting a proper geometrical nozzle orifice and MQL nozzle. Chapter 5 discusses advancements in metalworking fluid for sustainable manufacturing. Chapter 6 covers laser micro-texturing of silicon for reduced reflectivity. Chapter 7 discusses the advancement in AI and ML-based techniques to improve energy efficiency in the legacy machine tool. Chapter 8 explores the challenges in achieving sustainability during manufacturing and also suggests various ways to achieve sustainability in industries. Chapter 9 discusses the future trends for sustainability in manufacturing.

Part II of this book discusses sustainable materials and processing. Chapter 10 discusses the role of materials and manufacturing processes in sustainability. Chapter 11 explores the use of additives and nanomaterials for the sustainable production of biofuels. Chapter 12 discusses the modification of stainless steel 316 with the assistance of a high velocity oxy fuel (HVOF) process to upsurge its sustainability. Chapter 13 covers the effects of HVOF-sprayed TiC+25%CuNi-Cr coatings on the sustainability and cavitation erosion resistance of stainless steel 316. Chapter 14 deals with acid and alkali-treated hierarchical silicoaluminophosphate molecular sieves for sustainable industrial applications. Chapter 15 discusses a case study on recycling waste carpets into a sustainable composite. Chapter 16 lists the current status and explores the idea of using waste plastic for additive manufacturing to fabricate components.

This book will work as a reference book for researchers, practicing materials scientists, and manufacturing engineers and managers. This book can also be used as a textbook for the postgraduate courses and as an elective course book at undergraduate level. Sustainable materials and manufacturing technologies provide a foundational link to more specialized research work in the domain of sustainable manufacturing.

Acknowledgments

We would like to thank all the authors of various chapters for their contributions. Our heartfelt gratitude to CRC Press (Taylor & Francis Group) and the editorial team for their support during the completion of this book. We are sincerely grateful to reviewers for their suggestions and illuminating views on each book chapter presented in the book *Sustainable Materials and Manufacturing Technologies*.

Introduction

Earth is a finite source. It is a well-known fact. The burden of humanity on earth is increasing day by day, and it seems that we would be reaching the limits of growth due to planetary boundaries. Since the 19th century, humanity has been exploiting earth's resources in the name of growth and economic output. Our natural equilibria have a limit to which they can store carbon-based emissions. Also, the consumption of earth resources at an alarming rate is causing difficulty in timely replenishments. These are a few unresolved challenges of these times.

Even though the idea of sustainable development is quite popular and has received attention from the government and industrialists, the challenges and costs to achieve sustainability are enormous. Sustainability is based on three pillars – environment, economy, and society. As materials and manufacturing are the major sources of a country's growth, both industries and the government play an important role in sustainability. Sustainable materials and manufacturing techniques are a part of green manufacturing, which will help industries and researchers to increase efforts in the right direction to reduce the detrimental impacts of conventional/traditional techniques.

Sustainability can be achieved by implementing the 6R concept in most of the categories. 6R implies reduce, reuse, recover, redesign, remanufacture, and recycle. Processes and techniques can be improved by reducing energy consumption, reusing resources, recovering losses, redesigning processes, and improving efficiencies, as well as remanufacturing and recycling as much as possible for sustainable growth.

To introduce these multidimensional sustainable materials and manufacturing techniques in the industries as well as in the curriculum of future generation researchers and engineers is the ambition behind this book. This goal can be balanced by developing adequate economic, environmental, and social criteria, with analysis of their independencies and application of that analysis for guiding technological innovation in respective economic, environmental, and societal frameworks.

This book will provide a lucid way for readers to understand the processing of sustainable materials and manufacturing technologies. The book comprises 16 chapters divided into two parts. The chapters in Part I discuss the challenges faced by industries for sustainable drilling as well as machining under various lubrication for the machining of difficult-to-cut material. This book also covers topics such as optimization study, computation fluid dynamic analysis for various geometrical nozzle orifice, sustainable metalworking fluids, laser texturing of silicon, and artificial intelligence (AI) and machine learning (ML) based techniques to improve energy efficiency in the legacy machine tool. This book explores challenges in achieving sustainability during manufacturing as well as future trends for sustainability in manufacturing.

Part II further focuses on sustainable materials and processing and discusses topics such as the role of materials and manufacturing process in sustainability, the use of additives and nanomaterials for sustainable production of biofuels, and modification of stainless steel using high velocity oxy fuel (HVOF) process to upsurge its sustainability. This book also covers acid- and alkali-treated hierarchical silicoaluminophosphate molecular sieves for sustainable industrial applications, recycling waste carpets into a sustainable composite, and the idea of using waste plastic for additive manufacturing to fabricate components.

This book addresses the challenges and solutions for sustainable materials and manufacturing processes. It discusses prevailing trends and suggests research findings for industries to move towards sustainable development by improving economic and social perspectives as well as reducing detrimental effects on the environment. Overall, the aim of this book is to catalog the latest achievements in the modern processing of sustainable materials and manufacturing technologies that can be helpful for future generations.

Part I

Sustainable manufacturing technologies

Chapter 1

Cryogenic assisted drilling of Ti-6Al-4V

An industry-supported study

Darshit Shah, R. A. Rahman Rashid, Muhammad Jamil, Salman Pervaiz, Kishor Kumar Gajrani, M. Azizur Rahman, and Navneet Khanna

Contents

1.1 Introduction

Titanium is known for its mechanical properties, fracture resistance, good strength-to-weight ratio, and excellent corrosion resistance (Jamil et al. 2020). Titanium alloys have numerous applications in the aerospace, defense, automotive, and biomedical industries to replace steel and other metals to save space and to increase the efficiency of the system. Ti-6Al-4V is a widely utilized titanium alloy comprising an $\alpha+\beta$ microstructure. This alloy is used in the medical field, such as in dental implants, bone fracture fixation, cardiovascular implants, hip implants, knee implants, etc. (Brunette 2001). It possesses good fabricability and high-temperature strength (Pederson 2002).

Even though Ti alloys are demandable and desirable, the fabrication and machinability of these alloys have been very difficult due to low elastic modulus, propensity to strain hardening, low thermal conductivity, and high chemical reactivity (Liu and Shin 2019). Titanium alloys are considered to be one of the hard-to-machine materials (Machado 1990; Siju et al. 2021). These factors lead to rapid tool wear and low surface integrity (Khanna et al. 2017), thus making titanium a difficult-to-cut material by traditional machining techniques (Rahman et al. 2021, Yuan et al. 2021).

Though many experiments of improving the traditional machining techniques, such as milling and turning, of titanium alloys have been conducted, very little attention is given to drilling, which is one of the most widely used post-machining operations. In aerospace industries, 25% of the manufacturing time is utilized in drilling operations where rapid tool wear of the drill bit is a common occurrence (Lopez de Lacalle et al. 2000). There are four

DOI: 10.1201/9781003291961-2

major contributors to drill wear, viz., heat, pressure, stress, and friction. Generally, wear starts from the cutting edge corner and propagates to the chisel edge and margin of the drill (Sartori et al. 2016). At high cutting speeds, attrition and diffusion wear are the dominant mechanisms of tool wear. The diffusive wear was due to the temperature and chemical affinity in the cutting zone between the tool and the chip (Sharif et al. 2012).

The feed, cutting speed, cooling technologies, and drilling techniques affect machinability (Rahim et al. 2008). Built-up edge (BUE) is a dominant factor in the reduction of tool life while machining titanium, which directly influences the surface integrity of the component (Rahman Rashid et al. 2017). Therefore, the drill bit geometry is also considered a major factor that affects the machining quality (Furness et al. 1996; Khanna and Sangwan 2013). Moreover, many researchers have studied the machinability with different tool coatings, like TiN, TiAlN, AlCrN, TiCrN, CrN, and TiCN. These coatings can improve oxidation resistance and tool life (Kumar et al. 2014; Agrawal et al. 2022). AlCrN coating can be used in aerospace and biomedical industries for increased productivity with optimized process parameters and other machining conditions (Cadena et al. 2013). Likewise, many studies have been carried out on-chip formation as it is important to understand the machinability of the materials (Zang et al. 2018; Gajrani 2020; Shetty et al. 2014; Gajrani et al. 2021a). Gajrani et al. (2021b) studied the effect of drills with peripheral slits and concluded that the modified drill bit was successfully able to reduce the burr formation in the drilling of titanium alloy. Compared to turning and milling operations, tool engagement with workpiece material in a drilling operation is more complex, which makes the investigation of chip morphology difficult (Zhu et al. 2018). Serrated chip formation is common phenomenon in Ti-6Al-4V machining, which makes it different from the other metal alloys. The main reason for the serrated chip formation is the inferior thermal properties (Hua and Shivpuri 2004). In comparison with other materials, saw-tooth type chip formation takes place at a relatively lower cutting speed, which makes machinability poor (Upadhyay et al. 2014). In some cases, chips are intertwined with the cutting tool, which reduces the machining quality and increases tool wear (Ahsan et al. 2016).

During the machining of Ti alloy, the temperature rises up to 1000°C. The extreme cutting temperature in dry machining causes tool wear and tool breakage, thus limiting the cutting speed, which can reduce productivity. Therefore, temperature reduction is achieved by using the "cutting fluids" that act both as cooling and lubricating agent (Mia et al. 2022; Bambam et al. 2022; Gajrani et al. 2022). The use of coolant at the tool-chip interface reduces the cutting temperature and improves productivity (Junior et al. 2017). Minimum quantity lubrication (MQL) has emerged as a sustainable solution. MQL techniques significantly reduce the amount of cutting fluids; however, the effectiveness is somewhat limited (Jawahir et al. 2016). Due to its non-toxic and environment-friendly nature (Khanna et al. 2019), cryogenic machining has become an emerging sustainable machining technique compared to other machining processes, as illustrated in Figure 1.1 (Jawahir et al. 2016).

Figure 1.1 Sustainable machining options as machining with dry, MQL, cryogenic cooling (Jawahir et al. 2016).

Bordin et al. (2015) investigated that dry and cryogenic machining are better options compared with other machining conditions for medical equipment. The application of cryogenic coolant minimizes the cutting force that helps increase cutting speed. Cryogenic machining improves chip morphology that shows steady cutting (Impero et al. 2018) (Shokrani et. al. (2016a)). Cryogenic LN_2 improves chip breakability (Ahmed and Kumar 2016). Hong and Ding (2001) reported that LN_2 decreases the temperature at the cutting zone drastically and improves the tool life as cryogenic coolant is an excellent medium for heat dissipation. Moreover, it affects the chip formation and removal mechanisms, thereby improving machinability (Aramcharoen 2016).

Rahim and Sharif (2006) investigated the tool life and surface integrity of titanium alloy in drilling operations with water-soluble coolant as cutting fluid. They reported non-uniform flank wear and micro-chipping to be the dominant tool wear. Suhaimi et al. (2018) investigated indirect cryogenic cooling with flood cooling, minimum quantity lubrication (MQL), and other conventional cooling techniques. They found that the cutting force and tool life both improved by 54% and 90%, respectively, in cryogenic conditions. Shokrani et al. (2016c) reported a 40% reduction in surface roughness and a threefold improvement in tool life in cryogenic cooling compared to flood cooling in CNC milling operations.

This chapter focuses on the machinability of Ti-6Al-4V during drilling operations under dry and cryogenic conditions. The important machining parameters, including thrust force, tool wear, chip morphology, and the surface roughness of the drilled holes, were studied and compared.

1.2 Machining of Ti-6Al-4V: a case study

1.2.1 Materials and methods

A Ti-6Al-4V plate with dimensions of 256 mm × 256 mm × 5.6 mm is used for the investigation. The coated carbide drill bit of 5 mm diameter is used for machining operation. The drill tool is coated with a 3-μm AlCrN (Alcrona) coating. The drill bit specifications are mentioned in Table 1.1.

1.2.2 Experimental setup

The experiments are conducted on a 3-axes VMC–V544 with a Mitshubishi E70 controller having a maximum spindle speed of 8000 RPM. Figure 1.2 shows the cryogenic machining experimental setup. A cryogenic LN_2 cylinder of 200 liters capacity from the INOX Devar is used for the experiments. A special fixture is designed for the application of LN_2 coolant. This arrangement is designed in such a way as to give a steady and vibration-free flow of LN_2 during operation. The cryogenic LN_2 was delivered using a 2-mm diameter nozzle. Experiments are carried out at the feed of 0.06 mm/rev and cutting speed of 80 m/min, keeping rpm at 5000. The tool wear is investigated using Mitutoyo's Tool Makers microscope TM generation B series Tm505B 176–818E having *X* and *Y* travel lengths of 50 mm each and equipped with a Mitutoyo 2D data processing unit QM–Data 200. The surface roughness of the hole is measured using a Taylor–Hobson Sutronix S128. The Kistler dynamometer was configured

Table 1.1 Tool Specifications

Tool diameter	*Material*	*Coating*	*Helix angle*	*Point angle*	*Flutes*
5 mm	Solid carbide	AlCrN	40°	140°	2

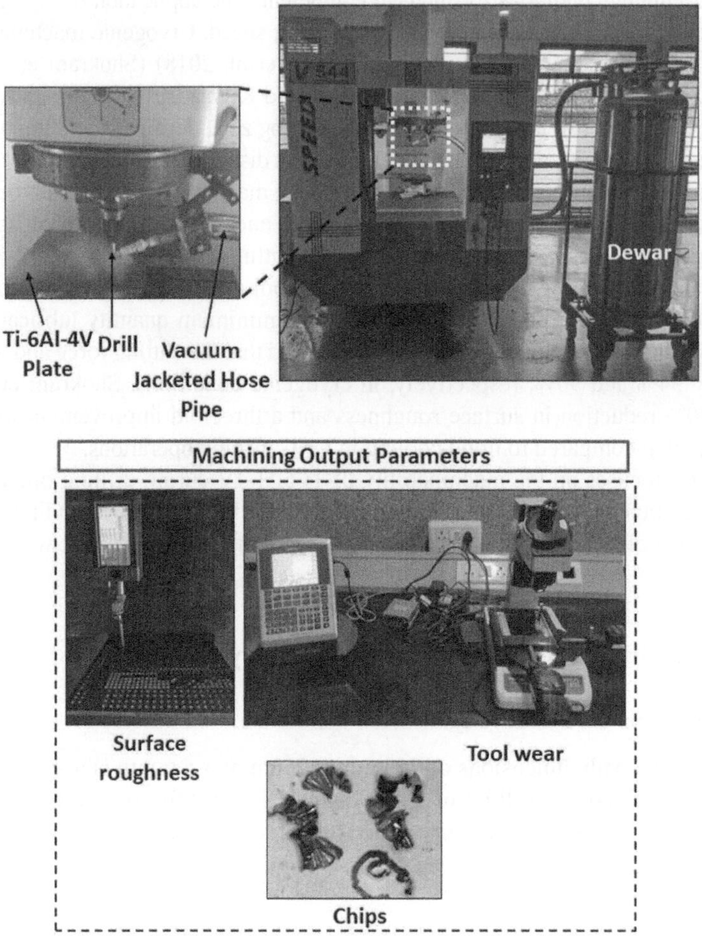

Figure 1.2 Overview of drilling experimental setup under cryogenic environment.

with DynoWare software and used for measuring thrust forces. The machinability of the titanium is evaluated under dry and cryogenic high-speed cutting conditions.

1.3 Results and discussion

1.3.1 Thrust force

Figure 1.3 illustrates the variation of thrust force (F_t) with the number of drilled holes for dry and cryogenic cooling environments. Results show that the thrust forces for cryogenic machining are higher compared to dry machining, as previously reported by Hong et al. (2001). When LN_2 is used as the coolant, the cutting zone temperature is considerably reduced, which contributes to the hardening of the workpiece material in the vicinity of the

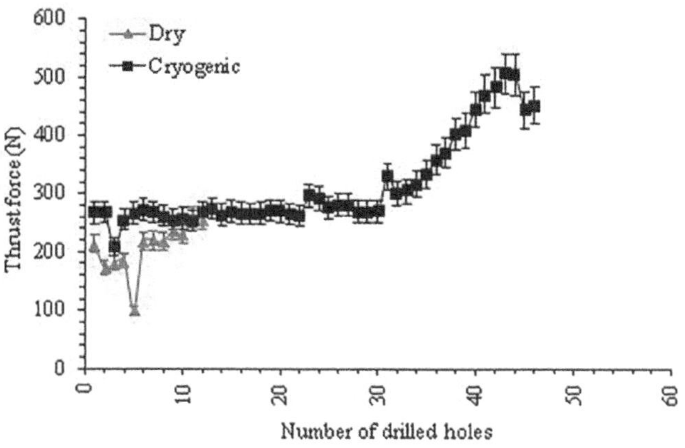

Figure 1.3 Variation of thrust forces with respect to the number of drilled holes under dry and cryogenic cooling environment.

cutting zone. This may increase the thrust forces (Zhao and Hong 1992). Thermal softening of the workpiece material gives consistent thrust forces in the middle runs. However, in a cryogenic cooling environment, after drilling 30 holes, a rapid increase in the thrust forces was recorded due to the flank face wear of the carbide drill bit.

1.3.2 Tool wear

Tool wear is one of the critical parameters for the machinability of workpiece material (Ti-6Al-4V). In the case of dry machining, a relatively higher amount of heat is generated due to friction between tool-chip and tool-workpiece interfaces. Therefore, significant adhesion of workpiece material (Ti-6Al-4V) was observed on the tool surface after drilling of the 2nd hole, as shown in Figure 1.4 (a). In contrast, little-to-none workpiece adhesion was observed under the cryogenic environment at the same machining conditions as shown in Figure 1.4 (b). This is attributed to the pressurized flow of the LN_2 coolant towards the cutting zone, which assisted in the easy removal of the chips from the tool flank face due to the better heat dissipation ability of LN_2.

With an increase in the number of drilled holes, the adhesion of workpiece material on the tool surface becomes more dominant. Under both machining environments, the adhesion of workpiece material can be seen on the cutting lip and the chisel edge. This adhered material protects the tool from excess wear (Rahim et al. 2008). But in dry drilling conditions, an extreme temperature rise occurs at the cutting zone leading to the welding of the chip on the cutting lip, as shown in Figure 1.5 (a). When drilling the 12th hole under dry conditions, the chips started to burn and sparks were seen, as illustrated in Figure 1.6.

Figure 1.5 (b) shows the surface morphology of the drill used under a cryogenic environment after drilling 46 holes. It is clear that the whole cutting edge is covered with a welded chip, and adhesion is spread on the flank face (Shokrani and Newman 2019; Bermingham et al. 2012). As the drilling progressed, the flow of chips started breaking the BUE, which removed grains of the tool material. Following this, a new BUE is formed. This is known as attrition wear (Rahman Rashid et al. 2017). For prolonged drilling operation, attrition wear is a dominant tool wear mechanism (Rahim et al. 2008; Shokrani and Newman 2019).

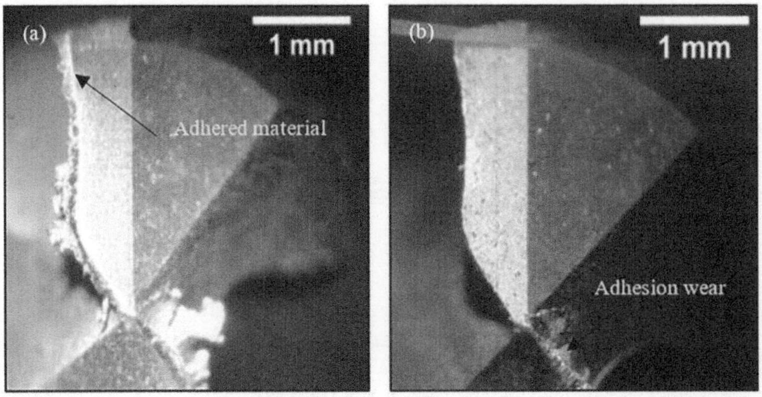

Figure 1.4 Morphology of tool surface after drilling two holes under (a) dry environment and (b) cryogenic environment.

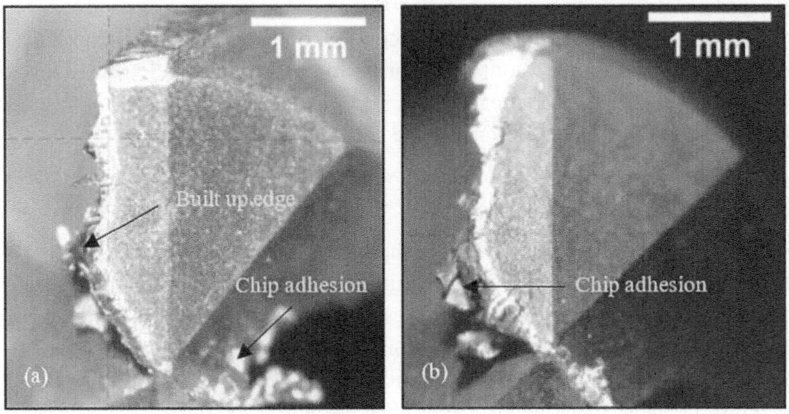

Figure 1.5 Morphology of tool surface after drilling (a) 12 holes under dry environment and (b) 46 holes under cryogenic environment.

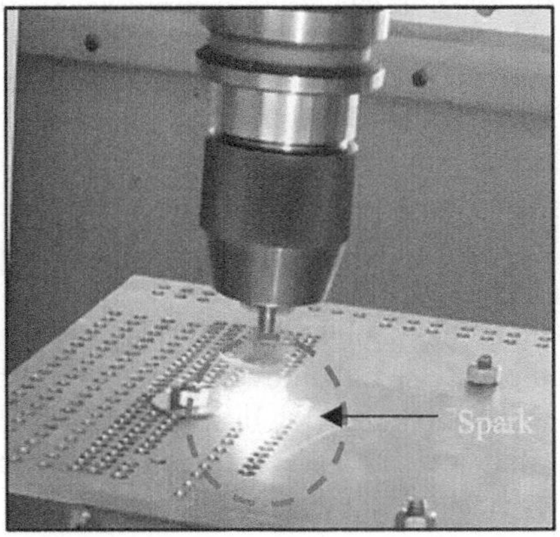

Figure 1.6 Sparks from the cutting zone at the 12th hole during dry machining.

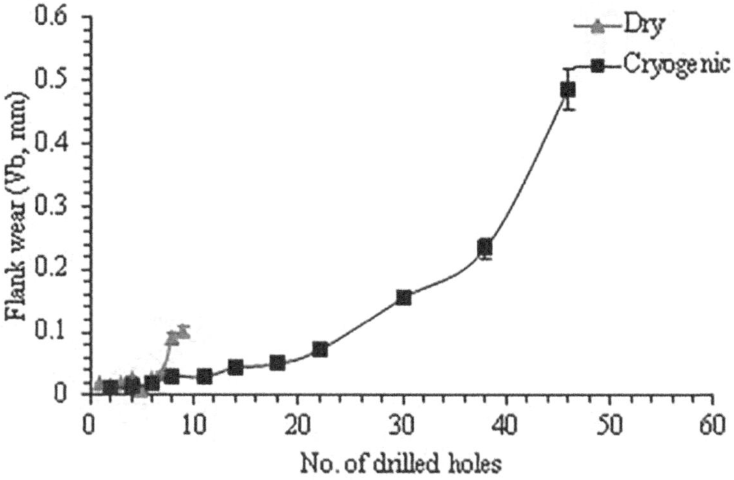

Figure 1.7 Variation of flank wear (Vb, mm) with respect to the number of drilled holes under dry and cryogenic cooling environment.

Figure 1.7 shows the comparison of average flank wear with respect to the number of drilled holes under dry and cryogenic environments. After drilling 46 holes under a cryogenic environment, flank wear exceeded its criteria of 0.3 mm. However, for dry drilling, only 12 holes were drilled before tool failure. The cryogenic environment improved the tool life by more than 350%.

1.3.3 Surface roughness

Machining processes change the surface integrity by producing roughness, crack, and tensile residual stress on the surface (Li et al. 2012). Surface roughness quantifies the microscopic asperities of a surface. Average surface roughness (R_a) is considered for the evaluation of the surface quality of the drilled holes. Figure 1.8 shows R_a values for respective holes after drilling under dry and cryogenic environments. This phenomenon can be explained by the mechanism of BUE formation and removal. As marked in Figure 1.8, under both environments, there is a convex section that shows a smoother surface finish at the end. This is due to adhered workpiece material on the flank face or deformed flank face (Che-Haron and Jawaid 2005). In cryogenic drilling, the flank face deformation occurred after 30 holes, which relates to the pressure of the convex section in Figure 1.8. The average R_a for dry machining is 1.11 µm and for cryogenic machining is 0.91 µm. Cryogenic drilling condition improves surface quality compared to dry condition. This is understandable as cryogenic LN_2 dissipates the accumulated heat from the tool-workpiece interface, thereby reducing the tool wear and assisting in steady drilling operation (Shokrani et al. 2016c).

1.3.4 Chip morphology

The chip geometry provides fundamental information on the cutting process. Thus, the qualitative assessment criterion was used to study the machined surface integrity through the chip morphology (Jawahir and Luttervelt 1993). The chips obtained are of saw-tooth type (serrated)

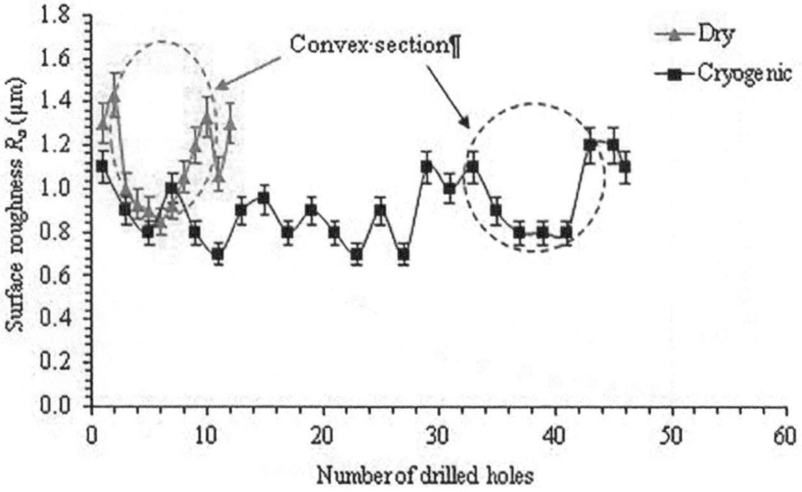

Figure 1.8 Variation of surface roughness with respect to the number of drilled holes under dry and cryogenic cooling environment.

for the dry machining, as shown in Figure 1.9 (a). In dry drilling, tooth height and segmentation frequency are high. This is attributed to the result of high friction force between the tool and workpiece. Due to high serration and segmentation, cyclic forces are increased, which leads to rapid tool wear and poor surface finish (Upadhyay et al. 2014). Figure 1.9 (b) shows the chip formed after drilling 12 holes, which is heavily oxidized. The presence of thick BUE after drilling 12 holes results in heat accumulation at the tool-chip interface. Figure 1.9 (c-d) shows the chip formed during machining under cryogenic machining. Cryogenic machining shows smaller serrated chip formation. This is attributed to the result of more heat dissipation under cryogenic environment, which allows smooth chip flow through the tool. Initially, the curvature of the chips is wide, as shown in Figure 1.9 (c). In the final stages of drilling, the curve bends become narrow, as shown in Figure 1.9 (d) (Ahmed and Kumar 2016; Hong and Ding 2001). Burning marks are seen at the bottom part of the chips. This shows that the tool wear is increasing as the heat accumulation is diminishing the cooling effect of LN_2. Even if the tool life criterion of 300 μm was not achieved, the chips were completely burned, as observed during dry drilling. This shows that cryogenic LN_2 gives better machinability compared to dry machining.

In both machining conditions, adhesion wear was found to be the dominant tool wear mechanism. Similar results were observed by Shokrani et al. (2016b). When using the cryogenic coolant, LN_2 is distributed to both the cutting zone and the tool-workpiece interface. This results in the hardening of the workpiece material around the cutting region (Shokrani and Newman 2019).

Continuous segmented chip formation was observed in both dry and cryogenic drilling conditions (S. Berger et al. 2021). Segmentation degree in cryogenic machining was lesser compared to dry machining. Also, the chip formed in the cryogenic was smoother and had smaller serration. Cryogenic reduces the cutting temperature, and workpiece material becomes harder, which improves chip formation and hence improves machinability. Figure 1.10 shows the chip for different holes in both the machining environment. From

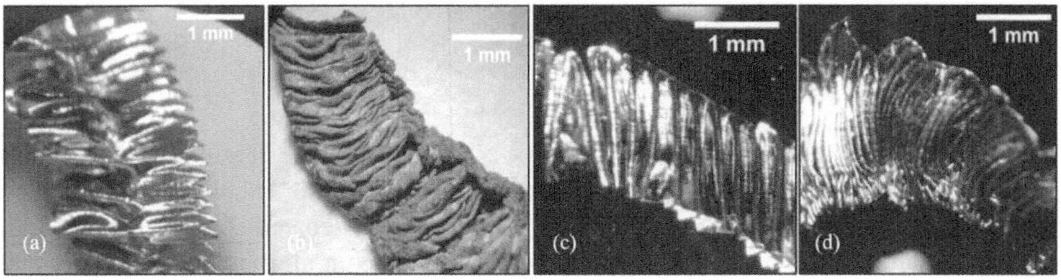

Figure 1.9 (a) Chip formed during drilling of the 1st hole under dry environment. (b) Chip formed during drilling of the 12th hole under dry environment. (c) Chip formed during drilling of the 6th hole under cryogenic environment. (d) Chip formed during drilling of the 38th hole under cryogenic environment.

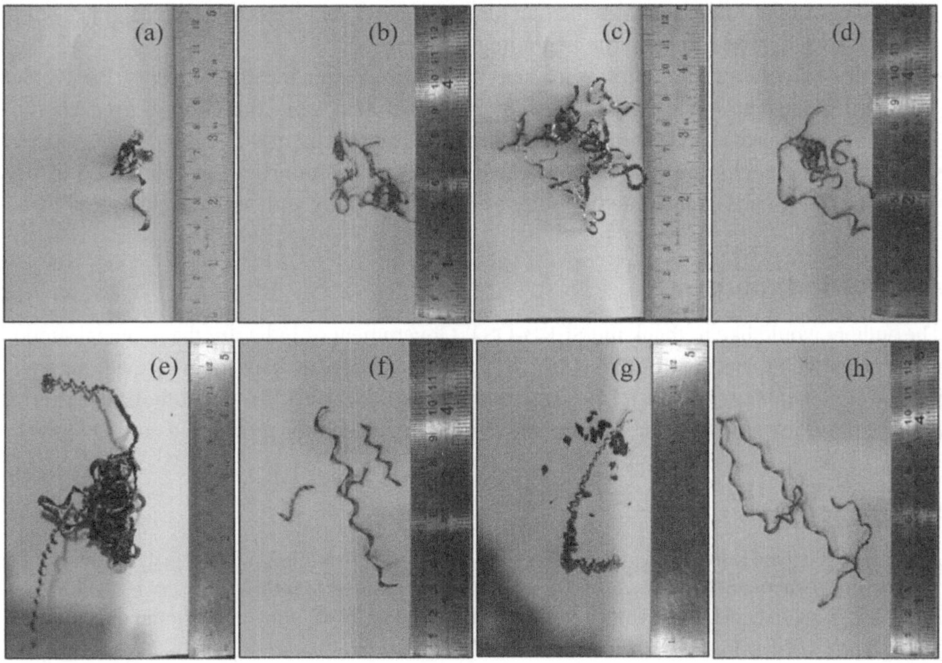

Figure 1.10 Chip formed during drilling of (a) 1st hole under dry environment, (b) 1st hole under cryogenic environment, (c) 6th hole under dry environment, (d) 6th hole under cryogenic environment, (e) 10th hole under dry environment, (f) 38th hole under cryogenic environment, (g) 12th hole under dry environment, and (h) 46th hole under cryogenic environment.

the comparison, it is clear that dry machining has produced more continuous chips than cryogenic. Effective evacuation and breakability of the chips are the major bottlenecks of the dry drilling environment. Figures 1.10 (f) and (h) show the increased length of the chip, which is due to increased tool wear in cryogenic drilling conditions. In cryogenic machining, pressurized LN_2 flow at the cutting zone and heat dissipation at the cutting zone improve the chip breakability and effective removal of chips. This shows the improved machinability in cryogenic machining conditions.

1.4 Summary

This chapter focuses on the overview of work carried out for the feasibility of an indigenously developed machining facility for cryogenic drilling. The tribological aspects are discussed for the machinability of Ti-6Al-4V with the AlCrN-coated solid carbide drill bit in terms of thrust force, tool wear, surface roughness, and chip morphology under dry and cryogenic environments. The salient findings from the experimental work are as follows:

- Thrust forces were greater in cryogenic conditions. After the tool flank wear of the drill bit, the effect of cryogenic cooling is reduced, which leads to high thrust forces.
- Attrition wear is the dominant wear type in a cryogenic machining environment. In dry machining, the welding of the chip reduces the machinability of the workpiece.
- In the initial drill pass, cryogenic machining showed no adhesion, whereas dry machining showed adhesion from the beginning.
- In dry machining, the chips were heavily oxidized after the formation of the welded chip. Cryogenic machining efficiently dissipates heat from the cutting zone, thereby resulting in prolonged tool life (up to threefold improvement).
- Cryogenic machining leads to favorable chip formation during the machining of titanium and improves the machinability of Ti-6Al-4V alloy and, subsequently, tool life.

Acknowledgments

The authors would like to thank the SERB-DST, Government of India, for the financial support given under the Project (ECR/2016/000735). The authors are also thankful to Elsevier for granting permission to reuse Figure 1.1 (License number: 5340240100558). The authors also express their gratitude towards Kenstar Tooltech Pvt. Ltd., Ahmedabad for sponsoring cutting tools.

References

Agrawal, V., Gajrani, K. K., Mote, R. G., Barshilia, H. C., Joshi, S. S. (2022). Wear analysis and tool life modeling in micro drilling of Inconel 718 superalloy. *ASME Journal of Tribology*, *144*(10), 101706.

Ahmed, L. S., Kumar, M. P. (2016). Cryogenic drilling of Ti-6Al-4V alloy under liquid nitrogen cooling. *Materials and Manufacturing Processes*, *31*(7), 951–959.

Ahsan, K. B., Mazid, A. M., Pang, G. K. (2016). Morphological investigation of Ti-6Al-4V chips produced by conventional turning. *International Journal of Machining and Machinability of Materials*, *18*(1–2), 138–154.

Aramcharoen, A. (2016). Influence of cryogenic cooling on tool wear and chip formation in turning of titanium alloy. *Procedia CIRP*, *46*, 83–86.

Bambam, A. K., Dhanola, A., Gajrani, K. K. (2022). Machining of titanium alloys using phosphonium-based halogen-free ionic liquid as lubricant additives. *Industrial Lubrication and Tribology*, *74*(6), 722–728. https://doi.org/10.1108/ILT-03-2022-0083.

Berger, S., Brock, G., Biermann, D. (2021). Simulative design of constraints for targeted restriction of chip thickness deviations when machining titanium alloy Ti6Al4V. *Procedia CIRP*, *102*, 85–90.

Bermingham, M. J., Palanisamy, S., Kent, D., Dargusch, M. S. (2012). A comparison of cryogenic and high pressure emulsion cooling technologies on tool life and chip morphology in Ti-6Al-4V cutting. *Journal of Materials Processing Technology*, *212*(4), 752–765.

Bordin, A., Bruschi, S., Ghiotti, A., Bariani, P. F. (2015). Analysis of tool wear in cryogenic machining of additive manufactured Ti6Al4V alloy. *Wear, 328*, 89–99.

Brunette, D. M. (2001). Principles of cell behavior on titanium surfaces and their application to implanted devices. In: *Titanium in medicine*. Engineering Materials. Springer-Verlag.

Cadena, N. L., Cue-Sampedro, R., Siller, H. R., Arizmendi-Morquecho, A. M., Rivera-Solorio, C. I., Di-Nardo, S. (2013). Study of PVD AlCrN coating for reducing carbide cutting tool deterioration in the machining of titanium alloys. *Materials*, *6*(6), 2143–2154.

Che-Haron, C. H., Jawaid, A. (2005). The effect of machining on surface integrity of titanium alloy Ti–6% Al–4% V. *Journal of Materials Processing Technology*, *166*(2), 188–192.

De Lacalle, L. L., Perez-Bilbatua, J., Sánchez, J. A., Llorente, J. I., Gutierrez, A., Albóniga, J. (2000). Using high pressure coolant in the drilling and turning of low machinability alloys. *The International Journal of Advanced Manufacturing Technology*, *16*(2), 85–91.

Furness, R. J., Wu, C. L., Ulsoy, A. G. (1996). *Statistical analysis of the effects of feed, speed, and wear on hole quality in drilling*. ASME Journal of Manufacturing Science and Engineering, *118*(3): 367–375.

Gajrani, K. K. (2020). Assessment of cryo-MQL environment for machining of Ti-6Al-4V. *Journal of Manufacturing Processes*, *60*, 494–502.

Gajrani, K. K., Divse, V., Joshi, S. S. (2021b). Burr reduction in drilling titanium using drills with peripheral slits. *Transactions of the Indian Institute of Metals*, *74*(5), 1155–1172.

Gajrani, K. K., Prasad, A., Kumar, A. (Eds.). (2022). *Advances in sustainable machining and manufacturing processes*. CRC Press. ISBN: 9781003284574. https://doi.org/10.1201/9781003284574.

Gajrani, K. K., Suvin, P. S., Kailas, S. V., Rajurkar, K. P., Sankar, M. R. (2021a). Machining of hard materials using textured tool with minimum quantity nano-green cutting fluid. *CIRP Journal of Manufacturing Science and Technology*, *35*, 410–421.Hong, S. Y., Ding, Y. (2001). Cooling approaches and cutting temperatures in cryogenic machining of Ti-6Al-4V. *International Journal of Machine Tools and Manufacture*, *41*(10), 1417–1437.

Hong, S. Y., Ding, Y., Jeong, W. C. (2001). Friction and cutting forces in cryogenic machining of Ti-6Al-4V. *International Journal of Machine Tools and Manufacture*, *41*(15), 2271–2285.

Hua, J., Shivpuri, R. (2004). Prediction of chip morphology and segmentation during the machining of titanium alloys. *Journal of materials processing technology*, *150*(1–2), 124–133.

Impero, F., Dix, M., Squillace, A., Prisco, U., Palumbo, B., Tagliaferri, F. (2018). A comparison between wet and cryogenic drilling of CFRP/Ti stacks. *Materials and Manufacturing Processes*, *33*(12), 1354–1360.

Jamil, M., Khan, A. M., Gupta, M. K., Mia, M., He, N., Li, L., Sivalingam, V. (2020). Influence of CO2-snow and subzero MQL on thermal aspects in the machining of Ti-6Al-4V. *Applied Thermal Engineering*, *177*, 115480.

Jawahir, I. S., Attia, H., Biermann, D., Duflou, J., Klocke, F., Meyer, D., Newman, S. T., Pusavec, F., Putz, M., Rech, J., Schulze, V., Umbrello, D. (2016). Cryogenic manufacturing processes. *CIRP Annals*, *65*(2), 713–736.

Jawahir, I. S., Luttervelt, C. A. (1993). Recent developments in chip control research and applications. *CIRP Annals—Manufacturing Technology*, *43*(2), 659–693.

Junior, A. S. Araujo Sales, W. F., Silva, R. B. Da, Costa, E. S., Machado, A. Rocha. (2017). Lubricooling and tribological behavior of vegetable oils during milling of AISI 1045 steel focusing on sustainable manufacturing. *Journal of Cleaner Production*, *156*, 635–647.

Khanna, N., Rashid, R. R., Palanisamy, S. (2017). Experimental evaluation of the effect of workpiece heat treatments and cutting parameters on the machinability of Ti-10V-2Fe-3Al β titanium alloy using Taguchi's design of experiments. *International Journal of Machining and Machinability of Materials*, *19*(4), 374–393.

Khanna, N., Sangwan, K. S. (2013). Interrupted machining analysis for Ti6Al4V and Ti5553 titanium alloys using physical vapor deposition (PVD): Coated carbide inserts. *Proceedings of the Institution of Mechanical Engineers, Part B: Journal of Engineering Manufacture*, *227*(3), 465–470.

Khanna, N., Suri, N. M., Agrawal, C., Shah, P., Krolczyk, G. M. (2019). Effect of hybrid machining techniques on machining performance of in-house developed Mg-PMMC. *Transactions of the Indian Institute of Metals*, *72*(7), 1799–1807.

Kumar, T. S., Prabu, S. B., Manivasagam, G., Padmanabhan, K. A. (2014). Comparison of TiAlN, AlCrN, and AlCrN/TiAlN coatings for cutting-tool applications. *International Journal of Minerals, Metallurgy, and Materials, 21*(8), 796–805.

Li, F. L., Xia, W., Zhou, Z. Y., Zhao, J., Tang, Z. Q. (2012). Analytical prediction and experimental verification of surface roughness during the burnishing process. *International Journal of Machine Tools & Manufacture, 62*, 67–75.

Liu, S., Shin, Y. C. (2019). Additive manufacturing of Ti6Al4V alloy: A review. *Materials & Design, 164*, 107552.

Machado, A. R., Wallbank, J. (1990). Machining of titanium and its alloys: A review. *Proceedings of the Institution of Mechanical Engineers, Part B: Management and Engineering Manufacture, 204*, 53–60.

Mia, M., Rahman, M. A., Gupta, M. K., Sharma, N., Danish, M., Prakash, C. (2022). Chapter 3 – advanced cooling-lubrication technologies in metal machining. Editor: Alokesh Pramanik. In *Processes, surfaces, coolants, and modeling*. Elsevier, 67–89.

Pederson, R. (2002). *Microstructure and phase transformation of Ti-6Al-4V* (Doctoral dissertation, Luleå tekniska universitet).

Rahim, E. A., Kamdani, K., Sharif, S. (2008). Performance evaluation of uncoated carbide tool in high speed drilling of Ti6Al4V. *Journal of Advanced Mechanical Design, Systems, and Manufacturing, 2*(4), 522–531.

Rahim, E. A., Sharif, S. (2006). Investigation on tool life and surface integrity when drilling Ti-6Al-4V and Ti-5Al-4V-Mo/Fe. *JSME International Journal Series C Mechanical Systems, Machine Elements and Manufacturing, 49*(2), 340–345.

Rahman, M. A., Mia, M., Ahmed, A. (2021). Chapter 3 – trends in electrical discharge machining of Ti- and Ni-based superalloys: Macro-micro-compound arc/spark/melt process. Editors: Tanveer Saleh, Mohamed Sultan, Mohamed Ali, Kenichi Takahata, In *Micro and nano technologies, micro electro-fabrication*. Elsevier, 63–87.

Rashid, R. R., Palanisamy, S., Sun, S., Dargusch, M. S. (2017). A case-study on the mechanism of flank wear during laser-assisted machining of a titanium alloy. *International Journal of Machining and Machinability of Materials, 19*(6), 538–553.

Sartori, S., Bordin, A., Moro, L., Ghiotti, A., Bruschi, S. (2016). The influence of material properties on the tool crater wear when machining Ti6Al4 v produced by additive manufacturing technologies. *Procedia Cirp, 46*, 587–590.

Sharif, S., Rahim, E. A., Sasahara, H. (2012). Machinability of titanium alloys in drilling. *Titanium Alloys-Towards Achieving Enhanced Properties for Diversified Applications, 3*, 117–137.

Shetty, P. K., Shetty, R., Shetty, D., Rehaman, N. F., Jose, T. K. (2014). Machinability study on dry drilling of titanium alloy Ti-6Al-4V using L9 orthogonal array. *Procedia Materials Science, 5*, 2605–2614.

Shokrani, A., Dhokia, V., Newman, S. T. (2016a). Comparative investigation on using cryogenic machining in CNC milling of Ti-6Al-4V titanium alloy. *Machining Science and Technology, 20*(3), 475–494.

Shokrani, A., Dhokia, V., Newman, S. T. (2016b). Investigation of the effects of cryogenic machining on surface integrity in CNC end milling of Ti-6Al-4V titanium alloy. *Journal of Manufacturing Processes, 21*, 172–179.

Shokrani, A., Huibin, S., Dhokia, V., Newman, S. T. (2016c). High speed cryogenic drilling of grade 5 ELI titanium alloy. In *26th International conference on flexible automation and intelligent manufacturing, Seoul, Korea, Republic of* (Vol. 27, pp. 6–30). https://researchportal.bath.ac.uk/en/publications/high-speed-cryogenic-drilling-of-grade-5-eli-titanium-alloy

Shokrani, A., Newman, S. T. (2019). A new cutting tool design for cryogenic machining of Ti-6Al-4V titanium alloy. *Materials, 12*(3), 477.

Siju, A. S., Gajrani, K. K., Joshi, S. S. (2021). Dual textured carbide tools for dry machining of titanium alloys. *International Journal of Refractory Metals and Hard Materials, 94*, 105403.

Suhaimi, M. A., Yang, G. D., Park, K. H., Hisam, M. J., Sharif, S., Kim, D. W. (2018). Effect of cryogenic machining for titanium alloy based on indirect, internal and external spray system. *Procedia Manufacturing*, *17*, 158–165.

Upadhyay, V., Jain, P. K., Mehta, N. K. (2014). Comprehensive study of chip morphology in turning of Ti-6Al-4V. In *5th Int. 26th all India manufacturing technology, design and research conference* (pp. 2–7). https://www.iitg.ac.in/aimtdr2014/PROCEEDINGS/papers/262.pdf

Yuan, C. G., Pramanik, A., Basak, A. K., Prakash, C., Shankar, S. (2021). Drilling of titanium alloy (Ti6Al4V): A review. *Machining Science and Technology*, *25*(4), 637–702.

Zang, J., Zhao, J., Li, A., Pang, J. (2018). Serrated chip formation mechanism analysis for machining of titanium alloy Ti-6Al-4V based on thermal property. *The International Journal of Advanced Manufacturing Technology*, *98*(1), 119–127.

Zhao, Z., Hong, S. Y. (1992). Cooling strategies for cryogenic machining from a materials viewpoint. *Journal of Materials Engineering and Performance*, *1*(5), 669–678.

Zhu, Z., Guo, K., Sun, J., Li, J., Liu, Y., Chen, L., Zheng, Y. (2018). Evolution of 3D chip morphology and phase transformation in dry drilling Ti6Al4V alloys. *Journal of Manufacturing Processes*, *34*, 531–539.

Chapter 2

Energy and economic assessment of machining Ti-6Al-4V in cryo-MQL environment

Nagella Lahari, Navneet Khanna, and Kishor Kumar Gajrani

Contents

2.1 Introduction

Titanium alloys have numerous applications in the biomedical, aerospace, sports, and automotive industries. Despite various advantages and applications, the machining of titanium alloys possesses various challenges due to its low thermal conductivity, which hinders the dissipation of heat from the machining zone [1]. Another challenge is its low elastic modulus, which causes heavy chatter and vibration during high-speed machining leading to a high surface roughness of the machined workpiece [2]. These result in high tool wear, higher machining forces, and enormous energy consumption, causing high carbon emissions. The higher energy consumption can be correlated with machined workpiece surface finish and tool wear. Higher energy requirements may be due to more tool wear and workpiece surface roughness [3].

Nowadays, government agencies are pushing industries to cut down/limit energy consumption due to global warming by applying various restrictions and sanctions. Electricity cost is increasing day by day. Consequently, industries and researchers are trying to find a solution

DOI: 10.1201/9781003291961-3

to reduce total energy consumption and associated costs without sacrificing product quality. Therefore, various lubrication and cooling strategies are explored in the literature, such as minimum quantity lubrication (MQL) [4–9], NMQL [10], cryogenic [11], and hybrid cryo-MQL [12], to reduce machining forces and overall energy consumption during machining.

Lei and Liu (2002) conducted the experiments during the machining of titanium alloy without any lubrication (dry condition). Assessment of results showed that the tool wear was high due to the high temperature and chemical affinity of titanium towards the tool material for dry machining [13]. In 2004, Weinert et al. compared dry machining and MQL. They discussed the issues and problems related to flood cooling and proposed MQL as a viable solution, and categorized it as a future of machining [14]. Further, MQL also has some limitations when machining hard materials, like titanium alloys, attributed to their low thermal conductivity. Therefore, cryogenic fluids are also popular for dissipating heat and cool the machining zone of hard materials. Cryogenic fluids can absorb or dissipate heat and can control the rising temperature. Cryogenic fluids also help to reduce machining forces and workpiece surface roughness [15]. However, either MQL or cryogenic alone is not efficient when machining hard-to-machine material at high speed. Consequently, the combination of both cryogenic cooling and MQL (cryo-MQL) is recommended for better cooling and lubrication while machining hard-to-machine materials at high speed [16].

In a nutshell, the literature suggests that the machining performance of hard-to-machine materials is already studied and reported extensively for dry, MQL, cryogenic, and hybrid cryo-MQL machining environments. However, the productivity and economics of machining are not adequately studied, which are equally important. Very limited studies are available in the literature related to the economic assessment of environmentally friendly machining processes. For a minimum cost per product, an economic analysis of hybrid cryo-MQL is necessary due to the use of additional coolant and its application techniques. In this work, an economic evaluation of a hybrid cryo-MQL environment for machining is carried out. The productivity and total machining cost per product, including energy consumption cost, overhead cost, coolant cost, cutting insert cost, and coolant disposal cost, are evaluated using microeconomics cost modeling. Energy consumption is another parameter that is studied and analyzed. Further, the surface roughness of the machined surface is measured after turning experiments. For comparison, economic assessment and turning experiments are also carried out for dry and MQL environments. This study is an extension of the authors' previously published work [12].

2.2 Materials and methods

In this study, the turning experiments were performed on Ti-6Al-4V titanium alloy having very high tensile strength, resistance to corrosion, and toughness. Cylindrical bars of 50 mm diameter and 250 mm length were used as specimens. The properties of Ti-6Al-4V are tabulated in Table 2.1 [12]. Turning tests were conducted on an EMCO turn 345 II CNC turning center. Carbide inserts TNMA 220412 were chosen as tool inserts having a nose radius of 1.2 mm and a rake angle of −6°. Tests were conducted under three lubrication/cooling environments: *(i)* Cryo-MQL, *(ii)* MQL, and *(iii)* dry. UNIST Coolubricator™ commercial MQL setup and semi-ester-based water-miscible coolant were used for the MQL environment. The properties of the cutting fluid are shown in Table 2.2 [12]. MQL and liquid nitrogen (LN_2) were applied to the machining zone. MQL and cryo-MQL input parameters were optimized in our previous study [12]. Figure 2.1 shows the schematic of the cryo-MQL experimental setup.

Table 2.1 Composition and Mechanical Properties of Ti-6Al-4V [12]

Composition (wt.%)							
V	*Al*	*O*	*N*	*H*	*C*	*Fe*	*Ti*
4.1	5.72	0.1	0.022	0.005	0.02	0.004	Balance

Mechanical Properties

Shear Modulus	*Young's Modulus*	*Hardness*	*Ultimate Tensile Strength*
40 GPa	110 GPa	348 HV	1020 MPa

Table 2.2 Properties of Cutting Fluid [12]

Kinematic viscosity at 40°C (mm²/s)	*Density (kg/m³)*	*Thermal conductivity (W/mK)*	*Viscosity index, VI*	*Flash point (°C)*
12	872	0.1792	198.04	242

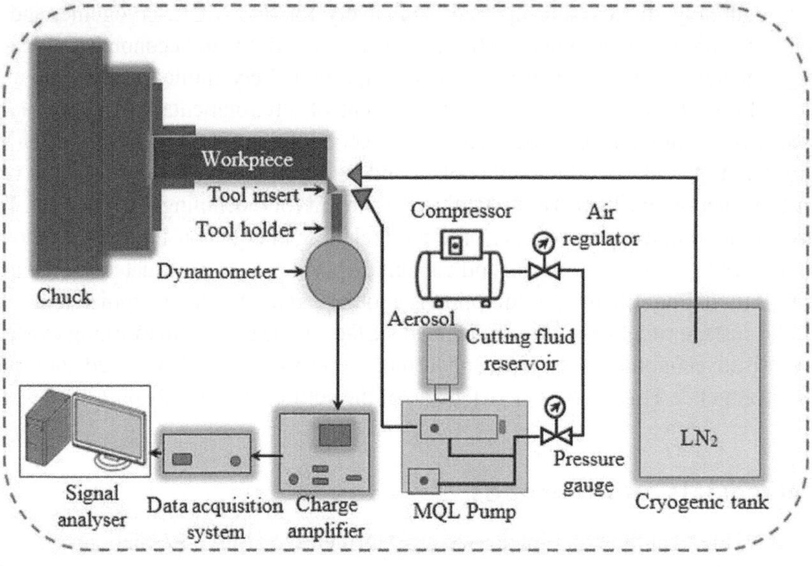

Figure 2.1 Schematic of the cryo-MQL experimental setup.

A piezoelectric dynamometer (Kistler make 9257A model) was used for determining machining forces. Specific cutting forces were calculated using measured machining forces for this study. Experimental machining conditions are tabulated in Table 2.3, and the experimental plan is tabulated in Table 2.4. The workpiece surface roughness was evaluated using the contact type Mitutoyo Surftest SJ-410. Each experiment was performed thrice, and average values were reported.

Table 2.3 Machining Environment and Conditions [12]

Input parameters	*Values*
Lubrication/cooling environments	Dry, MQL, and cryo-MQL
Turning conditions	
Cutting speed (*v*)	80 m/min and 120 m/min
Feed (*f*)	0.01 mm/rev, 0.15 mm/rev, and 0.2 mm/rev
Depth of cut (α)	1 mm
Length of cut	120 mm
Machining allowance	4 mm

Table 2.4 Experimental Plan

Experiment no.	*Cutting speed (m/min)*	*Feed (mm/rev)*	*Depth of cut (mm)*
1	80	0.1	1
2	80	0.15	1
3	80	0.2	1
4	120	0.1	1
5	120	0.15	1
6	120	0.2	1

2.3 Economic model

To study the economic behavior of any process, a simplified model designed to check the hypotheses about economic behavior is known as an economic model. In this study, an economic model is used to assess the lubrication/cooling environment while machining Ti-6Al-4V alloy. Total machining cost and energy consumption were computed as well as workpiece surface roughness were measured for three different lubrication/cooling environments.

2.3.1 *Total cycle time*

The efficiency and productivity of the machining process was based on total cycle time (T_{CT}). It is the summation of time required for one complete cycle of machining one product. Total cycle time includes tool air travel time (T_{air}), cutting time (T_{cut}), tool standby time (T_{sb}), and per cut tool change time (T_{ct}), as given in equation (2.1) [16]:

$$T_{CT} = T_{air} + T_{cut} + T_{sb} + T_{ct} \tag{2.1}$$

Tool changing time and standby time depend upon the operator's experience and skill. In this study, only one operator conducted all the experiments. Therefore, a constant tool changing time of 16 s and standby time of 35 s is considered based on measurement during experiments. Overall, productivity is calculated from total cycle time using equation (2.2) [16]:

$$Productivity\ (8\text{-}h\ shift) = 28000/T_{CT} \tag{2.2}$$

2.3.2 *Total machining cost*

Sustainable production includes economic production. This can be achieved by selecting appropriate machining conditions, which require an in-depth analysis of cost under various conditions. The total machining cost (C_T) is calculated using equation (2.3) [16, 17]:

$$C_T = C_t + C_e + C_o + C_c + C_d \tag{2.3}$$

where C_t is cutting insert cost, C_e is energy consumption cost, C_o is an overhead cost, C_c is coolant cost, and C_d is coolant disposal cost.

2.3.2.1 *Tool cost*

The tooling process is an essential part of manufacturing. During the machining process, tool wear is a common occurrence, and frequent tool change is required for mass production. In mass production, tool insert cost is a significant share of the total cost [18]. To decrease this cost, a few actions, such as optimizing the input parameters and usage of coolant, are preferred. Therefore, the study of overall tool cost is important for any sustainable product development. In this study, each cutting insert price was 20.54 USD. The tool cost is calculated as follows:

$$C_t = \frac{insert\ price}{tool\ life} \times machining\ time \tag{2.4}$$

2.3.2.2 *Energy consumption cost*

Within the industrial sector, manufacturing accounts for the largest share of annual industrial energy consumption. Machine tools employed in these manufacturing industries consume a significant portion of total energy consumption. Total energy consumption comprises energy for cutting workpieces and stand-by energy, which is calculated using equations (2.5–2.8).

$$C_e = C_{cut} + C_{ncut} \tag{2.5}$$

$$C_e = \frac{Unit\ enegry\ cost}{\eta \times 60 \times 1000} * t_m * (cutting\ power + stand\ by\ power) \tag{2.6}$$

where C_{cut} is cutting energy cost and C_{ncut} is machine stand-by cost. In this study, the unit energy cost was 0.15 USD/kWh, efficiency (η) was 88%, stand-by power is 1422 kW, and t_m is machining time. Machining time and cutting power can be calculated as follows:

$$t_m = \frac{length\ of\ cut + machining\ allowance}{feed \times rpm} \tag{2.7}$$

$$Cutting\ power = \frac{\alpha f V_c K_c}{60 \times 10^3 \times \eta} (kW) \tag{2.8}$$

where α is the depth of cut, V_c is the cutting speed, f is the feed, and K_c is the specific cutting force.

2.3.2.3 *Overhead cost*

Overhead cost is the sum of all the indirect costs incurred while manufacturing a product. It can be calculated using equations (2.9–2.11):

$$C_o = C_{op} + C_L + C_m \tag{2.9}$$

where C_{op} is operator cost (2.74 USD/h), C_L is lighting cost (0.4 kW), and C_m is machine depreciation cost (assumed to be zero for this study). C_{op} and C_L are calculated as follows:

$$C_{op} = operator\ labour\ \mathrm{cos}\,t \times t_m \times number\ of\ operators \tag{2.10}$$

$$C_L = \mathrm{cos}\,t\ of\ lighting \times unit\ energy\ \mathrm{cos}\,t \times t_m \tag{2.11}$$

2.3.2.4 *Coolant/lubricant cost*

Generally, the flood cooling technique is used for machining operations. In such operations, the share of cutting fluid costs is 15–17% of the final cost [19]. In such cases, the cost of cutting fluids is higher compared to the cost of tools. Therefore, it's desired to minimize cutting fluid costs to ensure economic machining. The coolant cost can be calculated using equation (2.12).

$$C_c = coolant\,/\,\mathrm{lub}\,ricant\ \mathrm{cos}\,t \times flow\,rate \times t_m \tag{2.12}$$

In this study, the lubricant cost and flow rate of MQL were 4.11 USD/L and 35 mL/h, respectively. The coolant cost and flow rate of LN_2 were 0.68 USD/L and 12 L/h, respectively.

2.3.2.5 *Chip and coolant/lubricant disposal cost*

Reused cutting fluids contain worn-out chip debris and various contaminants. Therefore, proper disposal of cutting fluids is important to protect society and the environment from the negative impacts of cutting fluids. However, in MQL or cryo-MQL machining environments, the overall amount of cutting fluid used is minimum, and always fresh cutting fluid is used. The chips produced under MQL remain almost dry, so they can be recycled easily without any need for further cleaning. In the case of LN_2, since it evaporates at ambient temperature, there is no concern about disposal as the workpiece and chip remain dry. Hence, dry, MQL, and cryo-MQL cutting have no associated cost for disposal.

2.4 Energy consumption analysis

Investigation of energy consumption is important to a company or organization more than that of electrical energy consumption. It's vital to reduce overall energy consumption to achieve sustainable production [20]. Energy consumption (E_c) can be calculated as given in equation 2.13:

$$E_c = F_c V_c t_m \tag{2.13}$$

where V_c is cutting speed, F_c is the main cutting force, and t_m is machining time. In our previous study, the measured cutting forces are reported [12], and the same are used for energy consumption analysis.

2.5 Results and discussion

Here, the economic assessment of the cryo-MQL lubrication/cooling environment is compared with that of dry and MQL machining environment results, followed by the discussion. Workpiece surface roughness is also reported and discussed.

2.5.1 *Total cycle time*

Total cycle time is an important factor for sustainable production as it directly affects overall productivity. Figure 2.2 shows the variation of total cycle time under three different lubricating/cooling environments for different machining parameters, as shown in Table 2.4.

Productivity depends on total cycle time. Using equation (2.2) and Figure 2.2, productivity under three different lubricating/cooling environments for different machining parameters was calculated and illustrated in Figure 2.3. From Figure 2.2 and Figure 2.3, it can be observed that a decrease in total cycle time resulted in an increase in the productivity of the number of parts in an 8-h shift.

After keeping the feed constant and increasing the cutting speed from 80 m/min to 120 m/min in all three machining environments, it was observed that the total cycle time had reduced by 16.06%, 18.30%, and 15.98% for cryo-MQL, MQL, and dry machining, respectively. Similarly, with an increase in the feed from 0.1 mm/rev to 0.15 mm/rev at a constant cutting speed (80 m/min), it was observed that the total cycle time had reduced by 29.93%, 26.80%, and 18.56% in cryo-MQL, MQL, and dry environment, respectively. The total

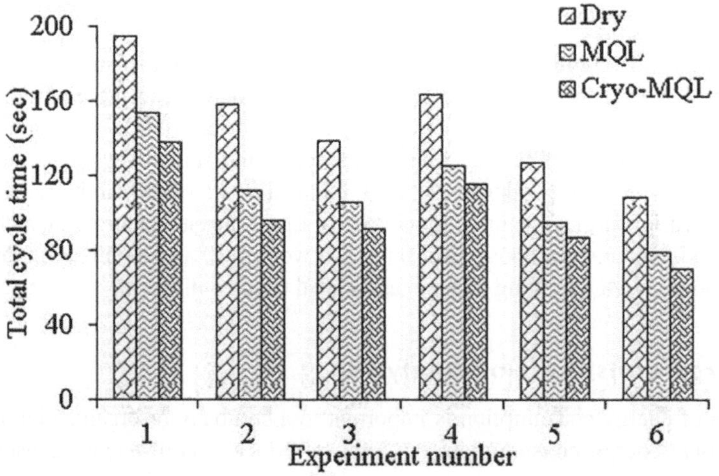

Figure 2.2 Total cycle time with respect to each experimental run under three different lubrication/cooling environments.

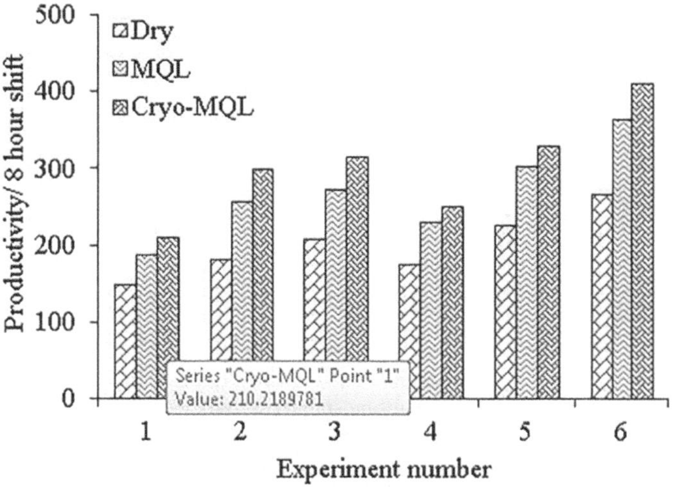

Figure 2.3 Influence on productivity corresponding to various experimental trials under three different lubrication/cooling environments.

cycle time decreased by 21.13% and 29.38% in MQL and cryo-MQL, respectively, when compared with dry environments. This result can be attributed to the reduction in tool wear under MQL and cryo-MQL environments, resulting in shorter tool handling time compared to dry environments.

2.5.2 *Economic and energy consumption assessment of cryo-MQL environment*

Economic assessment is important to estimate the probable costs and estimated benefits from a planned project, policy, or program in any industry. Economic evaluation is one of the parts which leads to sustainable manufacturing. Figure 2.4 illustrates the influence of cutting speed on total machining cost under three different lubricating/cooling environments. Figure 2.4 shows that under the same experimental design, the total machining cost in MQL and cryo-MQL had reduced by 16.17% and 19.47%, respectively, compared with the dry environment.

Figure 2.5 shows the cost sharing by various components, such as energy consumption cost, overhead cost, tool cost, and coolant cost, under three different lubricating/cooling environments. Based on economic model discussions on dry, MQL, and cryo-MQL machining in previous sections, it can be observed that the cryo-MQL machining environment is economical. It can be observed that the total machining cost is the lowest for the cryo-MQL environment (see Figure 2.4). This is due to a longer tool life, which leads to lower tool costs and low cutting power, due to which energy consumption cost reduces (see Figure 2.5). Also, the coolant/lubrication cost is low when compared with traditional flood coolants because the requirement of cryo-MQL coolant and lubricant is very low.

Figure 2.6 illustrates the influence of cutting speed on the consumption of total energy under three different lubricating/cooling environments. It was observed that by changing

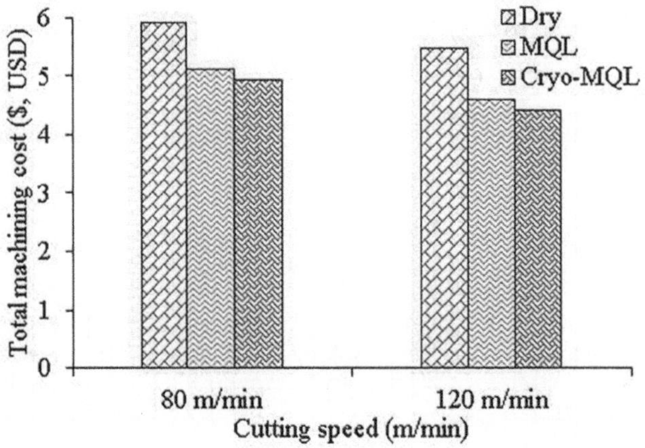

Figure 2.4 Influence of cutting speed on total machining cost under three different lubrication/cooling environments.

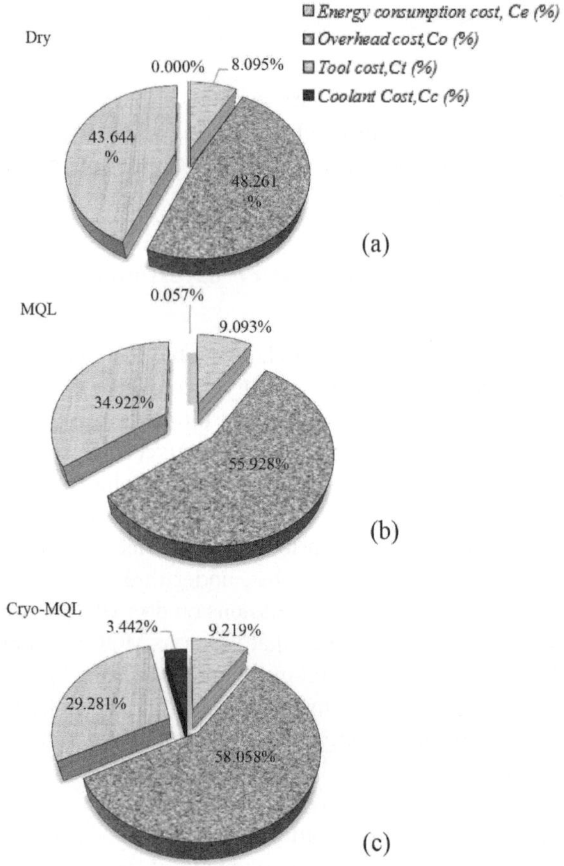

Figure 2.5 Cost sharing by various heads for three lubrication/cooling conditions (a) dry, (b) MQL, and (c) cryo-MQL.

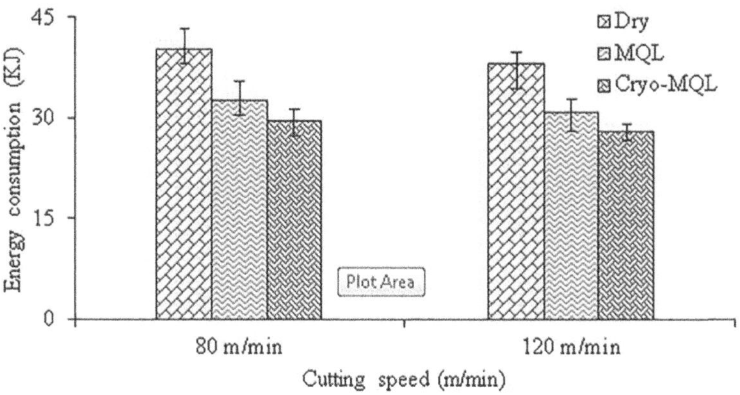

Figure 2.6 Consumption of total energy during machining under three different lubrication/cooling conditions.

the speed with constant feed, the energy consumption had reduced by 5.56%, 5.65%, and 4.62% under dry, MQL, and cryo-MQL machining, respectively. This is due to a reduction in cutting force by increasing cutting speed [10]. At the same cutting speed (120 m/min), it was observed that the energy consumption had reduced by 18.93% and 26.09% in MQL and cryo-MQL, respectively, as compared to dry conditions. It was attributed to a reduction in frictional force due to easy penetration of lubricant between chip and workpiece in MQL. In cryo-MQL, it is reduced further because of a fall in temperature due to coolant, which reduces adhesion forces on the tool chip interface, due to which material becomes harder and less sticky in addition to reduced frictional force. This results in the reduction of power consumption during machining. Therefore, cryo-MQL produces a hybrid effect, due to which a higher reduction in energy consumption is observed.

2.5.3 *Surface roughness*

The lubrication/cooling environment significantly affects the workpiece surface roughness (R_a) [21, 22]. Figure 2.7 illustrates the variation of workpiece R_a after machining under three different lubrication/cooling environments. From Figure 2.7, it is observed that the workpiece R_a reduces as the cutting speed increases. This was attributed to the reduction of machining forces with increasing speed leading towards lower tool vibration and hence reduced R_a. The workpiece R_a of 0.92 μm, 1.32 μm, and 1.71 μm was obtained while machining under cryo-MQL, MQL, and dry environments, respectively, for a cutting speed of 80 m/min. In the case of cryo-MQL, an overall 46.19% improvement in the surface finish was observed over a dry environment. Also, at the cutting speed of 120 m/min, overall R_a improved by 32.08% over a dry environment. It may be due to the lubricant from MQL, which will form a thin layer between the tool-workpiece interface, causing a reduction in friction leading towards lower tool wear and tool vibration and better surface finish. Moreover, applying LN_2 at the tool-workpiece interface might increase the cutting tool hardness attributed to the quenching effect of liquid nitrogen, which helps reduce the workpiece R_a.

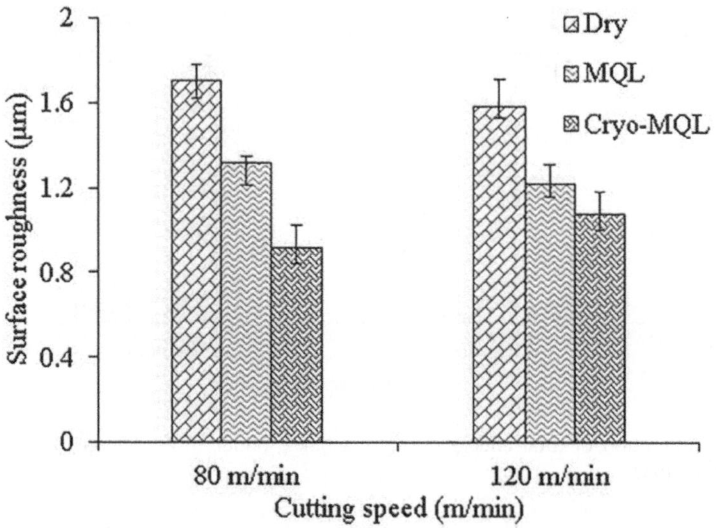

Figure 2.7 Influence of cutting speed on the workpiece surface roughness under three different lubrication/cooling conditions.

2.6 Conclusions

This study focuses on the economic assessment of the combined effect of MQL and cryogenic lubrication/coolant while machining Ti-6Al-4V. For comparison, experiments were also performed in MQL and dry environments. Responses obtained are total cycle time, productivity, total machining cost, energy consumption, and workpiece surface roughness (R_a). The salient findings are given below:

- Cryo-MQL environment reduces total cycle time by 29.38% compared to dry environment (V_c = 80 m/min). This is due to higher tool life.
- Cryo-MQL environment increases productivity by 49.97% on average over dry environment. This is due to the low total cycle time.
- Total machining cost in cryo-MQL machining is reduced by 19.47% compared to dry environment. This is due to lower tool cost attributed to shorter tool changing time and less energy consumption cost attributed to low cutting power required.
- Energy consumption in cryo-MQL environment is reduced by 26.09% compared to dry machining environment. This is attributed to a decrease in frictional force due to easy penetration of MQL between chip and workpiece and a reduction in temperature due to a reduction of adhesive forces on the tool-chip interface because of coolant (LN_2). Due to these reasons, the power required for machining decreases, and hence, energy consumption is also reduced.
- In the cryo-MQL environment, workpiece R_a is reduced by 46.19% and 22.80% over a dry environment at the cutting speed of 80 m/min, feed of 0.2 mm/rev, and depth of cut of 1 mm. This is attributed to a reduction in machining and frictional forces, which causes less vibration and chatter by the application of lubricant. Also, the surface roughness

is reduced, which is attributed to an increase in the hardness of the tool because of the quenching effect of liquid nitrogen.

Acknowledgments

This work was supported by the Institute Seed Grant from the Indian Institute of Information Technology, Design and Manufacturing, Kancheepuram, India (No. IIITDM/ISG/2022/ME/02). The authors acknowledge Elsevier for granting copyright permission to reuse Tables 2.1, 2.2, and 2.3 published in our previous study. The license number for copyright permission is 5335301041809. The authors are also thankful to Miss Hemalatha B for her effort to suggest improvement in this manuscript.

References

[1] N. Khanna, P. Shah, L.N.L. de Lacalle, A. Rodríguez, O. Pereira, 2021. In pursuit of sustainable cutting fluid strategy for machining Ti-6Al-4V using life cycle analysis. *Sustainable Materials and Technologies*, e00301.

[2] A.C. Hoyne, C. Nath, S.G. Kapoor, 2015. On cutting temperature measurement during titanium machining with an atomization-based cutting fluid spray system. *ASME Journal of Manufacturing Science and Engineering*, 137 (2), 024502(1-6).

[3] T. Kitagawa, A. Kubo, K. Maekawa, 1997. Temperature and wear of cutting tools in high-speed machining of Inconel 718 and Ti-6Al-6V-2Sn. *Wear*, 202 (2), 142–148.

[4] K.K. Gajrani, D. Ram, M.R. Sankar, U.S. Dixit, P.S. Suvin, S.V. Kailas, 2017. Machining of hardened AISI H-13 steel using minimum quantity eco-friendly cutting fluid. *International Journal of Additive and Subtractive Materials Manufacturing*, 1(3–4), 240–256. doi:10.1504/IJASMM.2017.10010905

[5] M. Bhuyan, A. Sarmah, K.K. Gajrani, A. Pandey, T.G. Thulkar, M.R. Sankar, 2018. State of art on minimum quantity lubrication in grinding process. *Materials Today: Proceedings*, 5 (9), 19638–19647.

[6] K.K. Gajrani, D. Ram, M.R. Sankar, 2017. Biodegradation and hard machining performance comparison of eco-friendly cutting fluid and mineral oil using flood cooling and minimum quantity cutting fluid techniques. *Journal of Cleaner Production*, 165, 1420–1435.

[7] M.S. Srinivas, K.K. Gajrani, A. Udayakumarc, M.R. Sankar, 2021. Sustainable machining of Cf/SiC ceramic matrix composite using green cutting fluids. *Procedia CIRP*, 98, 151–156.

[8] K.K. Gajrani, P.S. Suvin, S.V. Kailas, M.R. Sankar, 2019. Hard machining performance of indigenously developed green cutting fluid using flood cooling and minimum quantity cutting fluid. *Journal of Cleaner Production*, 206, 108–123.

[9] K.K. Gajrani, P.S. Suvin, S.V. Kailas, M.R. Sankar, 2019. Thermal, rheological, wettability and hard machining performance of MoS_2 and CaF_2 based minimum quantity hybrid nano-green cutting fluids. *Journal of Materials Processing Technology*, 266, 125–139.

[10] K.K. Gajrani, P.S. Suvin, S.V. Kailas, K.P. Rajurkar, M.R. Sankar, 2021. Machining of hard materials using textured tool with minimum quantity nano-green cutting fluid. *CIRP Journal of Manufacturing Science and Technology*, 35, 410–421.

[11] N. Khanna, P. Shah, 2020. Comparative analysis of dry, flood, MQL and cryogenic CO_2 techniques during the machining of 15–5-PH SS alloy. *Tribology International*, 146, 106196.

[12] K.K. Gajrani, 2020. Assessment of cryo-MQL environment for machining of Ti-6Al-4V. *Journal of Manufacturing Processes*, 60, 494–502.

[13] S. Lei, W. Liu, 2002. High-speed machining of titanium alloys using the driven rotary tool. *International Journal of Machine Tools and Manufacture*, 42 (6), 653–661.

[14] K. Weinert, I. Inasaki, J.W. Sutherland, T. Wakabayashi, 2004. Dry machining and minimum quantity lubrication. *CIRP Annals*, 53 (2) 511–537.

[15] A. Shokrani, V. Dhokia, P. Muñoz-Escalona, S.T. Newman, 2013. State-of-the-art cryogenic machining and processing. *International Journal of Computer Integrated Manufacturing*, 26 (7), 616–648.

[16] M.K. Gupta, Q. Song, Z. Liu, M. Sarikaya, M. Jamil, M. Mia, V. Kushvaha, A. Singla, Z. Li, 2020. Ecological, economical and technological perspectives based sustainability assessment in hybrid-cooling assisted machining of Ti-6Al-4V alloy. *Sustainable Materials and Technologies*, 26, e002018.

[17] C. Agrawal, J. Wadhwa, A. Pitroda, C.I. Pruncu, M. Sarikaya, N. Khanna, 2021. Comprehensive analysis of tool wear, tool life, surface roughness, costing and carbon emissions in turning Ti-6Al-4V titanium alloy: Cryogenic versus wet machining. *Tribology International*, 153, 106597.

[18] N. Khanna, C. Agrawal, M.K. Gupta, Q. Song, A.K. Singla, 2020. Sustainability and machinability improvement of Nimonic-90 using indigenously developed green hybrid machining technology. *Journal of Cleaner Production*, 263, 121402.

[19] K.K. Gajrani, M.R. Sankar, 2020. Role of eco-friendly cutting fluids and cooling techniques in machining. *Materials forming, machining and post processing*, 159–181. Springer, Cham. ISBN: 978-3-030-18853-5. doi:10.1007/978-3-030-18854-2_7

[20] N. Khanna, P. Shah, J. Wadhwa, A. Pitroda, J. Schoop, F. Pusavec, 2021. Energy consumption and lifecycle assessment comparison of cutting fluids for drilling titanium alloy. *Procedia CIRP*, 98, 175–180.

[21] A.K. Bambam, A. Dhanola, K.K. Gajrani, 2022. Machining of titanium alloys using phosphonium-based halogen-free ionic liquid as lubricant additives. *Industrial Lubrication and Tribology*, 74 (6), 722–728. doi:10.1108/ILT-03-2022-0083

[22] K.K. Gajrani, A. Prasad, A. Kumar (Eds.), 2022. *Advances in sustainable machining and manufacturing processes*. CRC Press, Boca Raton and London. ISBN: 9781003284574. doi:10.1201/9781003284574

Chapter 3

Optimization of sustainable manufacturing processes

A case study during drilling of laminated nanocomposites

Jogendra Kumar, Rajneesh Kumar Singh, and Jinyang Xu

Contents

Abbreviations

GRA: Grey Relation Analysis
MOORA: Multi-Objective Optimization on the basis of Ratio Analysis
MCDM: Multi-Criteria Decision Making
PCA: Principal Component Analysis
WPCA: Weighted Principal Component Analysis
RSM: Response Surface Methodology
BBD: Box Behnken Design
TOPSIS: Technique for Order Performance by Similarity to Ideal Solution
WASPAS: Weighted Aggregated Sum Product Assessment

3.1 Introduction

Manufacturers of epoxy composites encounter crucial problems like matrix crack propagation, fiber pull-out, delamination, plastic deformation, and so on during the machining process. Producing and assembling epoxy composites is more complicated than metal materials (Günay and Meral 2020; König and Graß 1989; J. Kumar et al. 2021). Due to epoxy composite's non-homogeneity

DOI: 10.1201/9781003291961-4

and anisotropic behavior, their machining characteristics are significantly different than those of metals and their alloys (John et al. 2019; Arul et al. 2006; Vinayagamoorthy 2018). Machinability and machining performance optimization is essential to properly utilizing any new composite. Eminent researchers have made significant contributions to polymer machining throughout the last three decades (Lachaud et al. 2001; Sorrentino et al. 2018; Ogawa et al. 1997). However, available data reveal very little research into modeling and optimizing nanocomposite machining characteristics (Xu et al. 2020; J. Kumar et al. 2020; J. Kumar et al. 2020). In addition, only a few papers have been found on multi-objective optimization of machining performance. Details on the influence of cutting circumstances and parameters on machinability were examined. Using proper modeling and optimization techniques makes it possible to optimize the machining process. Hence, an appropriate optimization tool for machining conditions is required to control the process parameters. Experts optimized the machining responses of laminated polymer nanocomposites by utilizing several modules. During multi-response optimization, the aggregation of process responses into a single objective function is done by various optimization modules, such as grey, analytic hierarchy process (AHP), utility functions, desirability functions, and technique for order performance by similarity to ideal solution (TOPSIS).

AHP is the most used for decision-making (DM) studies, but it works on the subjective judgment decision requirements and give preferential consistency. When aggregating conflicting responses, most studies assume equal weight and negligible response relation (Mohan et al. 2005; Palanikumar et al. 2006). They produce skewed, ambiguous, and riddled with the mistake. The exploration of the PCA module competently confronts these fundamental and traditional assumptions. PCA tackles multi-criteria optimization by converting associated output functions into independent indices. PCA works to estimate the proven statistical weight of the process response. The optimization method is commonly used during conventional manufacturing methods to evaluate the best possible sequences of machine limitations. Experts typically utilize it in advanced production processes, like welding (J. Kumar et al. 2022), milling (Uzun 2019; K. Kumar et al. 2021; J. Kumar and Verma 2021c), turning (Gok 2015), etc. PCA is a valuable statistical technique for assessing response priority weights with contradictory properties. It is feasible through parametric appraisal to improve the machining efficacy of the proposed material and reduce damages and waste (Sorrentino et al. 2017). Also, the existing literature on the machining behavior of nanocomposites is limited. It needs more attention from academia and industry. Some prominent academics have thoroughly investigated the theories of machining-induced defects have been thoroughly investigated by some prominent academics (Xu et al. 2018, 2013; Qiu et al. 2018). However, the manufacturing industry is still excessively concerned about the induced damages, which directly impact quality and performance. Robust tools and customization approaches are needed in this complicated challenge in laminated polymer nanocomposites. It is possible to optimize multiple criteria using the multi-criteria optimization modules. The improvement of quality and productivity indices in machining procedures is made possible by optimizing numerous machine restrictions. Fiber-reinforced polymers (FRP) are generally used in aircraft, automobile, naval, and building component manufacturing industries to achieve weight savings, precision, close tolerances, and generalization for precise component assembly.

3.1.1 Problem formulation

Polymer companies in fast-growing manufacturing sectors manage fast, innovative, and efficient manufacturing environments to adapt to changing consumer demands. Any industry or organization that deals with products or processes must prioritize ensuring their customers are satisfied. It is possible to meet the rapidly changing needs of customers through the use of a computerized regulated production procedure. These precise machines require skilled operators and large setup/installation costs. It is quite difficult for a manufacturer to recover the costs associated with these types of systems. It is necessary to balance the machining responses based on the experimental domain to study exact constraints to optimize them. Current techniques have many shortcomings and downsides, such as weight allocation responses mostly dependent on decision-makers (DM) and a lack of significant connection between responses. The previous literature review reveals that fewer tests on drilling polymer nanocomposites were conducted (Palanikumar 2011; Baraheni and Amini 2019; D. Kumar and Singh 2019). For example, researchers researched tool wear, fiber pullout, and matrix debonding and produced robust optimization techniques, like the Grey concept, TOPSIS, MOORA, and WPCA (Rajkumar et al. 2017). There is a lack of research on hybrids of multiple optimization modules during epoxy nanocomposite machining because most studies focus on single optimization modules. With hybridization modules, appropriate parametric parameters may improve machining qualities. The organized chapter are illustrated in Figure 3.1.

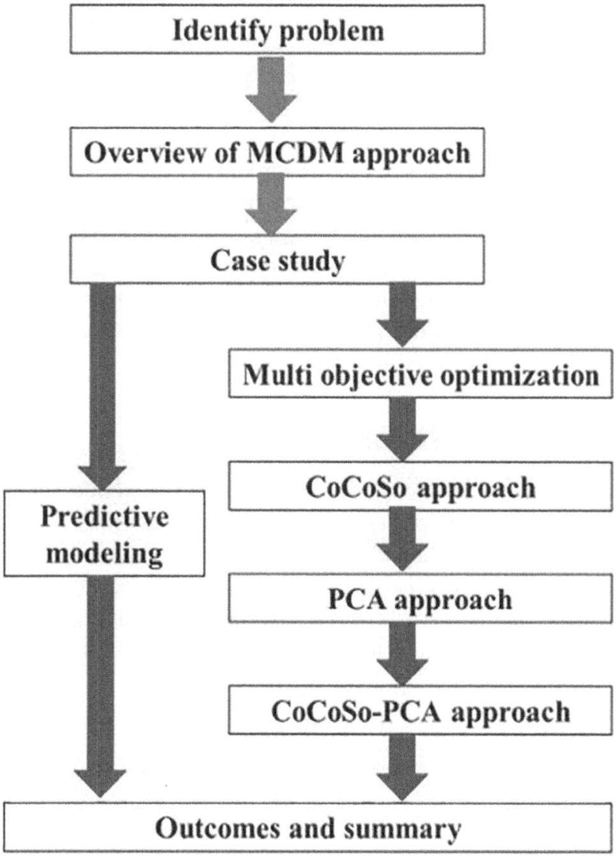

Figure 3.1 Flow chart for proposed module.

3.2 Multi-criteria decision making (MCDM) techniques

Conventional optimization tools consider the varying constraint of one factor while keeping the other factors constant. It is very time-consuming and costly. Still, the investigational technique of statistical strategy provides a more relaxed and efficient approach to optimizing several operational variables. The efficiently used test design method includes various evolutionary optimization modules and design-based methods (Islam and Pramanik 2016; Sampath Kumar et al. 2017; Kulkarni et al. 2020). RSM array is a statistically balanced investigational array that diminishes the test run efficiently. It is a highly acclaimed experimental pattern preferred in machining test design and multi-criteria evaluation. Thus, the current study optimizes polymer nanocomposite drilling using RSM. The design of the experiment has been carried out using the BBD concept for computing the correlation among the input parameters that affect a process parameter and the output response.

The machining of FRP composites employing GRA, TOPSIS, utility theory, etc., has been extensively studied. Most surveys assume little function correlation and equal weight during multiple response aggregation. These assertions result in haziness, inaccuracy, error, and imprecision in the real conclusion, which can cause the efficiency of the quality and production monitoring system to deviate from its original goal. PCA has the potential to address such essential challenges effectively, but current research indicates that this developing technology is only being used in a limited number of situations. Using a smaller collection of variables, PCA diminishes the data set (sample) without losing any model information, which helps it achieve its goal of reducing error. As seen by the association matrix between the real variable quantity, there is a lot of variety in the sample. Principal components are the terms used to describe these new, uncorrelated variables. Because the final objective value was calculated, it incorporates deviation (SD), covariance, eigenvectors, and eigenvalues, as well as the deviation (SD). In the sophisticated combined compromise solution (CoCoSo) module, this weight assignment is generated during the aggregated analysis of several responses. This chapter presents the hybrid optimization tool results, demonstrating its applicability in the machining environment.

3.2.1 *Box behnken design (BBD)*

The BBD considers all variables/factors simultaneously at maximum and minimum levels. It can easily avoid extreme conditions and test for feasible outcomes. A quadratic regression model was constructed using RSM, and the relationships among various inputs and outputs were established. As a result, compared to the conventional design of the experiment, it greatly diminishes trials and improves cost-effectiveness. The quadratic model second-order polynomial regression equation is given below (Gajrani et al. 2019).

$$F(X) = \beta_0 + \sum_{j=1}^{n} \beta_j X_j + \sum_{j-1}^{n} \beta_{jj} X_j^2 + \sum_{i=1}^{n=1} \sum_{j=2}^{n} \beta_{ij} X_i X_j \tag{3.1}$$

Here, β_j = Linear terms

β_{jj} = Quadratic terms

β_{ij} = Interaction terms

3.2.2 Combined compromise solution (CoCoSo) and principal component analysis (PCA)

In the beginning, Yazdani et al. (Yazdani et al. 2019) recommended this method to assess several management criteria in France (2018). Power control optimization, industrial energy concerns, and manufacturing trades currently use it (Wen et al. 2019; Mi and Liao 2020).

The CoCoSo approach normalizes the results following the intended quality characteristics as beneficial/cost-effective. The WASPAS approach was used to compute every possible weighted comparability order and drilling performance. Later, every option is referred to by three group operatives (K_{ia}, K_{ib}, and K_{ic}), comprising three compromise rankings. After considering the choices and criteria, the CoCoSo takes the following stages to solve the multifunctional decision issue (Kharwar et al. 2020).

Step-1: Decision matrix (DM) array is expressed as:

$$x_{ij} = \begin{bmatrix} x_{11} & x_{12} & \cdots & x_{1n} \\ x_{21} & x_{22} & \cdots & x_{2n} \\ \cdots & \cdots & \cdots & \cdots \\ x_{1m} & x_{2m} & \cdots & x_{mn} \end{bmatrix} \tag{3.2}$$

Here, x_{ij} = DM and m = 1, 2, 3 & n = 1, 2, 3.

Step-2: Data pre-processing involves normalizing measured quantities to reduce uncertainty. The cost criteria used in data pre-processing are primarily determined by the preferred characteristic criteria, referred to as "cost criteria".

$$r_{ij} = \frac{\min x_{ij} - x_{ij}}{\max x_{ij} - \min x_{ij}} \tag{3.3}$$

where x_{ij} = present data of array; $\min x_{ij}$, $\max x_{ij}$ = lowest and highest values of the array, respectively.

Step-3: Calculate the correlation coefficient.

A basic statistical methodology for evaluating the primary principal component, PCA, was invented by Karl Pearson in 1901 (PC) (Karl Pearson F. R. S. 1901). When substantially statistically independent components, known as the main components, are changed, the orthogonal transformation is used (PCs). The PCA-CoCoSo module controls quality and productivity concerns in industrial operations (Abhishek et al. 2014). This analysis assesses the particular response weight and the correlation factor among them.

$$R_{ij} = \frac{Cov\{X_i(k), X_i(l)\}}{\sigma_{(xi)}(k) \times \sigma_{(xi)}(l)} \tag{3.4}$$

where i = 1, 2, 3, 4. . . n and j = 1, 2, 3, 4. . . n, $X_i(j)$ is normalized for each performance, $Cov\{X_i(k), X_i(l)\}$ is the covariance of output performance, and $\sigma_{(xi)}(k) \times \sigma_{(xi)}(l)$ is the SD of the output response.

Step-4: Ascertain the eigenvalues.

$$(R_{ij} - \gamma_k I_m)V_{ik} \tag{3.5}$$

where γ_k = eigen values and $k = 1, 2, 3, 4 \ldots n$,

Step-5: Calculate the PC values.

$$Y_{mk} = \sum_{i=1}^{n} x_m(i)V_{ik} \tag{3.6}$$

where $Y_{m1}, Y_{m2}, \ldots$ are PC values.

Step-6: Calculate S_i and P_i.

The total weight comparability sequence is the sum of the proportionate component weighted at a normalized value for each alternative (S_i). The total comparability weight (P_i) is assessed for each alternate normalization value. The WASPAS method weighted multiplier of the product summation and exponential summation can attain the S_i and P_i values.

$$S_i = \sum_{j=1}^{n} (w_1 \times NR_a + w_2 \times NR_z + w_3 \times NC_e) \tag{3.7}$$

$$P_i = \sum_{j=1}^{n} \left((NR_a)^{w_1} + (NR_z)^{w_2} + (NC_e)^{w_3} \right) \tag{3.8}$$

Step-7: These are a few average strategies for determining the respective weights of the various alternatives. These techniques use three evaluation criteria to determine the relative importance of each possibility.

$$K_{ia} = \frac{P_i + S_i}{\sum_{i=1}^{m} (P_i + S_i)} \tag{3.9}$$

$$K_{ib} = \frac{S_i}{min\ S_i} + \frac{P_i}{min\ P_i} \tag{3.10}$$

$$K_{ic} = \frac{\lambda(S_i) + (1-\lambda)(P_i)}{(\lambda \max S_i + (1-\lambda)\max P_i)} \tag{3.11}$$

Step-8: Calculate the final aggregated K_i data and corresponding rank.

$$K_i = (k_{ia}k_{ib}k_{ic})^{\frac{1}{3}} + \frac{1}{3}(k_{ia} + k_{ib} + k_{ic}) \tag{3.12}$$

3.3 Case study: multi-objective optimization for drilling of laminated epoxy nanocomposites

3.3.1 *Selection of process parameters*

The CoCoSo-PCA method was utilized to evaluate and optimize the drilling data in this investigation. Directly and indirectly, several causes for machining are found but considered simultaneous, leading to limitations in the drilling process. It comprehends the machining environment; many process factors, such as cutter diameter, tool coating, tip angle, and coolant, could be selected (Sridharan et al. 2016; Fu et al. 2018; Gajrani et al. 2021; Agrawal et al. 2022). Additionally, these process characteristics were considered earlier in the drilling process (J. Kumar and Kumar Verma 2021; J. Kumar and Verma 2021a). Several other elements have been discovered in the literature to impact machining features. As a result of the experimental constraints in factor consideration simultaneously, it can affect machining time, experimental expense, and complexities. The proper RSM approach of 15 experiments is shown in Table 3.1. These three process constraints were chosen for this existing work study (J. Kumar and Verma 2021b; J. Kumar and Kumar Verma 2021). The responses are thrust, torque, delamination inlet, outlet, circularity error, and machining surface roughness (R_a and R_z). With respect to response aggregation, most pioneering scholars have assumed that all responses had the same weight. As a result of this notion, the results are ambiguous, reducing the module's overall effectiveness. During drilling testing, the proposed hybrid module improved the conflicting machining performance.

Table 3.1 Box Behnken Design With Observed Data

Exp. No.	*Input constraint*			*Output performance*						
	Drill speed (A)	*Feed (B)*	*GO (C)*	T_h	*T*	*IN*	*OUT*	R_a	R_z	C_e
1	0	+1	+1	64.98	0.1	1.0316	1.0436	1.69	4.2	0.366
2	+1	+1	0	53.94	0.06	1.012	1.0222	1.8	2.9	0.133
3	0	−1	−1	100.81	0.17	1.095	1.1118	1.99	5.8	0.52
4	0	−1	+1	61.59	0.16	1.0402	1.0618	2.1	4.43	0.287
5	−1	−1	0	74.07	0.13	1.0404	1.0664	1.75	5.09	0.42
6	+1	0	−1	57.17	0.1	1.0206	1.0408	1.84	3.66	0.169
7	+1	−1	0	95	0.15	1.0608	1.0912	1.87	5.7	0.41
8	−1	0	+1	61.29	0.12	1.0308	1.0534	2.09	3.76	0.21
9	+1	0	+1	54	0.07	1.0126	1.026	1.65	3.6	0.211
10	0	0	0	68.65	0.16	1.051	1.072	1.95	5	0.37
11	0	0	0	60.84	0.09	1.0306	1.0438	1.78	3.7	0.25
12	0	+1	−1	83.63	0.17	1.0718	1.118	2.09	5.9	0.401
13	−1	+1	0	67.27	0.13	1.0608	1.0818	1.88	4.4	0.32
14	0	0	0	70	0.12	1.0506	1.0642	1.89	4	0.301
15	−1	0	−1	72	0.11	1.036	1.064	1.9	4.3	0.31

Here, for drill speed (*A*) −1 = 800 rpm, 0 = 1600 rpm, and +1 = 2400 rpm; for Feed (*B*) −1 = 80 mm/min, 0 = 160 mm/min, and +1 = 240 mm/min; and for GO% (*C*) −1 = 1 wt.%, 0 = 2 wt.%, and +1 = 3 wt.%

Initially, all of the values were normalized to fall inside 0–1. The normalizing procedure is evaluated as a cost criterion using Eq. (3.3). It checks the correlation coefficient calculated using Eq. (3.4), as demonstrated in Table 3.2. Eigenvalues and PC values were calculated using Eqs. (3.5–3.6) and are tabulated in Tables 3.3 and 3.4. The Si and Pi values were determined using Eqs. (3.7–3.8) and organized in Table 3.5. Table 3.5 presents the results of three constant evaluations (Kia, Kib, and Kic) of each experiment, derived from Eqs. (3.9–3.11) to determine the relative weight criterion. The relative weight attribute combined the data to calculate the comparability sequence (S_i) with the selected weight (P_i) data. Table 3.5 shows the function values (K_i) calculated using Eq. (3.12).

The optimal drilling conditions are 2400 rpm drill speed, 80 mm/min feed, and 1% graphene oxide weight % (as shown in Figure 3.2). The value of the objective function is observed to be 38.065, which is greater than the value of the array valuation. Because of

Table 3.2 Correlation Coefficient

Sr. No.	*Response*	*Pearson value*	*P-value*	*Remarks*
1	T to T_h	0.730	0.002	<Significant
2	*IN* to T_h	0.889	0.000	
3	*IN* to T	0.852	0.000	
4	*OUT* to T_h	0.880	0.000	
5	*OUT* to T	0.891	0.000	
6	*OUT* to *IN*	0.956	0.000	
7	R_a to T_h	0.331	0.228	>Insignificant
8	R_a to T	0.719	0.003	<Significant
9	R_a to *IN*	0.516	0.049	
10	R_a to *OUT*	0.593	0.020	
11	R_z to T_h	0.897	0.000	
12	R_z to T	0.879	0.000	
13	R_z to *IN*	0.865	0.000	
14	R_z to *OUT*	0.923	0.000	
15	R_z to R_a	0.404	0.136	>Insignificant
16	C_e to T_h	0.878	0.000	<Significant
17	C_e to T	0.767	0.001	
18	C_e to *IN*	0.856	0.000	
19	C_e to *OUT*	0.824	0.000	
20	C_e to R_a	0.205	0.463	>Insignificant
21	C_e to R_z	0.917	0.000	<Significant

Table 3.3 Eigen Value

Eigenvalue	5.6164	0.9700	0.2011	0.1083	0.0713	0.0289	0.0040
Proportion	0.802	0.139	0.029	0.015	0.010	0.004	0.001
Cumulative	0.802	0.941	0.970	0.985	0.995	0.999	1.000

Table 3.4 PCA Analysis

Variable	*PC1*	*PC2*	*PC3*	*PC4*	*PC5*	*PC6*	*PC7*	*Contribution*
T_h	0.386	−0.264	0.507	−0.546	−0.353	−0.263	−0.179	0.149
T	0.392	0.261	−0.533	0.096	−0.057	−0.600	−0.348	0.154
IN	0.405	−0.028	0.403	0.640	0.021	−0.261	0.441	0.164
OUT	0.412	0.057	0.226	0.075	0.620	0.345	−0.516	0.17
R_a	0.239	0.828	0.105	−0.157	−0.293	0.326	0.172	0.057
R_z	0.404	−0.164	−0.354	−0.433	0.384	0.061	0.588	0.163
C_e	0.378	−0.382	−0.330	0.253	−0.504	0.523	−0.104	0.143

Table 3.5 Optimization Module

Exp. No.	S_i	P_i	K_a	K_b	K_c	K_i	*Rank*
1	0.6701	6.5912	0.0752	25.1556	0.9125	27.3431	5
2	0.9806	6.9771	0.0824	35.5519	1.0000	38.0654	1
3	0.0303	2.1249	0.0223	2.0000	0.2708	2.52273	15
4	0.5125	5.3333	0.0605	19.3752	0.7346	21.1221	9
5	0.4656	6.2272	0.0693	18.2527	0.8410	20.1841	11
6	0.8047	6.7772	0.0785	29.6724	0.9527	32.0084	3
7	0.2421	5.6130	0.0606	10.6104	0.7357	12.1863	13
8	0.6689	6.4745	0.0740	25.0602	0.8976	27.2171	6
9	0.9106	6.9031	0.0809	33.2163	0.9819	35.6612	2
10	0.4080	6.0553	0.0669	16.2771	0.8122	18.1164	12
11	0.7575	6.7268	0.0775	28.0941	0.9405	30.3822	4
12	0.1456	3.3219	0.0359	6.35668	0.4357	7.29175	14
13	0.4776	6.2827	0.0700	18.6728	0.8495	20.6281	10
14	0.5617	6.4419	0.0725	21.5165	0.8801	23.5809	8
15	0.5776	6.4675	0.0729	22.0520	0.8853	24.1355	7

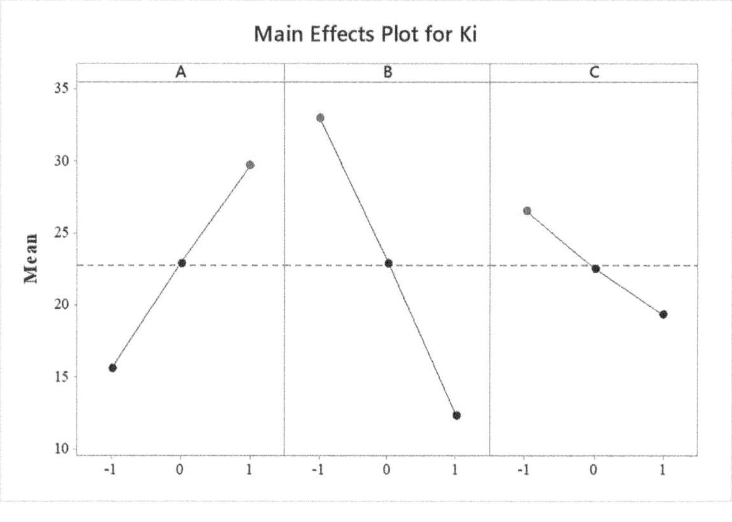

Figure 3.2 Optimum plot.

the intermittent cutting function of the drilling process, it is a more complicated process than the other operations. It ensures that the material is removed easily and set at high speed (2400 rpm), which could compromise the cutting tool features. In some cases, increasing the cutting speed raises the temperature at which cuts are being made. Aside from these advantages, a greater spindle speed and a higher temperature cause smoother material removal and, hence, lower surface roughness (Fu et al. 2018). Because of the increased interconnection strength of the laminated epoxy (polymer) composite, it is clear that the hole damage is at its lowest at a lower feed rate of 80 mm/min. Likewise, there is a reduced weight percent of GO drop in the fiber breaking and cracking (J. Kumar and Verma 2021a).

For low machining forces, T_h and T, hole damages *IN*, *OUT*, and C_e; machining surface roughness R_a and R_z; greater spindle speed; and low feed rate are preferred. Having a higher feed rate causes the drill to exert a greater bending strain on the job sample, leading to cracking (Wang et al. 2004; Köklü et al. 2019; Kyratsis et al. 2018). A lower cutting force is required because the fast spindle speed softens the epoxy phase, which raises the machining interface temperature (interface of the tool and the workpiece) (Kaybal et al. 2019; Tsao 2008). Drilling problems such as fractures, fiber breakage, and matrix debonding are reduced. A similar result and machining performance trend were noted in earlier investigations of distinguished researchers (Geier and Szalay 2017; Ragunath et al. 2017). Drilling-induced damage is highly susceptible to epoxy nanocomposite graphene oxide/carbon fiber laminates. The GO weight% can be added to the procedure to overcome this problem.

3.4 Predictive modeling

3.4.1 *Mathematical modeling of drilling parameters*

The effect of process constraints on the drilling performance of polymer nanocomposites was investigated. It was decided to conduct an ANOVA test at the 95% confidence level to determine the influence factor on machining performance (J. Kumar et al. 2021). When machining parameters are checked for maximum value, the Fisher test (F-value) determines which factors have the most impact on the response (Liu et al. 2010). On the other hand, the P-value implies that the process parameter is less than 0.05 times more significant than the component considered important. Table 3.6 contains the results of the variance analysis for the replies. According to Table 3.6, drilling force (T_h and T); hole quality (*IN*, *OUT*, C_e, R_a, and R_z) is a little error-prone and significant. The developed mathematical models for (T_h, T,

Table 3.6 Model Adequacy for the Response

Response	*Error*	*Contribution*	*F-value*	*P-value*
T_h	13.14%	86.86%	24.25	0.000
T	3.03%	96.97%	117.25	0.000
IN	13.64%	86.36%	23.22	0.000
OUT	8.62%	91.38%	38.85	0.000
R_a	3.04%	96.96%	116.89	0.000
R_z	6.07%	93.93%	56.76	0.000
C_e	1.51%	98.49%	239.57	0.000

IN, *OUT*, R_a, R_z, and C_e) are well satisfied as their characteristics (86.86%, 96.97%, 86.36%, 91.38%, 96.96%, 93.93%, and 98.49%) are very close to 1.

When the P-value for the model term is between 0.00 and 0.05, it is clear that the mean response of the machining output is affected. The model adequacy has been confirmed to be sufficient for further examination. The findings of this study's analysis of variance can predict the best possible environment for future research. Thrust, torque, delamination factor (*IN* & *OUT*), circularity error (C_e), and surface roughness (R_a & R_z) were found to be lower during drilling of GO/CF polymer nanocomposites.

Correlations between independent and dependent process factors can be generated using quadratic mathematical models. Eqs. (3.13–3.19) illustrate the regression analysis generated by the RSM method:

$$T_h = 25.62 - 0.00183 \times A + 0.3649 \times B + 13.04 \times C + 0.000004 \times A^2 - 0.000354 \times A^2 - 0.712 \times C^2 - 0.000110 \times A^2 - 0.00525 \times A \times C + 0.0254 \times B \times C \quad (3.13)$$

$$T = 0.0850 - 0.000048 \times A + 0.000406 \times B + 0.0037 \times C + 0.00250 \times C^2 + -0.000031 \times B \times C \quad (3.14)$$

$$IN = 0.9313 + 0.000028 \times A + 0.000497 \times B + 0.0407 \times C - 0.00709 \times C^2 - 0.000003 \times A \times C + 0.000009 \times B \times C \quad (3.15)$$

$$OUT = 0.9643 + 0.000035 \times A + 0.000563 \times B + 0.0086 \times C - 0.000001 \times B^2 - 0.00097 \times C^2 - 0.000004 \times A \times C + 0.000088 \times BC \quad (3.16)$$

$$R_a = 1.2897 + 0.000083 \times A + 0.001891 \times B + 0.08000 \times C \quad (3.17)$$

$$R_z = 3.829 - 0.000944 \times A + 0.01052 \times B + 0.2137 \times C \quad (3.18)$$

$$C_e = 0.3614 - 0.000143 \times A + 0.000966 \times B + 0.01263 \times C \quad (3.19)$$

3.5 Summary of the chapter

Laminated polymer nanocomposites are the subject of this chapter, which includes an investigation, statistical observations, and multi-criteria optimization experimental case study during drilling. To find the most cost-effective machining conditions, the process restrictions, such as drill speed (*A*), feed (*B*), and weight% of GO (*C*), were optimized. Because of this, CoCoSo-PCA has been able to identify the ideal parametric circumstances. The investigation results suggest that the proposed hybridization method may be implemented successfully. Based on the results collected, it is possible to draw the following conclusion:

- The previous optimization approach could not combine the competing machining performances into a single objective function K_i.
- In the case of machining responses, PCA was designed to give weights to them since, in most studies, the decision-makers assume that all responses have the same weight, which produces ambiguity, inaccuracy, and mistakes in the outcomes.

- Through CoCoSo-PCA, the optimal parametric combination is discovered as a useful strategy; test results show that the predicted values are in good accord with reality.
- The developed mathematical models for T_h, T, IN, OUT, R_a, R_z, and C_e are satisfactory as their characteristics are 86.86%, 96.97%, 86.36%, 91.38%, 96.96%, 93.93%, and 98.49%, respectively.
- The developed model has very less error for T_h (13.14%), T (3.03%), IN (13.64%), OUT (8.62%), R_a (3.04%), R_z (6.07%), and C_e (1.51%), which is effectively considerably.
- A hybridization approach can address concerns about offline process performance quality and manufacture.

It shows how to use robust hybridization to machine graphene oxide doped carbon fiber composites. Future research can study the effects of tool geometry and coated materials on the tool, chip structure, material removal, temperature behavior, and the protection of one health and safety. The suggested module is a thorough optimization tool that may be modified for different machining procedures. Further research into the machining environment of nanocomposites can include other parameters such as cutter diameter, tool coating, tip angle, coolant, etc. This chapter explains the weight assignment method, hybridization, and mathematical modeling of the claims presented in this chapter.

Conflict of response

The author has declared that there is no conflict of interest.

References

Abhishek, Kumar, Saurav Datta, and Siba Sankar Mahapatra. 2014. "Optimization of Thrust, Torque, Entry, and Exist Delamination Factor during Drilling of CFRP Composites." *International Journal of Advanced Manufacturing Technology* 76 (1–4): 401–16.

Arul, S., D. Samuel Raj, L. Vijayaraghavan, S. K. Malhotra, and R. Krishnamurthy. 2006. "Modeling and Optimization of Process Parameters for Defect Toleranced Drilling of GFRP Composites." *Materials and Manufacturing Processes* 21 (4): 357–65. doi:10.1080/10426910500411587.

Baraheni, Mohammad, and Saeid Amini. 2019. "Comprehensive Optimization of Process Parameters in Rotary Ultrasonic Drilling of CFRP Aimed at Minimizing Delamination." *International Journal of Lightweight Materials and Manufacture* 2 (4). Elsevier Ltd: 379–387.

Fu, Rao, Zhenyuan Jia, Fuji Wang, Yan Jin, Dan Sun, Lujia Yang, and De Cheng. 2018. "Drill-Exit Temperature Characteristics in Drilling of UD and MD CFRP Composites Based on Infrared Thermography." *International Journal of Machine Tools and Manufacture* 135. Elsevier Ltd: 24–37.

Gajrani, Kishor Kumar, Reddy, Rokkham Pavan Kumar, and Sankar, Mamilla Ravi. 2019. "Tribo-Mechanical, Surface Morphological Comparison of Un-Textured, Mechanical Micro-Textured (Mμt) And Coated-Mμt Cutting Tools During Machining." *Proceedings of the Institution of Mechanical Engineers, Part J: Journal of Engineering Tribology* 233 (1): 95–111. doi:10.1177/1350650118764975

Gajrani, Kishor Kumar, Vishwas Divse, and Suhas S. Joshi. (2021). "Burr Reduction in Drilling Titanium using Drills with Peripheral Slits." *Transactions of the Indian Institute of Metals*, 74 (5): 1155–1172.

Agrawal, Vipul, Kishor Kumar Gajrani, Rakesh G. Mote, Harish C. Barshilia, and Suhas S. Joshi. (2022). "Wear Analysis and Tool Life Modeling in Micro Drilling of Inconel 718 Superalloy". *Journal of Tribology, 144*(10), 101706.

Geier, Norbert, and Tibor Szalay. 2017. "Optimisation of Process Parameters for the Orbital and Conventional Drilling of Uni-Directional Carbon Fibre-Reinforced Polymers (UD-CFRP)." *Measurement: Journal of the International Measurement Confederation* 110: 319–334.

Gok, Arif. 2015. “A New Approach to Minimization of the Surface Roughness and Cutting Force via Fuzzy TOPSIS, Multi-Objective Grey Design and RSA.” *Measurement: Journal of the International Measurement Confederation* 70. Elsevier Ltd: 100–109.

Günay, Mustafa, and Tolga Meral. 2020. “Modelling and Multiresponse Optimization for Minimizing Burr Height, Thrust Force and Surface Roughness in Drilling of Ferritic Stainless Steel.” *Sadhana—Academy Proceedings in Engineering Sciences* 45 (1). doi:10.1007/s12046-020-01490-3

Islam, M. N., and A. Pramanik. 2016. “Comparison of Design of Experiments via Traditional and Taguchi Method.” *Journal of Advanced Manufacturing Systems* 15 (3): 151–160.

John, K. M., S. Thirumalai Kumaran, Rendi Kurniawan, Ki Moon Park, and J. H. Byeon. 2019. “Review on the Methodologies Adopted to Minimize the Material Damages in Drilling of Carbon Fiber Reinforced Plastic Composites.” *Journal of Reinforced Plastics and Composites* 38 (8): 351–68.

Karl Pearson F. R. S. 1901. “On Lines and Planes of Closest Fit to Systems of Points in Space.” *The London, Edinburgh, and Dublin Philosophical Magazine and Journal of Science* 2 (11): 559–572.

Kaybal, Halil Burak, Ali Ünüvar, Murat Koyunbakan, and Ahmet Avcı. 2019. “A Novelty Optimization Approach for Drilling of CFRP Nanocomposite Laminates.” *International Journal of Advanced Manufacturing Technology* 100 (9–12): 2995–3012.

Kharwar, Prakhar Kumar, Rajesh Kumar Verma, and Abhishek Singh. 2020. “Neural Network Modeling and Combined Compromise Solution (CoCoSo) Method for Optimization of Drilling Performances in Polymer Nanocomposites.” *Journal of Thermoplastic Composite Materials*, 1–28.

Köklü, Uğur, Murat Mayda, Sezer Morkavuk, Ahmet Avci, and Okan Demir. 2019. “Optimization and Prediction of Thrust Force, Vibration and Delamination in Drilling of Functionally Graded Composite Using Taguchi, ANOVA and ANN Analysis.” *Materials Research Express* 6 (8): 085335 (1–11).

König, W., and P. Graß. 1989. “Quality Definition and Assessment in Drilling of Fibre Reinforced Thermosets.” *CIRP Annals—Manufacturing Technology* 38 (1): 119–24. doi:10.1016/S0007-8506(07)62665-1

Kulkarni, Giridhar S., G. S. Shivashankar, R. Suresh, and N. G. Siddeshkumar. 2020. “Optimization of Drilling Parameters of GFRP with Liquid Silicone Rubber and Fine Silica Powder by Taguchi Approach.” *Silicon* 12 (7). Silicon: 1651–1666.

Kumar, Dhiraj, and K. K. Singh. 2019. “Investigation of Delamination and Surface Quality of Machined Holes in Drilling of Multiwalled Carbon Nanotube Doped Epoxy/Carbon Fiber Reinforced Polymer Nanocomposite.” *Proceedings of the Institution of Mechanical Engineers, Part L: Journal of Materials: Design and Applications* 233 (4): 647–663.

Kumar, Jogendra, and Rajesh Kumar Verma. 2021. “A Novel Methodology of Combined Compromise Solution and Principal Component Analysis (CoCoSo-PCA) for Machinability Investigation of Graphene Nanocomposites.” *CIRP Journal of Manufacturing Science and Technology* 33 (March). CIRP: 143–157. doi:10.1016/j.cirpj.2021.03.007

Kumar, Jogendra, Rajesh Kumar Verma, and Kishore Debnath. 2020. “A New Approach to Control the Delamination and Thrust Force during Drilling of Polymer Nanocomposites Reinforced by Graphene Oxide/Carbon Fiber.” *Composite Structures* 253 (August). Elsevier Ltd: 112786. doi:10.1016/j.compstruct.2020.112786

Kumar, Jogendra, Sujay Majumder, Arpan Kumar Mondal, and Rajesh Kumar Verma. 2022. “Influence of Rotation Speed, Transverse Speed, and Pin Length during Underwater Friction Stir Welding (UW-FSW) on Aluminum AA6063: A Novel Criterion for Parametric Control.” *International Journal of Lightweight Materials and Manufacture* 5 (3). The Authors: 295–305. doi:10.1016/j.ijlmm.2022.03.001

Kumar, Jogendra, Rajesh K. Verma, Arpan K. Mondal, and Vijay K. Singh. 2021. “A Hybrid Optimization Technique to Control the Machining Performance of Graphene/Carbon/Polymer (Epoxy) Nanocomposites.” *Polymers and Polymer Composites* 29 (9_suppl): S1168–S1180. doi:10.1177/09673911211046789

Kumar, Jogendra, and Rajesh Kumar Verma. 2021a. “A New Criterion for Drilling Machinability Evaluation of Nanocomposites Modified by Graphene/Carbon Fiber Epoxy Matrix and Optimization Using Combined Compromise Solution.” *Surface Review and Letters* 28 (9). doi:10.1142/S0218625X21500827

Kumar, Jogendra, and Rajesh Kumar Verma. 2021b. "Experimental Investigation for Machinability Aspects of Graphene Oxide/Carbon Fiber Reinforced Polymer Nanocomposites and Predictive Modeling Using Hybrid Approach." *Defence Technology* 17 (5). Elsevier Ltd: 1671–1686. doi:10.1016/j.dt.2020.09.009

Kumar, Jogendra, and Rajesh Kumar Verma. 2021c. "Multiple Response Optimization in Machining (Milling) of Graphene Oxide-Doped Epoxy/Cfrp Composite Using CoCoSo-PCA: A Novel Hybridization Approach." *Journal of Advanced Manufacturing Systems* 20 (2): 423–446. doi:10.1142/S0219686721500207

Kumar, Jogendra, Rajesh Kumar Verma, and Prateek Khare. 2021. "Graphene-Functionalized Carbon/Glass Fiber Reinforced Polymer Nanocomposites: Fabrication and Characterization for Manufacturing Applications." In *Handbook of Functionalized Nanomaterials*, edited by Chaudhery Mustansar Hussain and Vineet Kumar, 57–78. Netherlands: Elsevier. doi:10.1016/B978-0-12-822415-1.00011-1

Kumar, Jogendra, Rajesh Kumar Verma, and Arpan Kumar Mondal. 2020. "Predictive Modeling and Machining Performance Optimization during Drilling of Polymer Nanocomposites Reinforced by Graphene Oxide/Carbon Fiber." *Archive of Mechanical Engineering* 67 (2): 229–258. doi:10.24425/ame.2020.131692

Kumar, Kuldeep, Jogendra Kumar, Vijay Kumar Singh, and Rajesh Kumar Verma. 2021. "An Integrated Module for Machinability Evaluation and Correlated Response Optimization during Milling of Carbon Nanotube/Glass Fiber Modified Polymer Composites." *Multiscale and Multidisciplinary Modeling, Experiments and Design*, no. 0123456789. Springer International Publishing. doi:10.1007/s41939-021-00099-1

Kyratsis, Panagiotis, Angelos P. Markopoulos, Nikolaos Efkolidis, Vasileios Maliagkas, and Konstantinos Kakoulis. 2018. "Prediction of Thrust Force and Cutting Torque in Drilling Based on the Response Surface Methodology." *Machines* 6 (2): 24 (1–12).

Lachaud, Frédéric, Robert Piquet, Francis Collombet, and Laurent Surcin. 2001. "Drilling of Composite Structures." *Composite Structures* 52 (3–4): 511–516. doi:10.1016/S0263-8223(01)00040-X

Liu, Yung Tien, Wei Che Chang, and Yutaka Yamagata. 2010. "A Study on Optimal Compensation Cutting for an Aspheric Surface Using the Taguchi Method." *CIRP Journal of Manufacturing Science and Technology* 3 (1). CIRP: 40–48.

Mi, Xiaomei, and Huchang Liao. 2020. "Renewable Energy Investments by a Combined Compromise Solution Method with Stochastic Information." *Journal of Cleaner Production* 276. Elsevier Ltd: 123351 (1–13). doi:10.1016/j.jclepro.2020.123351

Mohan, N. S., A. Ramachandra, and S. M. Kulkarni. 2005. "Influence of Process Parameters on Cutting Force and Torque during Drilling of Glass-Fiber Polyester Reinforced Composites." *Composite Structures* 71 (3–4): 407–413.

Ogawa, K., E. Aoyama, H. Inoue, T. Hirogaki, H. Nobe, Y. Kitahara, T. Katayama, and M. Gunjima. 1997. "Investigation on Cutting Mechanism in Small Diameter Drilling for GFRP (Thrust Force and Surface Roughness at Drilled Hole Wall)." *Composite Structures* 38 (1–4): 343–350.

Palanikumar, K. 2011. "Experimental Investigation and Optimisation in Drilling of GFRP Composites." *Measurement: Journal of the International Measurement Confederation* 44 (10): 2138–2148.

Palanikumar, K., L. Karunamoorthy, and R. Karthikeyan. 2006. "Multiple Performance Optimization of Machining Parameters on the Machining of GFRP Composites Using Carbide (K10) Tool." *Materials and Manufacturing Processes* 21 (8): 846–852.

Qiu, Xinyi, Pengnan Li, Changping Li, Qiulin Niu, Anhua Chen, Puren Ouyang, and Tae Jo Ko. 2018. "Study on Chisel Edge Drilling Behavior and Step Drill Structure on Delamination in Drilling CFRP." *Composite Structures* 203: 404–413.

Ragunath, S., C. Velmurugan, and T. Kannan. 2017. "Optimization of Drilling Delamination Behavior of GFRP/Clay Nano-Composites Using RSM and GRA Methods." *Fibers and Polymers* 18 (12): 2400–2409.

Rajkumar, D., P. Ranjithkumar, and C. Sathiya Narayanan. 2017. "Optimization of Machining Parameters on Microdrilling of CFRP Composites by Taguchi Based Desirability Function Analysis." *Indian Journal of Engineering and Materials Sciences* 24 (5): 331–338.

Sampath Kumar, T., S. Balasivanandha Prabu, and T. Sorna Kumar. 2017. "Comparative Evaluation of Performances of TiAlN-, AlCrN- and AlCrN/TiAlN-Coated Carbide Cutting Tools and Uncoated Carbide Cutting Tools on Turning EN24 Alloy Steel." *Journal of Advanced Manufacturing Systems* 16 (3): 237–261.

Sorrentino, L., S. Turchetta, and C. Bellini. 2017. "In Process Monitoring of Cutting Temperature during the Drilling of FRP Laminate." *Composite Structures* 168: 549–561.

Sorrentino, L., S. Turchetta, and C. Bellini. 2018. "A New Method to Reduce Delaminations during Drilling of FRP Laminates by Feed Rate Control." *Composite Structures* 186: 154–164.

Sridharan, V., T. Raja, and N. Muthukrishnan. 2016. "Study of the Effect of Matrix, Fibre Treatment and Graphene on Delamination by Drilling Jute/Epoxy Nanohybrid Composite." *Arabian Journal for Science and Engineering* 41 (5): 1883–1894.

Tsao, C. C. 2008. "Thrust Force and Delamination of Core-Saw Drill during Drilling of Carbon Fiber Reinforced Plastics (CFRP)." *International Journal of Advanced Manufacturing Technology* 37 (1–2): 23–28.

Uzun, Gültekin. 2019. "Analysis of Grey Relational Method of the Effects on Machinability Performance on Austempered Vermicular Graphite Cast Irons." *Measurement: Journal of the International Measurement Confederation* 142: 122–130.

Vinayagamoorthy, R. 2018. "A Review on the Machining of Fiber-Reinforced Polymeric Laminates." *Journal of Reinforced Plastics and Composites* 37 (1): 49–59. doi:10.1177/0731684417731530

Wang, Xin, L. J. Wang, and J. P. Tao. 2004. "Investigation on Thrust in Vibration Drilling of Fiber-Reinforced Plastics." *Journal of Materials Processing Technology* 148 (2): 239–244.

Wen, Zhi, Huchang Liao, Edmundas Kazimieras Zavadskas, and Abdullah Al-Barakati. 2019. "Selection Third-Party Logistics Service Providers in Supply Chain Finance by a Hesitant Fuzzy Linguistic Combined Compromise Solution Method." *Economic Research-Ekonomska Istrazivanja* 32 (1). Routledge: 4033–4058.

Xu, Jinyang, Qinglong An, Xiaojiang Cai, and Ming Chen. 2013. "Drilling Machinability Evaluation on New Developed High-Strength T800S/250F CFRP Laminates." *International Journal of Precision Engineering and Manufacturing* 14 (10): 1687–1696.

Xu, Jinyang, Xianghui Huang, Ming Chen, and J. Paulo Davim. 2020. "Drilling Characteristics of Carbon/Epoxy and Carbon/Polyimide Composites." *Materials and Manufacturing Processes* 35 (15). Taylor & Francis: 1732–1740. doi:10.1080/10426914.2020.1784935

Xu, Jinyang, Chao Li, Sipei Mi, Qinglong An, and Ming Chen. 2018. "Study of Drilling-Induced Defects for CFRP Composites Using New Criteria." *Composite Structures* 201: 1076–1087.

Yazdani, Morteza, Pascale Zarate, Edmundas Kazimieras Zavadskas, and Zenonas Turskis. 2019. "A Combined Compromise Solution (CoCoSo) Method for Multi-Criteria Decision-Making Problems." *Management Decision* 57 (9): 2501–2519. doi:10.1108/MD-05-2017-0458

Chapter 4

Computational fluid dynamics analysis of MQL mist droplets characteristics with various geometrical nozzle orifice

Karuturi Manohar Sai Krishna, Satyam Dwivedi, Rajendra Soni, and Kishor Kumar Gajrani

Contents

Abbreviations

2D: 2-Dimensional
CFD: Computational Fluid Dynamics
DPM: Discrete Phase Model
MQL: Minimum Quantity Lubrication

4.1 Introduction

The efficiency of a machining process is greatly dependent on turning speed. During turning at high speed and feed rate, high heat is generated at the tool-chip interface, causing a rise in temperature, which leads to tool wear. Tool wear starts at a relatively faster rate due to the break-in wear caused by attrition and micro-chipping at the sharp cutting edges. Tool wear affects the surface roughness of machined parts, which is a significant design specification. The surface roughness of fabricated components considerably influences properties such as wear resistance, cleanability, assembly tolerances, coefficient of friction, wear rate, and corrosion resistance.

Usually, flood coolants are used to dissipate generated heat and reduce tool-chip interface temperature. However, flood coolants using petroleum-based mineral oil have long-term

DOI: 10.1201/9781003291961-5

detrimental effects on the environment [1]. Therefore, nowadays, minimum quantity lubrication (MQL) techniques are being used with environmentally friendly vegetable-based cutting fluids to avoid negative effects on the environment [2–4].

MQL is rapidly being used in metal cutting machining and has established itself as a viable alternative to traditional flood coolant in several areas. Unlike flood lubrication, minimal quantity lubrication requires only a few drops of lubricant in machining (about 5 ml to 50 ml per hour) [5]. Today, the enormous cost-saving potential resulting from doing almost entirely without metalworking fluids in machining production is recognized and implemented by many companies, primarily in the automotive industry. Compared to water-mixed metalworking fluids, MQL has several advantages. A significant benefit is the much-improved skin compatibility of the operators working in a machining environment for a prolonged time. Also, workpieces and chips are nearly dry due to the substantial drop in lubricant quantities. It significantly decreases the health risks posed by metalworking fluid emissions in the air and on employees' skin at work. Metalworking fluids do not spread all over the machine, resulting in a cleaner working environment. With the MQL technique, the costs associated with traditional flood lubrication (for example, maintenance, inspection, preparation, and disposal of metalworking fluids) are no longer a concern [6–8]. MQL systems have various components, such as fluid nozzle, pressure regulator, oil tank, air compressor, and solenoid valve. Of these, MQL nozzle's shapes and sizes significantly affect the lubrication characteristics.

The nozzle is the final component before the point of application, and it determines the shape and size of the spray pattern. Even the most exact lubricant applicator is useless if it is not pointed precisely at the target. Therefore, the nozzle is a key component of the entire system. Depending on the application, MQL nozzles have a variety of designs to control lubricant spray pattern, shape, and flow rate. Excessive lubricant will adversely affect the chip formation and be wasted, whereas inadequate lubricant oils reduce the tool's life. Moreover, although the oil mist of MQL has been proven to enhance the machining quality, the tiny droplets of oil mist also easily evaporate, thus weakening the lubrication effects during the machining process. Moreover, the appropriate droplet size of oil was also proven to increase the penetration ability of the oil mist through the cutting zone. Since the surface roughness in-between tool-chip and tool-work interfaces are relatively high, only oil with a smaller droplet size can successfully pass through the cutting zone [9]. Hence, the size of lubricant oil droplets under the MQL machining system must be fundamentally investigated to clarify the lubricating mechanism in the cutting zone. Nozzle angle, nozzle distance, oil flow rate, and air pressure are examples of factors affecting the droplet size of lubricant oil [3].

Nozzle angle with respect to tool feed direction is important because the droplets of oil mist that adhere to the insert are thrown away from the insert surface while the tool rotates to the contact zone. Different nozzle angles can affect the spray impingement capability and send the oil mists to drop into the target cutting zone and penetrate the tool edges. However, if the nozzle angle is not optimum, the other positions cannot allow the MQL lubrication to reach the cutting zone effectively. Also, the elevation of the nozzle and stand-off distance affect the overall efficiency of the MQL lubrication [3]. Apart from these parameters, nozzle oil, air orifice diameter, and the ratio of oil to air orifice significantly affect the MQL droplet velocity, droplet diameter, and tool-chip interface temperature [10–12].

In this work, the effect of oil and air orifice diameter in the nozzle and the ratio of oil-to-air orifice diameter was studied in terms of MQL droplet velocity and droplet diameter at various air pressure. A total of four different ratios at five different geometrical pairs are numerically studied, and their responses are presented and discussed for a better understanding of flow parameters, like particle velocities and diameter, under varying flow rates and pressure using ANSYS FLUENT® version 2021 R2.

4.2 Design and modelling of MQL nozzle for different orifice ratios

A discrete phase model (DPM) is used to atomize in turbulent conditions. The Euler Lagrange approach is used, where air is considered as the continuous phase and oil is considered as a discrete phase. Turbulence with viscosity constraints is described using the *k*-ε model. In this MQL model, oil makes up only 10% of the total volume of the mixture. Air is assumed as an ideal gas, implying that there will be no heat transfer between the air phase and the oil phase. To analyze the droplet dynamics, an injection model is created with DPM. The rate of injection of particles was kept constant at 1 μs to maintain the continuity of flow. Oil droplets are assumed to be fully spherical as per the spherical drag rule. For the breakup of particles, the Wave breakup model was used. The Rosin Rammler model was used to determine the droplet distribution. The solutions of the simulations were considered to be converging if the residues of all the variables fell below 10^{-4}. Depending upon the parameters used in the simulations, appropriate under-relaxation factors were chosen to tackle the non-linearity of the equations to get the desired convergence.

To study the effect of MQL nozzle orifices, they were designed in such a way that the ratio of oil orifice-to-air orifice is in the range of 1.14–1.25. Also, the same ratio of 1.25 was studied twice by changing the oil orifice and air orifice to study the effect of individual oil and air orifice parameters in MQL nozzles. The orifice ratio range corresponds to the normal range of 1.1–1.3 for air-atomizing nozzles. A total of four different ratios by using five different configurations of oil and air orifice were designed and modeled in ANSYS FLUENT® as listed in Table 4.1. The design of a two-dimensional (2D) nozzle model with 1.16 oil-to-air orifice diameter ratio, using details provided by the Spraying Systems Co., USA, is shown in Figure 4.1. This modeled nozzle has a 0.6 mm air aperture and a 0.7 mm liquid opening.

4.2.1 Meshing and boundary conditions

In Gambit 2.3.16, 2D views of the model shown in Figure 4.1 were used to construct a mesh. In all regions, the conventional quadrilateral element type was used for meshing due to its less computational time. Figure 4.2 shows the MQL atomizer nozzle domain within the atmosphere virtual box with different regions.

Table 4.1 Chosen Dimensions for Oil and Air Orifice Diameter and the Corresponding Ratio

S. No.	*Oil orifice diameter*	*Air orifice diameter*	*Ratio*
1	0.8	0.7	1.14
2	0.7	0.6	1.16
3	0.6	0.5	1.20
4	0.5	0.4	1.25a
5	0.75	0.6	1.25b

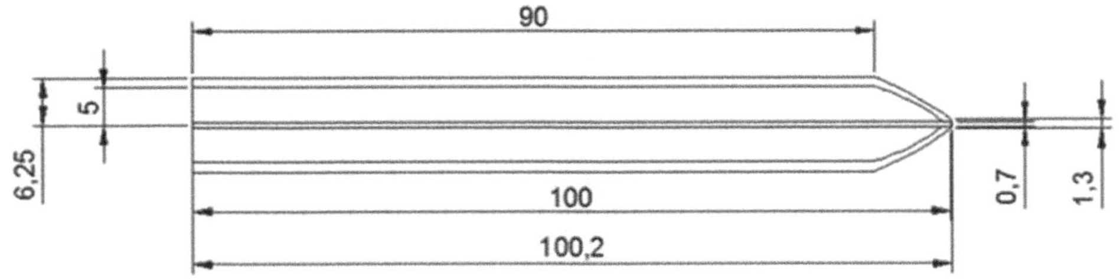

Figure 4.1 2D model of a nozzle with orifice ratio of 1.16.

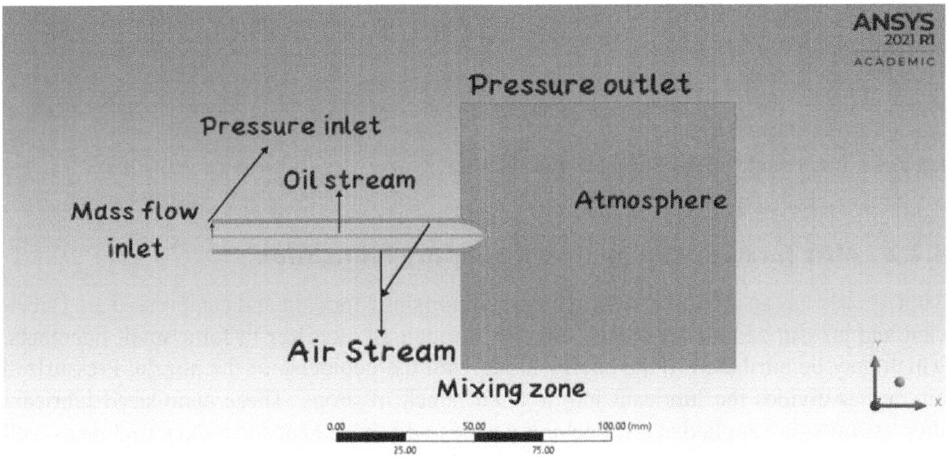

Figure 4.2 Design of the minimum quantity lubrication atomizer nozzle domain within the atmosphere virtual box with different regions.

Figure 4.3 shows the meshing of the entire domain. Fine mesh element size was chosen near the nozzle opening, and coarse mesh was used for the virtual atmosphere to reduce computational time. The element sizes of oil and air streams were around 0.5 mm, and particle properties were examined in the atmospheric domain having a finer resolution of 0.1 mm. As mass transfer occurs and severe turbulence is envisaged, the element size in the mixing zone is reduced to approximately 0.05 mm.

The boundary conditions for various regions were determined depending on the constant inputs provided for that region. As the air was supplied at constant pressure (2, 3, and 4 bar) from that boundary, the air stream input boundary was designated as a "pressure inlet". Similarly, the oil stream input and boundaries of the wall were selected as mass flow inlet and pressure outlet, respectively. Different domains, such as oil/air streams, mixing area, and nozzle outlet, were separated using the "interior" boundary condition.

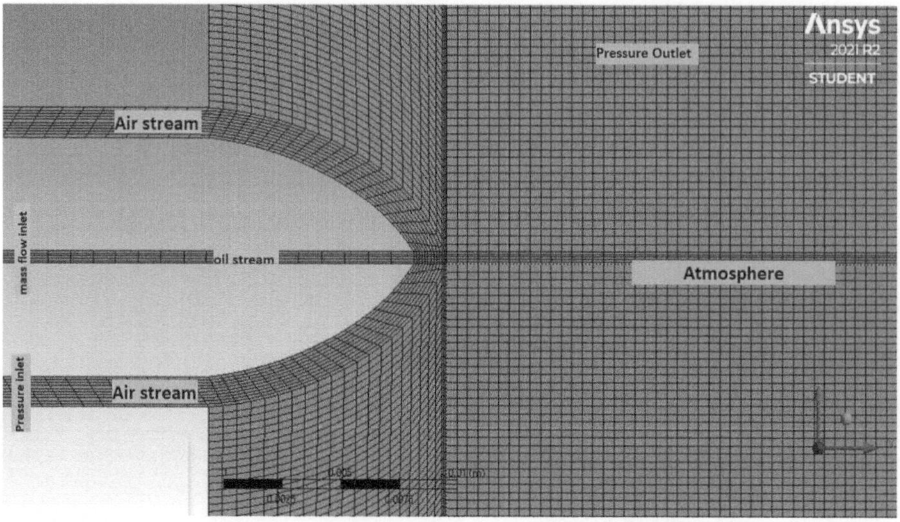

Figure 4.3 Mesh model of the designed MQL nozzle.

4.2.2 *Mist formation in minimum quantity lubrication*

MQL nozzle has an atomizer to form a mist comprising lubricant and compressed air. Lubricant and pressurized air are forced together through an atomizer to form small ligaments, which may be attributed to the energy along with the geometry of the nozzle. Pressurized air further divides the lubricant into a small bunch of drops. These nano-sized lubricant droplets/mists are applied at the machining zone to dissipate heat and reduce friction as well as temperature. As the amount of lubricant is minimum, this technique is known as MQL.

4.3 Results and discussions

4.3.1 *Effect of air pressure and oil mass flow rate on MQL mist droplet velocity*

MQL mist droplet velocity majorly depends on the diameter of the nozzle orifice. Mist droplet velocity directly impacts drift characteristics, interface friction, and surface roughness of the machined material. Mist droplets with different velocities lead to different scenarios. Figures 4.4–4.6 show the magnitude of MQL mist droplet velocity at varying lubricant mass flow rates and air pressures. Figure 4.4 (a-c) shows the mist droplet velocity at varying air pressures of 2, 3, and 4 bar with an oil mass flow rate of 60 mL/h and a constant nozzle orifice ratio of 1.16. Similarly, Figure 4.5 (a-c) and Figure 4.6 (a-c) show the mist droplet velocity at varying air pressures of 2, 3, and 4 bar with oil mass flow rates of 80 mL/h and 100 mL/h and a constant nozzle orifice ratio of 1.16.

From Figure 4.4 (a-c), it was observed that the mist droplet velocity increases with an increase in air pressure. At the air pressure of 2 bar and mass flow rate of 60 mL/h, the maximum mist droplet velocity was 2.81×10^2 m/s (see Figure 4.4 (a)). However, at the air pressure of 4 bar, the maximum observed droplet velocity was 3.90×10^2 m/s (see Figure 4.4 (c)). Figure 4.7 shows the effect of varying oil mass flow rate and air pressures at a constant nozzle orifice ratio of 1.16 on MQL mist droplet velocity.

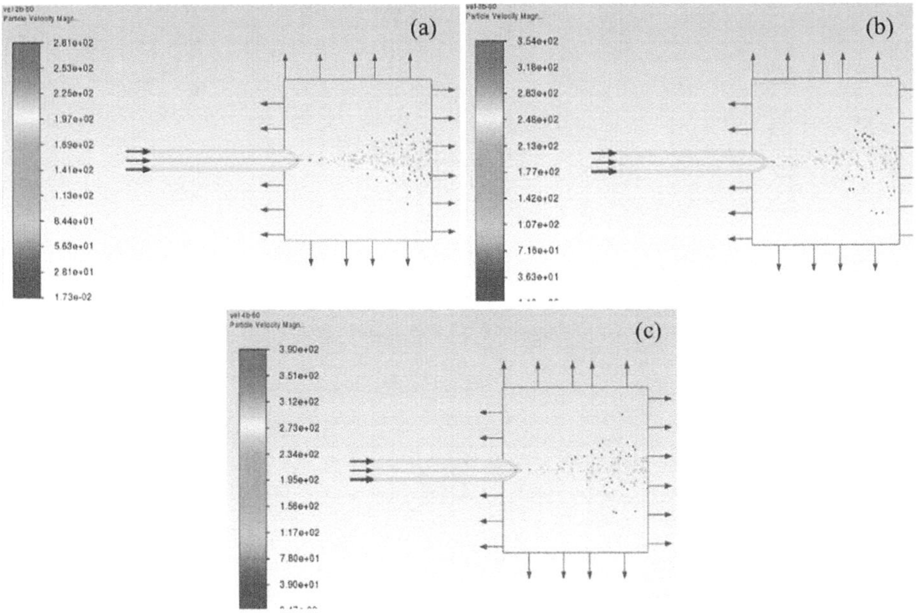

Figure 4.4 Particle track contours showing particle velocity of oil droplets at a constant mass flow rate of 60 mL/h as well as orifice ratio of 1.16 and varying air pressures of (a) 2 bar, (b) 3 bar, and (c) 4 bar.

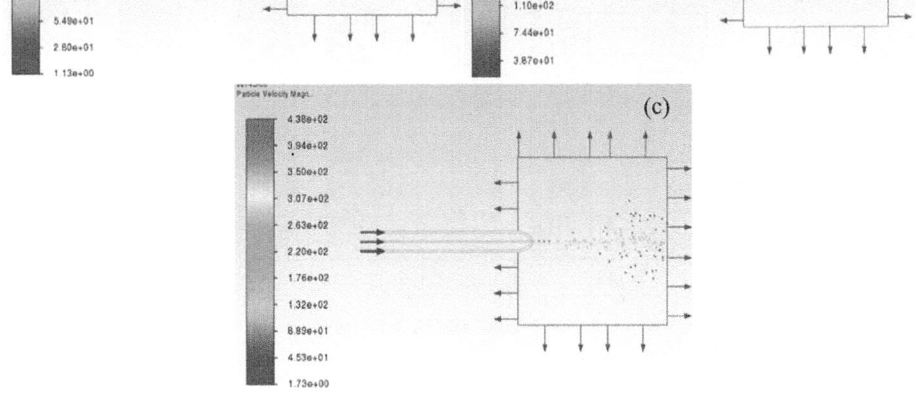

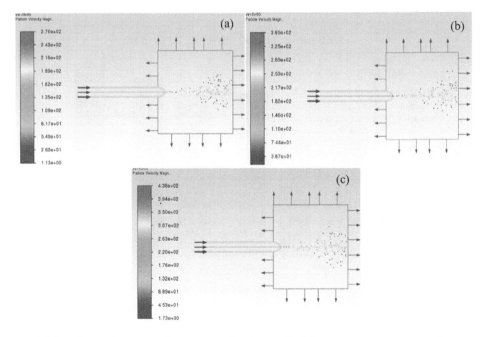

Figure 4.5 Particle track contours showing particle velocity of oil droplets at a constant mass flow rate of 80 mL/h as well as orifice ratio of 1.16 and varying air pressures of (a) 2 bar, (b) 3 bar, and (c) 4 bar.

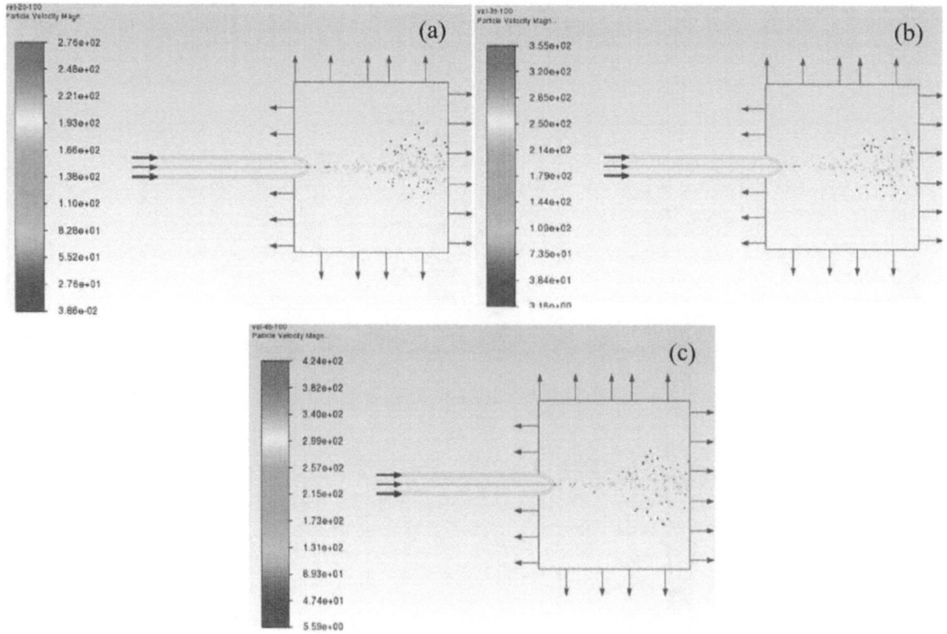

Figure 4.6 Particle track contours showing particle velocity of oil droplets at a constant mass flow rate of 100 mL/h as well as orifice ratio of 1.16 and varying air pressures of (a) 2 bar, (b) 3 bar, and (c) 4 bar.

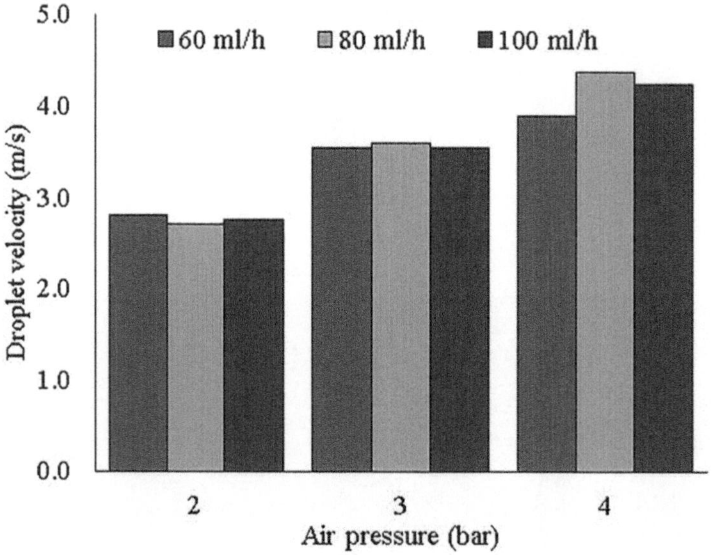

Figure 4.7 Influence on oil droplet velocity at varying oil mass flow rate and air pressures at a constant nozzle orifice ratio of 1.16.

From Figures 4.4–4.7, it was observed that in most cases, with an increase in oil mass flow rate, the oil droplet velocity reduces. However, with an increase in air pressure, the oil droplet velocity increases. At low air pressure, oil gets more time for atomization resulting in low droplet velocity. However, at a higher air pressure, fine atomization of oil droplets takes place, which leads to higher droplet velocity. Usually, low mist droplet velocity leads to a larger droplet diameter that faces difficulties in penetrating in-between the tool-chip interface. In contrast, high mist droplet velocity leads to a small droplet diameter, which may bounce off from the cutting zone. In both cases, machining performance may not be optimum. Therefore, optimum droplet velocity and droplet diameter are important to study.

4.3.2 Effect of nozzle orifice ratio on MQL mist droplet diameter

Figure 4.8 (a-c) shows the MQL mist droplet diameter at varying air pressures of 2, 3, and 4 bar with an oil mass flow rate of 60 mL/h and a constant nozzle orifice ratio of 1.16. Figures 4.9 (a-c) and 4.10 (a-c) show the mist droplet diameter at varying air pressures of 2, 3, and 4 bar, with oil mass flow rates of 80 mL/h and 100 mL/h and a constant nozzle orifice ratio of 1.16.

From Figure 4.8 (a-c), it was observed that the mist droplet diameter reduces with an increase in air pressure. At the air pressure of 2 bar and mass flow rate of 60 mL/h, the maximum mist droplet diameter was 3.81 μm (see Figure 4.8 (a)). However, at the air pressure of 4 bar, the maximum observed droplet velocity was 2.42 μm (see Figure 4.8 (c)). Figure 4.11 shows the effects of varying oil mass flow rates and air pressures at a constant nozzle orifice ratio of 1.16 on average MQL mist droplet diameter.

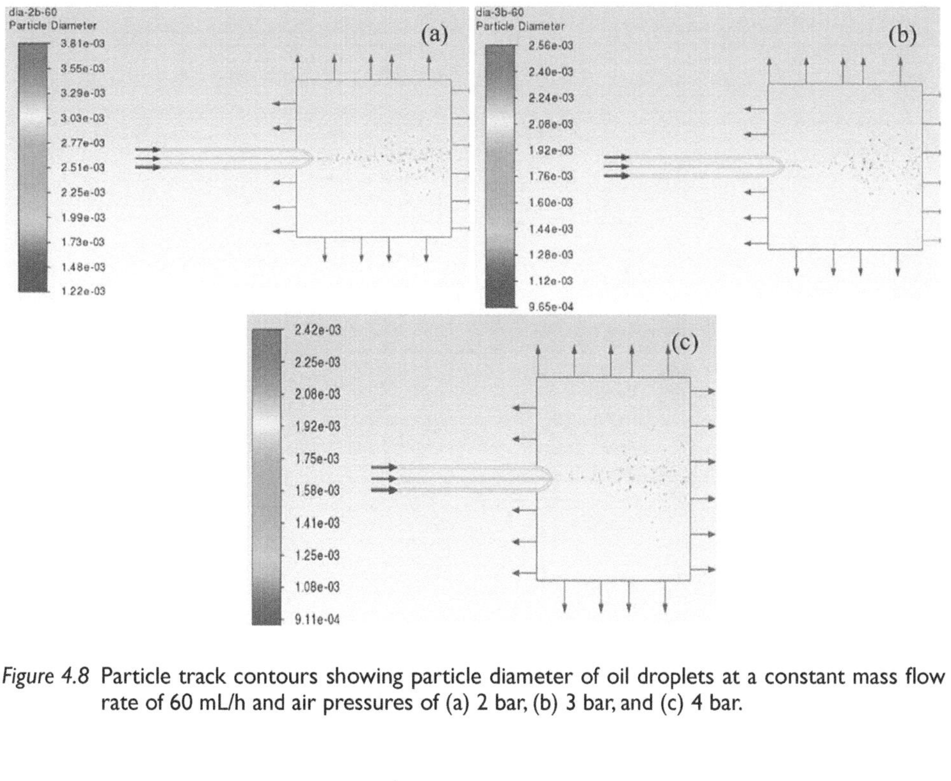

Figure 4.8 Particle track contours showing particle diameter of oil droplets at a constant mass flow rate of 60 mL/h and air pressures of (a) 2 bar, (b) 3 bar, and (c) 4 bar.

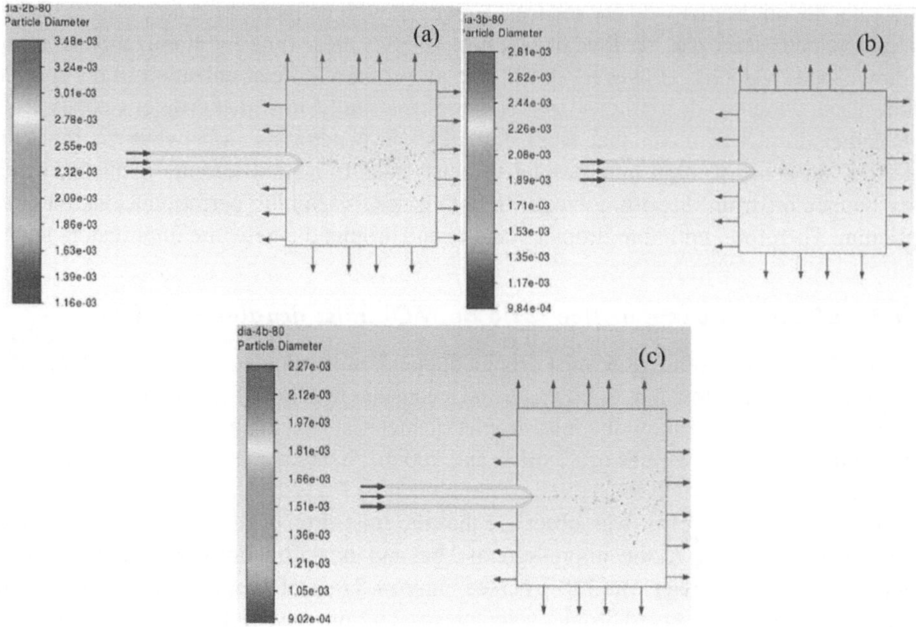

Figure 4.9 Particle track contours showing particle diameter of oil droplets at a constant mass flow rate of 80 mL/h and air pressures of (a) 2 bar, (b) 3 bar, and (c) 4 bar.

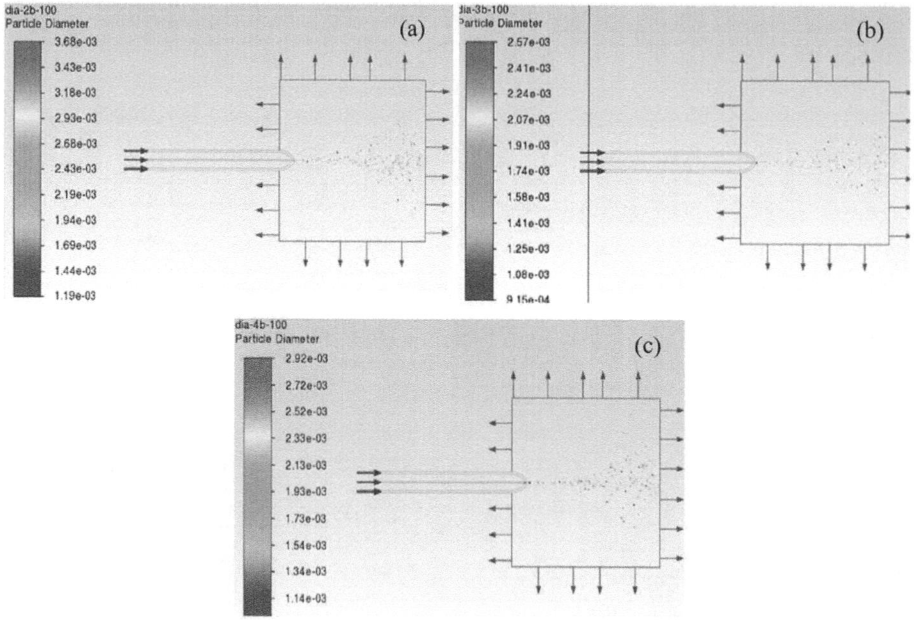

Figure 4.10 Particle track contours showing particle diameter of oil droplets at a constant mass flow rate of 100 mL/h and air pressures of (a) 2 bar, (b) 3 bar, and (c) 4 bar.

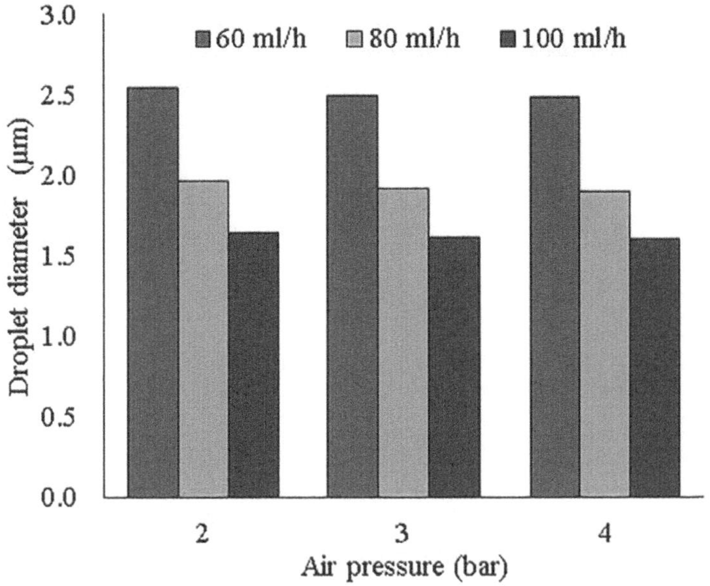

Figure 4.11 Effect of air pressure and oil mass flow rate on mist droplet diameter at a constant nozzle orifice ratio of 1.16.

It is observed that the air pressure significantly affects the mist droplet diameter, which, in turn, affects the volume-to-surface area ratio. Small droplet diameters result in poor penetration in-between the tool-chip interface, and high droplet diameters are difficult to carry up to the cutting zone. At the same time, the oil mass flow rate has an almost negligible effect in the tested range. For better machining efficiency, the objective is to optimize the mist oil droplet diameter as well as mist droplet velocity.

4.3.3 Effect of nozzle orifice ratio on MQL mist droplet velocity and droplet diameter

Figure 4.12 shows the MQL mist droplet velocity at a constant air pressure of 2 bar with four different nozzle orifice ratios and two different geometrical configurations for a ratio of 1.25. Figure 4.13 shows the contour of particle mist droplet velocity at an air pressure of 2 bar and an oil mass flow rate of 60 mL/h at five different nozzle orifice ratios.

Figure 4.14 shows the MQL mist droplet diameter at a constant air pressure of 2 bar with four different nozzle orifice ratios and two different geometrical configurations for a ratio of 1.25. Figure 4.15 shows the contour of particle mist droplet diameter at an air pressure of 2 bar and an oil mass flow rate of 60 mL/h at five different nozzle orifice ratios.

From Figures 4.12–4.15, it was observed that the nozzle orifice ratio significantly affects the MQL mist droplet velocity as well as droplet diameter. With an increase in the nozzle orifice ratio from 1.14 to 1.25, MQL mist droplet velocity reduces. At the same time, with an increase in nozzle orifice ratios, MQL mist droplet diameter increases. For the nozzle orifice

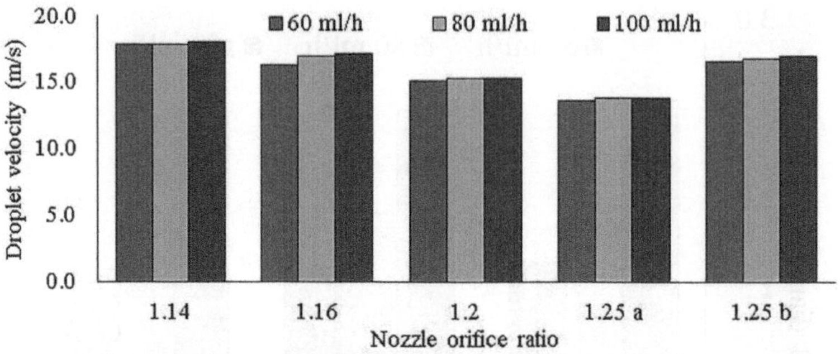

Figure 4.12 Influence of oil mass flow rate and nozzle orifice ratios on mist droplet velocity at a constant air pressure of 2 bar.

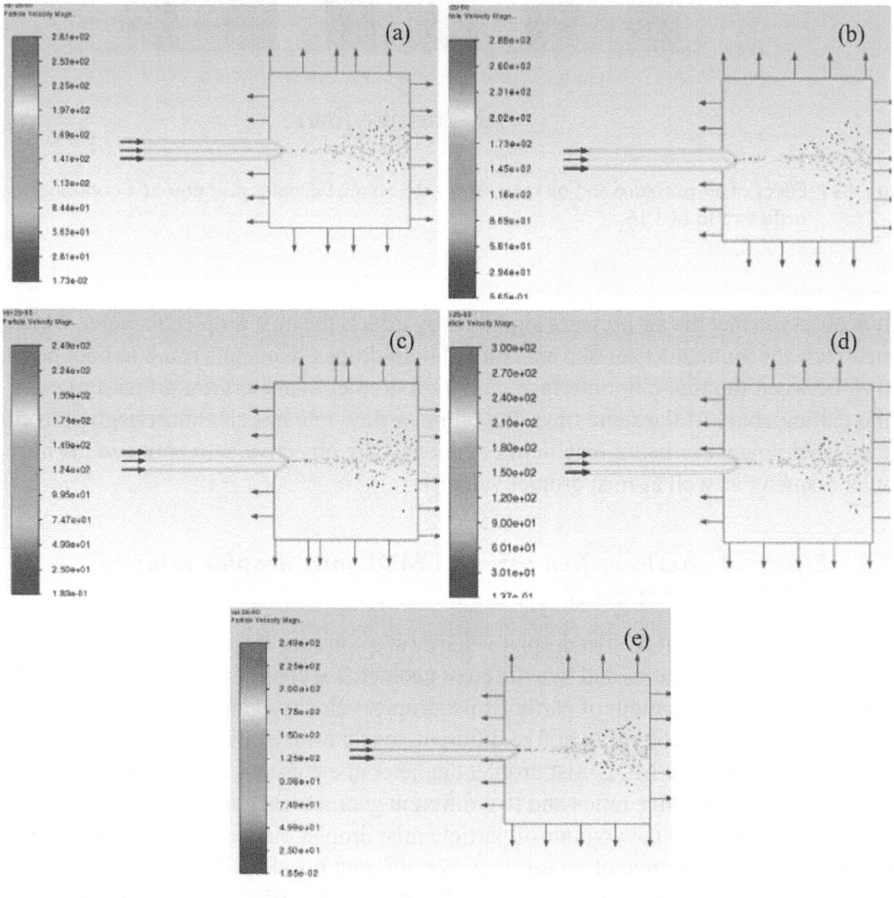

Figure 4.13 Particle track contours showing the velocity of oil droplets at a constant air pressure of 2 bar, oil mass flow rate of 60 mL/h, and varying nozzle orifice ratios of (a) 1.14, (b) 1.16, (c) 1.2, (d) 1.25a, and (e) 1.25b.

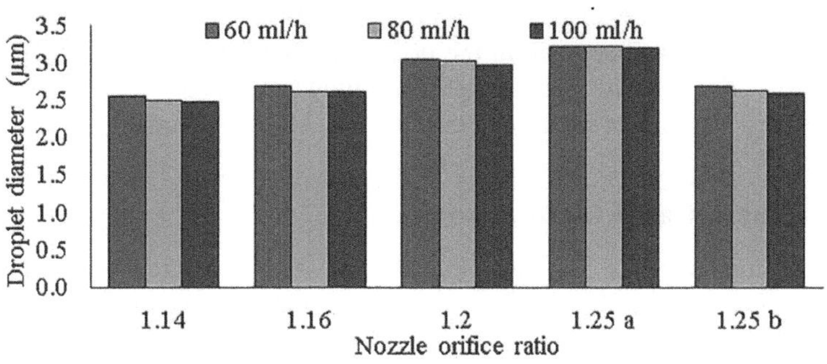

Figure 4.14 Influence of oil mass flow rate and nozzle orifice ratios on mist droplet diameter at a constant air pressure of 2 bar.

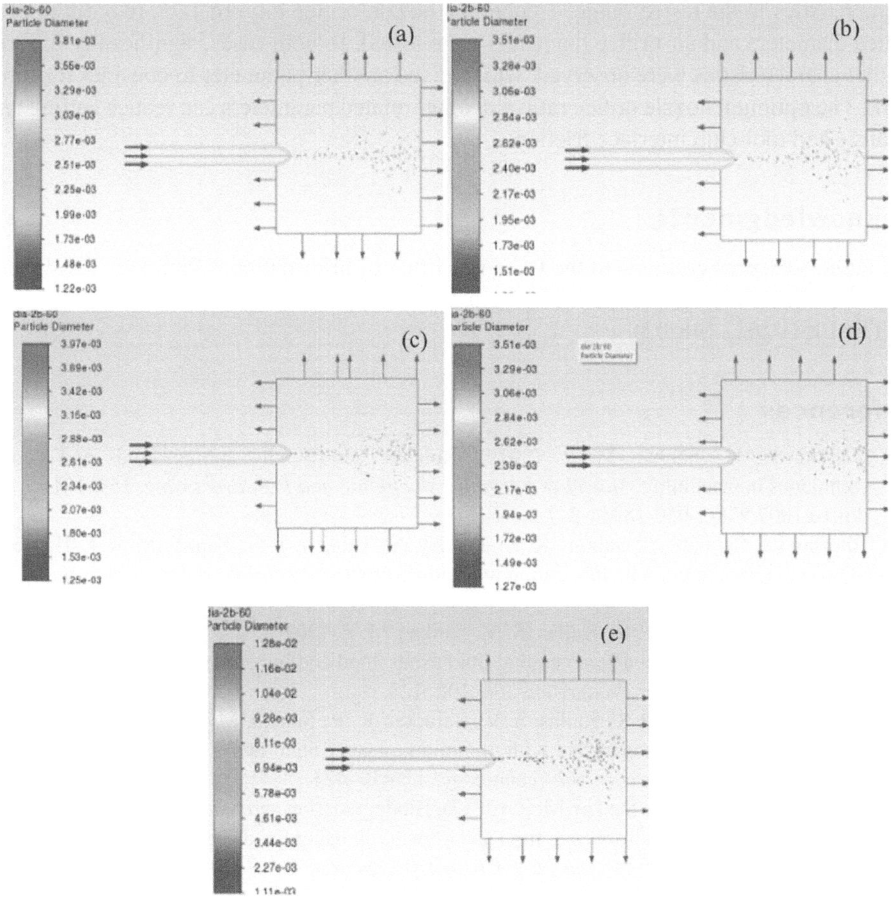

Figure 4.15 Particle track contours showing the diameter of oil droplets at a constant air pressure of 2 bar, oil mass flow rate of 60 mL/h, and varying nozzle orifice ratios of (a) 1.14, (b) 1.16, (c) 1.2, (d) 1.25a, and (e) 1.25b.

ratio of 1.25, two different oil orifice diameters and air orifice diameters were chosen, as listed in Table 4.1. In both cases, the MQL mist droplet velocity and droplet diameter were different, which shows that apart from the orifice ratios, oil and air orifice diameters also affect the characteristics of the mist droplet. This study can help select an optimum oil mass flow rate, air pressure, and nozzle orifice ratio for any MQL machining.

4.4 Conclusion and future work

In this chapter, computational fluid dynamic analysis of MQL mist droplets was carried out with varying air pressures, oil mass flow rates, and nozzle orifice ratios. A discrete phase model was developed in ANSYS FLUENT®, and meshing was done using Gambit 2.3.16. The effect of air pressure, oil mass flow rate, and nozzle orifice diameter on MQL mist droplet velocity and droplet diameter was studied using computational fluid dynamics. It was observed that the air pressure and nozzle orifice ratio significantly affect the characteristics of oil droplets. At the same time, the oil mass flow rate shows a negligible effect on droplet characteristics in the tested range. To obtain a nozzle orifice ratio of 1.25, two different oil orifice diameters and air orifice diameters were tested. In both cases, significantly different droplet characteristics were observed, which gives another parameter to consider for future work. The optimum nozzle orifice ratio and other related parameters can reduce cutting temperature and tool-chip interface friction.

Acknowledgments

The authors express gratitude to the Indian Institute of Information Technology, Design and Manufacturing, Kancheepuram, India for financial support through the Institute Seed Grant (IIITDM/ISG/2022/ME/02).

References

1. Gajrani, K. K., Sankar, M. R. (2020). Role of eco-friendly cutting fluids and cooling techniques in machining. *Materials Forming, Machining and Post Processing*, 159–181. doi:10.1007/978-3-030-18854-2_7
2. Bhuyan, M., Sarmah, A., Gajrani, K. K., Pandey, A., Thulkar, T. G., Sankar, M. R. (2018). State of art on minimum quantity lubrication in grinding process. *Materials Today: Proceedings*, *5*(9), 19638–19647.
3. Gajrani, K. K., Suvin, P. S., Kailas, S. V., Sankar, M. R. (2019). Hard machining performance of indigenously developed green cutting fluid using flood cooling and minimum quantity cutting fluid. *Journal of Cleaner Production*, *206*, 108–123.
4. Gajrani, K. K., Suvin, P. S., Kailas, S. V., Rajurkar, K. P., Sankar, M. R. (2021). Machining of hard materials using textured tool with minimum quantity nano-green cutting fluid. *CIRP Journal of Manufacturing Science and Technology*, *35*, 410–421.
5. Gajrani, K. K., Ram, D., Sankar, M. R. (2017). Biodegradation and hard machining performance comparison of eco-friendly cutting fluid and mineral oil using flood cooling and minimum quantity cutting fluid techniques. *Journal of Cleaner Production*, *165*, 1420–1435.
6. Sharma, V. S., Singh, G., Sørby, K. (2015). A review on minimum quantity lubrication for machining processes. *Materials and Manufacturing Processes*, *30*(8), 935–953.
7. Gajrani, K. K., Prasad, A., Kumar, A. (Eds.). (2022). *Advances in Sustainable Machining and Manufacturing Processes*. CRC Press, Boca Raton and London. ISBN: 9781003284574. doi:10.1201/9781003284574

8. Gajrani, K. K., Sankar, M. R. (2018). Sustainable cutting fluids: thermal, rheological, biodegradation, anti-corrosion, storage stability studies and its machining performance. *Reference Module in Materials Science and Materials Engineering, 1*, 839–852. doi:10.1016/B978-0-12-813195-4.11152-X
9. Dhar, N. R., Islam, M. W., Islam, S., Mithu, M. A. H. (2006). The influence of minimum quantity of lubrication (MQL) on cutting temperature, chip and dimensional accuracy in turning AISI-1040 steel. *Journal of Materials Processing Technology, 171*(1), 93–99.
10. Jadhav, P. A., Deivanathan, R. (2021). Numerical analysis of the effect of air pressure and oil flow rate on droplet size and tool temperature in MQL machining. *Materials Today: Proceedings, 38*, 2499–2505.
11. Duchosal, A., Serra, R., Leroy, R., Louste, C. (2015). Numerical steady state prediction of spitting effect for different internal canalization geometries used in MQL machining strategy. *Journal of Manufacturing Processes, 20*, 149–161.
12. El-Bouri, W., Deiab, I., Khanafer, K., Wahba, E. (2019). Numerical and experimental analysis of turbulent flow and heat transfer of minimum quantity lubrication in a turning process using discrete phase model. *International Communications in Heat and Mass Transfer, 104*, 23–32.

Chapter 5

Advancements in metalworking fluid for sustainable manufacturing

Pandi Jyothish, Upendra Maurya, and V. Vasu

Contents

Abbreviations

MWF: Metal Working Fluid
BF_4: Tetrafluoroborate
BMIM: 1-Butyl-3-methylimidazolium
BTMPP: bis(2,4,4- trimethylpentyl)phosphinate
DEHP: bis(2- ethylhexyl)phosphate
N_{1888}: Trioctylmethylammonium
EMIM: 1-Ethyl-3-methylimidazolium
NTf_2: bis(trifluoromethane)sulfonimide
PF_6: Hexafluorophosphate
P_{66614}: Tetradecyltrihexylphosphonium

5.1 Introduction

Machining involves different processes to remove excess material and shape modification to achieve desired product dimensions [1, 2]. It is necessary for the smooth operation of the modern world and is an essential part of manufacturing things. During machining, most energy generated due to plastic deformation of the workpiece gets converted into heat. Friction between the tool and the flowing chip interface also generates heat, which increases tool temperature. Exposers to the high heat negatively impact the tool (tool wear, thermal

DOI: 10.1201/9781003291961-6

cracking of tool), the workpiece (poor product quality, high surface roughness, dimensional inaccuracy), and overall productivity [3, 4]. The heat generated due to the breakage of the chemical bond of a workpiece (chip formation) cannot be avoided, and most of this heat is removed with the chip. While heat generated due to friction between the tool and chip can be reduced by cooling and lubrication with effective metalworking fluids (MWFs) [2]. The use of MWFs can be traced back to the beginning of human civilization. From inception till the second half of the 19th century, MWFs were composed of vegetable oils and animal fats. Due to various types of vegetation and seasons across the world, various types of vegetable oils are used. With the discovery of mineral oil in the second half of the 19th century, a paradigm shift occurred. Slowly, mineral oil replaced vegetable oil and animal fats in almost all the applications of MWFs. Mineral oil still dominates the MWFs industry (over 85%), and vegetable oil-derived MWFs are used in places where mineral oil is not suitable [4]. The majority of existing MWFs, particularly mineral-based, are not sustainable due to performance limitations and negative impacts on the environment (soil and water pollution due to leakage and disposal) and operator's health (cancer, skin irritation, and chronic respiratory problems) [2]. Current and future demand for MWFs not only depends on economic but also on environmental factors; the occupational health hazards of mineral oil-based MWFs and their performance limitations are the driving force for the development of environmentally benign MWFs. Scientists, MWFs manufacturers, and tribologists are relentlessly pursuing effective alternatives to replace mineral oil-based cutting fluids.

The primary focus has been to develop harder tools by tailoring tool compositions and/or introducing new tool materials [5, 6]. In addition, various high-temperature resistant multilayered coating is being explored to further enhance the tool's hardness [7–9]. The other approaches revolve around either softening the workpiece locally using thermally enhanced machining (e.g., laser-assisted machining) [10, 11], reducing cutting temperature using enhanced cooling techniques (e.g., cryogenic coolants) [12–15], or lubricating tool-workpiece interface [minimum quantity lubrication (MQL)] with an overall aim to reduce friction and enhance productivity [16–23]. The cryogenic cooling of the tool-workpiece interface is achieved by a cryogenic medium, such as liquid nitrogen (boiling point: −196°C). Cryogenic cooling is particularly applied for difficult-to-cut materials, such as Inconel, titanium, etc. [24–26], and it can further be used to enhance the machined surface quality [15, 27–29]. Few studies have reported the advantages of combining cryogenic methods with MQL and claimed improved performance [27, 30]. Another approach has been to develop new additives that impart specific properties to MWFs (see Figure 5.1). Room temperature ionic liquids (RTILs) and nanolubricants are promising candidates for effective lubrication of the tool-workpiece interface and an alternative to harmful additives [31, 32].

5.2 Health hazards of metal working fluids and need for sustainability

The demand for high-performance MWFs grew with rapid industrialization in the 20th century. The aviation and automotive industries promoted the relentless pursuit of efficient and high-performance MWFs to enhance productivity in mass production. Various chemical compounds were explored using the principle of "Trial and error", and it was observed that the addition of sulfur, chlorine, or boron and phosphorus compounds improved performance under high pressure [33–35]. A large number of common compounds earlier used in MWFs are no longer permitted due to their adverse effects on operator health and

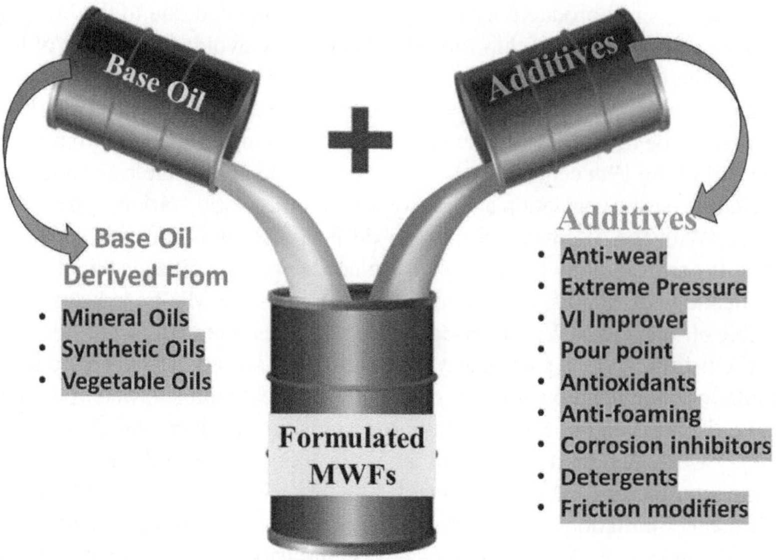

Figure 5.1 Composition of MWFs. The bulk of the MWF properties depend on the base oil, while the additive imparts specific ability depending on the service requirement.

the environment [2]. For instance, additive groups such as chlorinated compounds, amine groups, and boric acid are strictly prohibited from being used in MWFs due to their carcinogenic nature and other adverse health effects [36]. Animal studies on exposure to nitrosamine and a few polycyclic aromatic hydrocarbons showed their carcinogenic properties in animals; hence, these compounds are also no longer available for formulating MWFs [2, 37]. Currently, several regulatory organizations, such as Registration, Evaluation, Authorization, and Restriction of Chemicals (Reach), Environment Health and Safety (EHS), etc., have set stringent regulations and limited the use of harmful volatile organic compounds and biocides [2, 36–38]. Other than regulations, the cost and the availability of MWFs impact the application in the manufacturing process. Modern MWFs are composed of over a dozen types of performance additives, and a majority of oil-based emulsions are still derived from mineral oil. With the evolution of technology and developmental methods, new economic base fluids and additives with lesser environmental impact compared to mineral oil are relentlessly explored. Vegetable and synthetic easter-based MWFs are biodegradable and eco-friendly and have shown promising potential as an alternative to mineral oil [35]. Particularly, vegetable oils have shown better lubricity than mineral oil, but lower oxidation stability and microbial growth might occur during storage. This limitation can be overcome using additives such as antioxidants, biocides to limit microbial growth, corrosion inhibitors, anti-foam agents to limit foam formation, etc. [35, 39–41].

5.3 What are MWFs?

MWFs are supplied between the tool and workpiece interface, with a specific focus on reducing power consumption and enhancing product quality and overall productivity. Even with the same chemical composition, these MWFs, based on the machining process, may be loosely termed as cutting oil, forming oil, grinding oil, etc. German Institute for Standardization (DIN-51385) categorizes MWFs in two broad categories, that is, water-based and oil-based. Oil-based MWFs can further be classified as straight oils (neat oils), soluble or emulsifiable oils (30–85% refined petroleum oil + water), semi-synthetic oils (5–30% severely refined petroleum oil+30–50% water + other additives), and synthetic oils (detergent-like compounds and other additives). Commercial MWFs are formulated from base stocks (mineral, synthetic, or vegetable base oils) and might contain performance enhancers called additives.

5.3.1 Mineral oil and its application as MWFs

Since its inception, human civilization was dependent on vegetable oil till the drilling of the first oil well (the year 1859 at Titusville, Pennsylvania, USA) in the second half of the 19th century [35]. Due to superior stability and cost-effectiveness, mineral oils rapidly replaced vegetable oil, which facilitated rapid industrialization in the 20th century. Currently, the majority of MWF's market share comprises conventional (mineral-based) oils. The crude oil obtained from beneath the earth is a cheap source of many products (fuels, chemicals for industries, and lubricants) essential for the modern world [3]. Crude oil is composed of organic compounds (primarily hydrocarbons) and some residuum (contaminant of metal and metal salts), and their composition varies greatly attributed to changes in density, wax content, sulfur content, and geographical location [35]. The organic parts can further be categorized as hydrocarbons (paraffinic, naphthenic, and aromatic), asphaltenes, aromatics, hetero-organic compounds, and resins, with atomic nitrogen, sulfur, and oxygen in their structures. These complexities make it extremely difficult to know the exact chemical structure of mineral oil and hence are generally categorized based on viscosity, sulfur content, and degree of refining (API group 1, 2, 3) [35].

As discussed above, mineral oils used as cutting fluids are primarily composed of hydrocarbons (each molecule comprising almost 30 carbons) along with few impurities and might further consist of additives (phosphorous, sulfur, and chlorine) to improve extreme-pressure performance under severe working conditions [35]. Based on task-specific application, mineral oil is enriched with other additives and might further be categorized as straight, sulfurized mineral oil, fatty mineral oil, sulfurized fatty mineral oil, sulfochlorinated mineral oil, fatty mineral oil, and water-soluble oil [3]. Straight mineral oil does not contain any additives and provides poor tool protection. Hence, it is recommended for light-duty operation. Contrarily, sulfurized mineral oil contains sulfur and during cutting operation forms sulfide-rich protective film on the tool resulting in improved machining parameters (decreased built-up edge, friction, and heat generation). This type of MWF is suitable for tough tools and ductile workpiece materials. Fatty oils are known to provide excellent lubricity by increasing wettability and are generally preferred where surface finish and precision are paramount.

5.3.2 Synthetic oil and its application as MWFs

Contrary to mineral oil, synthetic oils have known molecular structures and give predictable behavior. Mineral oils, due to complexities, residuum, and unknown molecular structure, may not give predictable behavior [35]. Synthetic processes enable molecules to be built in a laboratory using chemistry (polymerization, esterification, and alkylation) from simpler substances to give the precise or desired oil properties [3, 35]. In this chapter, the term "synthetic" is exclusively used to refer to man-made compounds used for lubrication.

Even though mineral oils are economical and abundant, they have several disadvantages compared to synthetic oils, such as relatively poor oxidation stability, shear-thinning, lower flash and fire points, higher pour point, etc. [35, 42]. This makes synthetic oil valuable blending components for MWF formulation to operate under extreme service conditions (both high and low temperatures). Synthetic-based MWFs are preferred only for specialized applications where mineral oil cannot be used [3]. It is well known that MWFs derived from synthetic base oil inherit better thermos-oxidation and chemical stability than mineral oil. Synthetic oils can be classified into seven categories based on chemical compounds [3, 35]: (1) Synthetic hydrocarbon polymers (polybutenes, alkylated aromatics, and polyalphaolefins), (2) Phosphate esters, (3) Carboxylate esters (polyol esters and aliphatic esters), (4) Silicon compounds (silicate esters and silicones), (5) Poly-alkylene glycols, (6) Halogenated hydrocarbons, and (7) Poly-phenyl ethers. Based on their chemical structures, these synthetic oil groups possess distinct physicochemical properties and costs. MWFs formulated from a synthetic hydrocarbon can be economically as close to mineral oil and may give superior performance to mineral oil [3]. In addition, synthetic oils have low-temperature fluidity, stable at extreme temperatures and pressure but are economically restrictive and are in limited supply.

Polybutene is non-toxic synthetic oil and is generally used for metallic aluminum. Polybutene easily depolymerizes, resulting in negligible staining or deposition on the workpiece, even at elevated temperatures. It forms a stable emulsion and is resistant to biodegradability, making it less prone to microbial attack and hence longer operational life [3]. Another important synthetic oil is PAG (poly alkyl glycols), which is generally blended with water for lubrication while water acts as a coolant. PAG added in water does not affect the lubrication properties of PAG and also imparts other desirable properties such as hydrolytic stability, excellent penetrating and wetting, non-corrosiveness, resistance to microbial attacks, and non-toxicity [3, 35]. At elevated temperatures, PAG solubility decreases, and the released PAG from the solution is adsorbed on metallic surfaces (protective film), resulting in better lubrication.

5.3.3 *Vegetable oil and its application as MWFs*

Vegetable oils are renewable and inherit several desirable properties, such as extremely low toxicity, eco-friendliness, and sustainability, making them a popular alternative to mineral oil-based MWFs. Vegetable oils are obtained from plants and are composed of fatty acids and glycerols. Each molecule of vegetable oil is composed of glycerol linked to three long-chain fatty acid molecules, as shown in Figure 5.2 [4, 35]. Free fatty acids of vegetable oil consist of a polar part that has a great affinity towards metallic surfaces, forming an adhesive film, resulting in excellent lubricity. Further, vegetable oil has a high flash point, which not only eliminates fire and smoke risks but also enables operation at high temperatures [4].

Table 5.1 lists some of the popular vegetable oils explored for cooling and lubrication, along with important physicochemical properties. The inherent properties of vegetable oil, such as high flash point, high viscosity index, excellent lubricity, biodegradability, renewable nature, and disposability, make them a potential alternative to mineral oils. Vegetable oil-based MWFs are often very efficient and have found application in many machining processes, such as milling, drilling, grinding, and turning [43]. Table 5.2 lists sustainability studies of vegetable oil-based MWFs along with important observations.

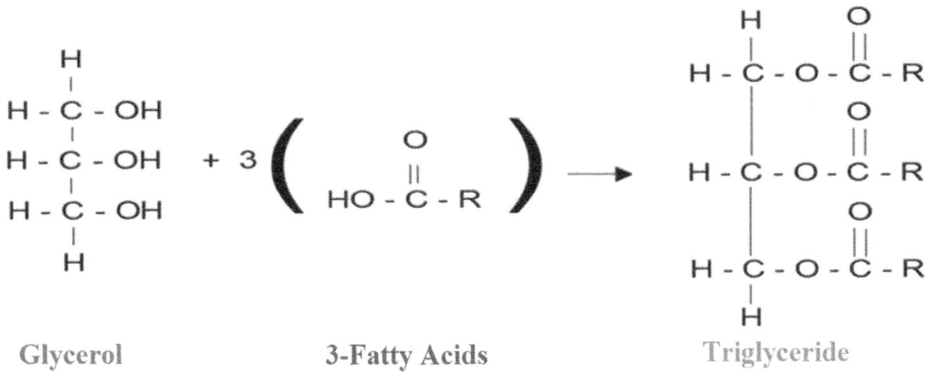

Figure 5.2 Molecular structure of vegetable oil.

Table 5.1 List of Popular Edible and Non-Edible Vegetable Oil Along With Their Respective Physico-chemical Properties

Sl. No.	*Oil*	*Kinematic viscosity at 313 K (cSt)*	*Density at 298 K (kg/m³)*	*Cloud point (K)*	*Flash point (K)*	*Oxidation stability at 383 K (h)*
			Edible oils			
1	Palm	5.72	875	286	438	4.0
2	Rice bran	4.95	886	273.3	591	0.5
3	Peanut	4.92	882	278	450	2.1
4	Olive	4.52	892	-----	591	3.4
5	Sunflower	4.45	878	276.42	525	0.9
6	Rape seed Oil	4.45	880	269.7	525	7.5
7	Soybean	4.05	885	274	598	2.1
8	Linseed	3.74	890	269.2	451	0.2
9	Coconut	2.75	805	273	598	35.4
			Non-edible oils			
1	Rubber seed oil	31.4	870.9 (at 313 K)	-----	-----	-----
2	Castor oil	15.25	898	259.5	533	1.2
3	Neem	5.20	885	287.5	317	7.2
4	Jatropha	4.82	878	275.75	409	2.3
5	Karanja	4.80	918	282	423	6.0
6	Tobacco	4.25	887	-----	439	0.8
7	Mahua	3.40	850	-----	483	-----

Table 5.3 Sustainability Studies of Vegetable Oil-Based MWFs Along With Important Observations

Sr No.	*MWFs compositions*	*Machining process*	*Workpiece and tool material*	*Main findings*	*References*
1	Rapeseed oil, ester oil, and meadowfoam oil, along with sulfur and phosphorus additives	Drilling	Workpiece - AISI 316L Tool - HSS-Co	All formulated vegetable oil outperformed commercial mineral oil and showed a maximum of 177% enhancement in tool life.	Belluco et al. [44]
2	Four vegetable oils, that is, groundnut, palm kernel oil, coconut oil, and shear butter oil	Turning	Workpiece - Mild steel, copper, and aluminum rods Tool - Tungsten carbide	The cutting force depends on materials. As per the highest reduction of cutting forces: Groundnut > shear butter oil > palm kernel oil > coconut oil	Ojol et al. [45]
3	Vegetable oil	Turning under dry, wet and VMQL	Workpiece AISI - 9310 alloy steel Tool - Carbide, TTS, SNMG	MQL decreased cutting zone temperature, resulting in improved chip formation mode, surface finish, and wear at a wide range of cutting speeds. Wet cooling performed poorly and severity increased with the increase in cutting speed.	Khan et al. [46]
4	Sunflower oil	Turning under dry; wet; and VMQL, NMQL, and N_2 cryogenic cooling.	Workpiece -Inconel-800 Tool - Carbide, CNMG120408	Compared to dry machining conditions, the N2 cooling conditions lowered overall machining cost by 9.3%, carbon emissions by 49.17%, and tool wear by 46.6%.	Gupta et al. [47]
5	Vegetable oil	Turning under dry, wet, and VMQL	Workpiece - Hastelloy C-276 Tool - CNMG120408	Vegetable oil using MQL machining exhibits lower cutting temperatures, lower chip reduction coefficient, and excellent surface quality.	Singh et al. [48]
6	Vegetable oil	Turning under wet, VMQL, and MQSL	Workpiece - AISI 4140 Tool - CNMG 120	MQSL and VMQL outperformed wet and dry machining, and VMQL in machining enhanced energy efficiency, sustainability, and product quality.	Makhesana et al. [49]
7	Canola oil	End milling under dry, wet, and VMQL	Workpiece - Aluminum 6061 Tool - TiAlN-coated carbide	MQL with vegetable oil gave the finest surface and lowest tool wear and was superior compared to dry and flood cooling conditions.	Shukla et al. [50]

8	A mixture of aloe vera and cottonseed oil	Turning under MQL	Workpiece - M2 Steel Tool - Carbide tool	Formulated vegetable oil outperformed mineral oil with respect to surface roughness and tool wear.	Agarwal et al. [51]
9	Three vegetable oils (Palm, coconut, and sunflower oils)	Turning under MQL	Workpiece - AISI 304 Tool - Carbide tool	Sunflower oil outperformed the other two vegetable oils in terms of chip compression ratio and surface roughness.	Majak et al. [52]
10	1). Jojoba 2). Mineral oil (LRT 30)	Turning under VMQL and NMQL	Workpiece - Ti-6Al-4V Tool - MEGACOAT PVD-coated carbide	MQL machining with jojoba oil results in a 35–47% reduction in surface roughness, cutting force, and tool wear. They claimed that jojoba oil with MoS_2 NP is a good sustainable option for machining.	Gaurav G et al. [53]
11	Vegetable Oil (Rapeseed) and Al_2O_3	Turning under VMQL and NMQL	Workpiece - Austempered ductile iron Tool - CNMG120416MR2	MQL-nano-fluids improved the interface adhesion between the workpiece and the tool, which, in turn, enhanced machinability.	Eltaggaz A et al. [54]
12	Coconut oil and soluble oil	Turning under dry, wet, and VMQL	Workpiece - AISI 1040 Tool - Tungsten carbide	The MQL technique with coconut oil as a base fluid resulted in a considerable reduction in coefficient of friction, tool wear, chip morphology, and surface finish.	Ajay B et al. [55]
13	Synthetic ester and palm oil	Drilling under air blow, wet, MQL, and VMQL	Workpiece - Ti-6Al-4V Tool - Uncoated tungsten carbide	This study concludes that the MQL technique can be used to replace the flood cooling technique in terms of reducing pollution and enhancing safety and health. Further, palm oil is a reasonable substitute for synthetic ester.	Rahim et al. [56]

* *MQL*: Minimum Quantity Lubrication, VMQL: MQL with vegetable oil, NMQL: MQL with nano-fluids

Belluco et al. conducted a comprehensive study exploring direct comparative performance among two commercial and four vegetable oil-based formulated MWFs enriched with surface and phosphorus additives for drilling the AISI 316L workpiece [44]. Their group quantitatively studied the performance of MWFs with respect to tool life, tool wear, thrust force, and tangling of chips. They observed that formulated MWFs outperformed commercial mineral oil in every parameter and the most significant improvement of 177% was for tool life and a 7% reduction of the thrust force. Ojolo et al. explored the effect of cutting force for various vegetable oil (groundnut, palm kernel oil, coconut oil, and shear butter oil) and workpiece materials (mild steel, copper, and aluminum rods) for turning operation [45]. They reported that cutting force is highly influenced by material properties and feed rate. For instance, at a lower feed rate, groundnut oil registered the lowest cutting force for aluminum, while at a higher feed rate, it performed better for turning copper. Overall, groundnut oil gave excellent product quality and showed the lowest cutting force compared to the other three oils. Khan et al. reported that cutting zone temperature increases due to friction between the tool-workpiece interface, which negatively influences product quality and needs to be minimized [46]. Higher material removal rate (MRR) and specific energy consumption increase cutting temperature, which impacts tool life, cutting force, and mode of chip formation. Generally, the majority of the heat generated in the primary shear zone is taken away by the chip, but the heat generated in the secondary shear zone due to friction between the tool and workpiece increases the cutting zone temperature and negatively impacts product quality. The conventional cooling method uses a large quantity of coolant and attempts to provide bulk cooling to the tool and workpiece interface but fails to provide effective lubrication and subsequently minimizes heat generation. This happens due to the inability of coolant to enter in bulk contact between the flowing chip and tool rake surface, which becomes more severe with the increase in cutting velocity. Therefore, a process that can minimize friction between the tool and workpiece interface becomes intuitive for enhanced machinability and product quality. MQL provides a protective lubricant film between the flowing chip and tool interface, resulting in an overall decrease in cutting zone temperature and, subsequently, productivity. They observed a 10% reduction in cutting temperature under MQL lubrication compared to conventional wet lubrication and concluded that this cutting temperature decrease significantly improved other machinability indices. Gupta et al. explored the sustainability of various cooling/lubrication methods, such as dry, nitrogen cooling, MQL with vegetable oil, and MQL with nanolubricant for machining Inconel-800 [47]. Their group estimated carbon emission, energy consumption, cost, etc., during turning operation and clubbed these parameters to obtain sustainable indicators. They published a surprising observation that nitrogen cooling is the most sustainable and outperformed other lubrication methods in all parameters. Nitrogen cooling gave the highest tool life, which directly improved machining time and overall efficiency compared to other cooling/lubrication methods. Following is their order of performance compared to the dry machining.

Nitrogen cooling (9.3%) > NMQL (6%) > MQL (2.3%)

The excellent performance of NMQL could be attributed to a decrease in friction between chip tool interfaces due to graphene nanoparticles resulting in lower cutting zone temperature. On the other hand, cryogenic cooling removed relatively more heat from the cutting zone resulting in lower energy consumption. The order of performance with respect to total energy consumption compared to the dry lubrication method is reported as follows.

Nitrogen cooling (11.3%) > NMQL (9.4%) > MQL (7.5%) @ cutting speed of 100 m/min
Nitrogen cooling (13.5%) > NMQL (11.3%) > MQL (5.5%) @ cutting speed of 200 m/min

Energy consumption is directly related to carbon emissions during the machining process. The lowest carbon emission was observed for nitrogen cooling, as given below.

Nitrogen cooling (49.17%) > NMQL (32.81%) > MQL (13.47%)

They further observed the lowest tool wear due to the ability of cryogenic cooling to remove a significant amount of heat, as given below.

Nitrogen cooling (46.6%) > NMQL (36.3%) > MQL (12.1%)

Finally, they concluded that ecological benefits and machining parameters make cryogenic cooling relatively more sustainable compared to other colling methods for Inconel-800 alloy.

Makhesana et al. compared VMQL (vegetable oil MQL) and MQSL (minimum quantity solid lubricants) with dry and wet lubrication for machining steel (AISI 4140) with a carbide tool [49]. Their group assessed these lubrication/cooling methods for machining performance (cutting temperature, surface roughness, and tool life) and sustainable parameters (energy consumption, machining efficiency, and carbon emission). They observed that MQSL and VMQL decreased surface roughness by 20% and 27%, respectively, compared to dry conditions at 100 m/min cutting speed. Interestingly, at a cutting speed of 170 m/min, MQSL and VMQL showed excellent improvement of 41% and 45%, respectively, which could be attributed to effective cooling and lubrication provided by MWFs, while in the absence of any additive, dry condition lacked any surface protection. Due to friction reduction and high heat dissipation capability of the VMQL, oil cutting temperature was reduced by 31% and 18% compared to dry and wet conditions, respectively. Carbon emission depends on total energy consumption, and they reported 35 and 36% carbon emission reduction (depends on total energy consumption) for MQSL and VMQL, respectively, while tool wear reduction was observed to be 64% and 70%, respectively. They further made a camera comparison of chips formed under different machining conditions (a) Dry, (b) Wet, (c) VMQL, and (d) MQSL. Under dry and wet machining conditions, snarled chips were formed due to severe plastic deformation of the workpiece, resulting in poor surface quality and accelerated tool flank wear. Under MQSL and VMQL lubrication, fragmented chips with thin and narrow curls were formed due to effective lubrication and cooling (by compressed air), resulting in improved surface finish and tool wear.

Shukla et al. investigated the effects of feed, cutting speed, and depth of cut on the surface roughness under three machining conditions, such as flood, dry, and MQL, for the machining of Aluminium alloy (Al 6061-T6) [50]. They observed that an increase in cutting speed decreased surface roughness for all machining conditions. Further, at higher feed and depth of cut and lower cutting speed, the MQL showed better improvement in surface roughness. Machining under dry conditions resulted in poor performance with respect to cutting forces, tool life & cutting temperature [56]. Rahim et al. observed that MQL palm oil recorded lower cutting temperature & cutting force than MQL with synthetic ester, which was closely approaching flood cooling [56]. This was attributed to the generation of a thin self-lubricating film, resulting in friction reduction and a subsequent decrease in heat. They further concluded that the MQL could be used to replace the flood cooling technique with respect to pollution reduction and improvement of the health and safety of the operator. At the same time, the inclusion of palm oil is not only advantageous in terms of biodegradability but could also improve cooling and lubricating ability compared to MQL with synthetic ester. Under MQL conditions, they claimed that palm oil could be a reasonable substitute for synthetic ester. Ajay Vardhaman et al. evaluated the influence of cutting fluids on friction coefficient, tool wear, chip morphology, and surface quality for machining AISI 1040 material under wet, dry, and MQL with coconut oil [55]. They reported excellent wettability of the carbide tool resulting in better lubrication and lower friction. They observed a 33.7° wettability angle for coconut oil which was significantly higher than conventional cutting fluid hence a better lower coefficient of friction, tool wear, and improved product quality.

5.4 Additives

The bulk of the MWFs properties depends on the base fluid (mineral, synthetic, and bio-based). This means if base oil is fire-resistant and biodegradable, then the formulated MWF is most likely to be fire-resistant and biodegradable. Neat base oil is generally not adequate for dynamic operating conditions and may require performance-enhancing compounds called additives (listed in Figure 5.1) to be blended into the base oil. Most of the additives used in commercial MWFs are not sustainable and negatively impact ecology. Ionic liquids and nanoparticles are green alternatives and possess the potential to replace harmful additives and, at the same time, meet the dynamic operating conditions of the future MWFs.

5.4.1 Ionic liquids and their performance as MWFs

Ionic liquids are entirely made up of ions (cation and anion) [57, 58]. Nowadays, in scientific literature and general use, the term "ionic liquids" refers to the ionic compounds that are in a liquid phase below 100°C, while the ionic compounds having a melting temperature above 100°C are loosely termed "molten salt". Ionic liquids are also called "designer solvents" for the possibility of obtaining desired physicochemical properties by altering cations and anions (tunability). For instance, by altering cations and anions or both, viscosity, water miscibility, corrosive resistance, and polarity can be tuned for task-specific applications [59]. Ionic liquids inherit several desirable properties, such as negligible vapor pressure or non-volatility at ambient conditions, noncorrosive, nonflammable, excellent thermal stability, tunable polarity, solubility, basicity, acidity, etc. [60, 61]. Compared to hazardous volatile organic compounds (VOCs), ionic liquids are often considered green solvents since they have an extremely low vapor pressure. Table 5.3 lists the ionic liquids used in MWFs formulation along with important findings.

The first study of ILs in machining was reported by Pham et al. in 2014 [62]. Their group used two ILs with different viscosity for machining aluminum workpieces as neat oil and observed better performance than conventional oil. A couple of other groups attempted to use ILs as an additive in metalworking fluids (MWFs), but most of the initial ILs were immiscible in the base oil, so their emulsion was used for machining purposes [31, 32, 69]. Just over half a dozen studies exploring an ionic liquid's potential for machining oil application have been reported to date, as shown in Table 5.1. The table lists the pair of cation

Table 5.3 List of Ionic Liquids Used in MWF Formulation, Along With Important Findings

Sr. No	*Cation*	*anion*	*MWF*	*% of ILs*	*Workpiece and tool material*	*Machining method*	*Important findings*	*References*
Neat Ionic Liquid as MWF								
1	EMIM BMIM	NTf_2 iodide	-----	100%	Aluminum & Tungsten carbide	Three-axis micro milling CNC machining	Performed better or equal to commercial conventional machining oils. Negligible evaporative loss (1%) compared to 75–85% of conventional oils.	Pham et al. 2014 [62]
Miscible and Immiscible Ionic Liquid-Enriched MWFs								
2	BMIM	PF_6	Deionized water (emulsion)	0.5%	Titanium and CBN	Turning operation Using machining center	Significant tool wear reduction. IL emulsion showed the smallest cutting forces, largest shear band spacing, and best surface finish.	Devis et al. 2015 [63]
3	BMIM BMIM BMIM	PF_6 BF_4 NTf_2	Canola oil (emulsion)	1 wt.%	Medium carbon steel and uncoated tungsten carbide	CNC vertical milling	**PF_6** & BF_4 cation-based ILs gave the smoothest surface and lowest cutting force. **In** contrast, NTf_2 cation-based IL showed performance similar to the vegetable oil and flood cooling system.	Goindi et al. 2017 [32]
4	BMIM BMIM P_{4449}	PF_6 BF_4 DEHP	Conola oil (emulsion) and PEG400 (miscible)	3 wt.% 1 wt.% 0.5 and 1 wt.%	Medium carbon steel and uncoated tungsten carbide	CNC vertical milling	**PF_6** and BF_4 cation-based ILs showed improvement only at higher cutting speeds. **P**hosphonium IL showed improvement only at a lower cutting speed.	Goindi et al. 2018 [31]
5	BMIM	BF_4	Neem oil (emulsion) and Neem oil + water (emulsion)	1 and 2 wt.%	Carbon steel and titanium nitride coated and uncoated tungsten carbide	Turning operation using a geared lathe	**S**urface roughness was improved in the range of 40–62% compared to neat Neem oil.	Panneer et al. 2018 [64]

(*Continued*)

Table 5.3 (Continued)

Sr. No	*Cation*	*anion*	*MWF*	*% of ILs*	*Workpiece and tool material*	*Machining method*	*Important findings*	*References*
6	N_{1888} P_{66614}	NTf_2 BTMPP	Modified Jatropha-based ester (miscible)	1, 5, and 10 wt.%	Medium carbon steel and uncoated cermet	Turning operation using NC lathe machine	**O**ptimum concentration of 10 wt.% and 1 wt.% for ammonium phosphonium ILs, respectively **O**verall reduction of 4–5% in cutting force, 7–10% in cutting energy, 2–3% in friction coefficient, 8–11% in tool chip contact, 22–25% in chip thickness, and 1–2% in friction angle over synthetic esters	Abdul Sani et al. 2019 [65]
6	N_{1888} P_{66614}	NTf_2 BTMPP	Modified Jatropha-based ester (miscible)	1, 5, and 10 wt.%	Medium carbon steel and uncoated cermet	Turning operation using NC lathe machine	**10** wt.% ammonium IL concentration reduced cutting force up to 12%, cutting temperature by 9%, and surface roughness by 7%, and improved tool life by up to 50%. **1** wt.% phosphonium IL concentration reduced cutting force by up to 11%, cutting temperature by 7%, surface roughness by 4%, and improved tool life by up to 40%.	Abdul Sani. et al. 2019 [66]
7	P_{66614}	Cl	Coconut oil (miscible)	1 wt.%	Steel and multilayer-coated cermet	Turning operation using CNC lathe	**E**xcellent surface finish **N**o significant benefit for flank wear was observed for 1 wt.% IL concentration. **R**ecommends use of higher IL concentration (above wt.%)	Pandey et al. 2020 [67]
8	BMIM	PF_6	Deionized water and coconut oil	0.5 wt.%	Inconel 825 Steel and carbide	Turning operation using CNC Lathe	**R**eduction of surface roughness by 88%, cutting temperature by 74%, chip thickness by 89%, and tool wear minimized.	M. Naresh Babu 2021 [68]

and anion, type of base oil, blend/emulsion concentration, tool and workpiece material, and machining methods, along with important findings. These reported studies show an addition of ILs at even a small concentration (0.5–10 wt.%) can enhance machining performance.

Pham et al. compared two imidazolium ILs (i.e., [BMIM][I] and [EMIM][TFSI]) with other lubrication conditions, such as dry, distilled water, and two conventional commercial machine cutting fluid (i.e., ST501 and TC#1) [62]. As expected, dry and distilled water machining conditions gave the highest cutting force, while both the ILs performed similarly to TC#1. Viscosity influences the lubricant film formation and higher viscosity may form a thicker film on the surface interface, but despite higher viscosity of [BMIM][I] (viscosity of 401 cP at 25°C as compared to 32 of [EMIM][TFSI]) both ILs performed nearly similar. The similar performance of both the ILs was attributed to the reduction of viscosity of [BMIM][I] due to the rise in temperature and the increase in viscosity of [EMIM][TFSI] due to dispersion of nano/micro-sized chip particles during machining. Cation and anion both significantly affect the machining parameters, and altering any can change performance. Goindi et al. synthesized three unique ILs with a common cation (BMIM) and different fluorine-containing anions (PF_6, BF_4, and NTf_2). Emulsion in canola oil of PF_6 and BF_4 anion-based ILs gave the lowest cutting forces at both smooth and rough machining conditions [31, 32]. While [BMIM][NTf_2] in canola oil performed in the same range of flood cooling and MQL with neat canola oil. They reported a similar trend of peak and mean cutting forces for different lubrication methods. Their order of performance with cutting forces was dry machining> flood cooling> MQL with neat vegetable oil > MQL with vegetable oil + IL emulsion. Cutting speed, depth of cut, and feed rate influence the performance of the lubricants. Some ionic liquids can perform better at light operating conditions, while others are more suited for heavy operating conditions.

Goindi et al. observed vigorous scouring on the workpiece surfaces due to a relatively higher adhesion rate on the surfaces lubricated under dry, flood cooling, and MQL with vegetable oil [31, 32]. ILs showed very low adhesive scouring, probably due to protective film formation, on the tool flank and workpiece surfaces, resulting in a smoother surface. Sani et al. reported a 4–7% improvement in surface roughness performance over synthetic easter for both PIL and AIL at optimum concentrations. They also attributed this improved surface roughness to the non-flammable, thermally stable, and non-volatile properties of the ILs, which enhanced the adsorption of a protective layer on the tool-workpiece interface, subsequently reducing friction wear and peak temperature [67, 69, 70]. Pandey et al. observed excellent surface roughness improvements by blending imidazolium-based ILs in coconut oil. They also made similar conclusions about the protective film formation of the tool-workpiece interface.

Most conventional machining oils evaporate at higher operating temperatures. Pham et al. measured the evaporative weight loss of lubricant before and after the machining. They reported both the commercial machining oils evaporated by 75–89% while negligible evaporative losses (<1%) of ILs were recorded, which can be attributed to its extremely low vapor pressure. This eliminates evaporative losses and fumes formation during machining. Considering the vastness of unique available ILs very limited studies have been conducted exploring the potential of ILs in the field of machining. Moreover, a significant proportion of these published studies involve non-miscible Ils. With the first study of ILs for machining reported in the year 2014 by Pham et al., this field is still very new, with a lot of potential to be explored. Significant numbers of oil-miscible ILs have already been reported in the field of tribology for their friction and wear-reduction properties [61, 71]. These ILs may be explored in the field of machining using MQL and/or other MWF supplying methods.

5.4.2 Nanoparticles

Due to their tiny size (<100 nm), nanoparticles have to potential to enter the contact zone and convert sliding motion into a combination of sliding a rolling motion [72]. Jamil et al. explored a hybrid nano additive composed of multi-walled carbon nanotubes and Al_2O_3 nanoparticles dispersed in vegetable oil for machining Titanium-based alloy (Ti-6Al-4V) under MQL lubrication [73]. They further compared the machining performance of a hybrid nanofluid with CO_2 cryogenic cooling. They reported hybrid nano-fluid outperformed cryogenic cooling and reduced cutting force, surface roughness, and tool life by 11.8%, 8.72%, and 23%, respectively. It was interesting to note that cryogenic cooling reduced significant heat from the tool-workpiece interface and showed an 11.2% lower cutting temperature compared to hybrid nano-fluid. Additionally, hybrid nanofluid provided superior surface quality and lower particle deposition, which could be attributed ability of hybrid nanoparticles to decrease friction between the tool and chip interface. Shen et al. [74] analyzed the performance of nano-fluid for grinding cast iron under various lubrication conditions. Grinding performance was analyzed with respect to surface roughness, grinding forces, and grinding temperature, and compared with pure water. Nanoparticle dispersion was found to be crucial, particularly when a slurry layer was produced. The slurry layer resulted in a greater G-ratio (volume ratio of material removed and tool wear), reduced grinding forces, and improved surface quality. Table 5.6 lists the nanoparticles used in MWFs formulation along with important findings.

Tran Minh et al. conducted a novel comparative study between MQL and MQCL (minimum quantity cooling lubrication) for Al_2O_3-enriched nano-fluid, neat oil, and emulsion [76]. They observed superior microstructure and surface roughness performance under MQCL compared to the MQL method for Al_2O_3-enriched emulsion. Further, they claimed the tool lasted 4.5 times longer compared to dry conditions for drilling of Hardox 500 material (difficult-to-cut material). Padmini et al. combined both boric and MoS_2 micro and nanoparticles in coconut oil and sesame oil for the cutting fluid [43]. The outcome of this study was that nano-fluids performed much better than micro-fluids in terms of cutting forces and cutting temperature minimization during machining.

Virdi et al. evaluated machining output parameters, such as G-ratio, Surface roughness, and grinding energy, under various lubrication conditions using rice bran, sunflower (SF) with CuO nanoparticle at 0.5 wt.% and 1 wt.% concentration nanofluid [78]. The experimental results show that sunflower oil's cooling and lubrication performance under nanofluid conditions is superior to that of pure lubrication and flood lubrication. This is because sunflower oil has the ability to produce a viscous film in the grinding region, which increases both the heat-carrying and the wetting ability. The nanofluid under the MQL method showed a superior lubrication effect than that of flood cooling and pure oil. The higher heat transfer rate that results from the adsorption of nanoparticles on the surface makes effective lubrication possible. The presence of nanoparticles changes the sliding friction to a combination of sliding and rolling friction resulting in reduced friction and subsequent heat generation. They attributed lower surface roughness to the polishing action provided by nanoparticles. Triglycerides in vegetable oils produce a viscous layer that improves the lubrication effect, whereas the deficiency of lubrication during traditional cooling leads to greater surface

Table 5.7 List of Nanoparticles Used in MWF Formulation, Along With Important Findings

Sr No.	*MWF compositions*	*Machining process*	*Workpiece and tool materials*	*Main findings*	*References*
1	Vegetable oil with hybrid nanoadditive (Al_2O_3 + CNT) and cryogenic (CO_2) cooling	Turning under NMQL and cryogenic cooling.	Workpiece - Ti-6Al-4V Tool - CNGA 120408 T01020 WG	Al_2O_3-carbon nanotube-based MQL significantly reduced cutting force and surface roughness compared to cryogenic cooling. Moreover, under hybrid-nanofluids, tool life was extended due to a lower cutting force.	Jamil et al. [73]
2	Ethylene glycol with graphene	Turning under dry, wet, and NMQL	Workpiece - D3 tool steel Tool - CNGA 120408 T01020 WG.	They claimed that graphene nanofluids with an MQL cooling are a better alternative technique for lowering the cutting temperature, tool wear, and surface roughness.	Naresh Babu et al. [75]
3	Water with diamond and Al_2O_3	Grinding under NMQL conditions	Workpiece - Ductile iron Tool - Aluminum oxide grinder wheel	Significant reduction of grinding force, temperature, and surface roughness due to nanoparticles in MWF	Shen et al. [74]
4	Emulsion and rice bran oil with Al_2O_3	Drilling under NMQL and MQL	Workpiece - Hardox 500 Tool - Coated carbide drill	MQL with Al_2O_3 nanofluid provided better machining performance, surface quality, and lower thrust force than neat emulsion.	Duc et al. [76]
5	Sunflower oil with MoS_2	Drilling under flood, dry, VMQL, and NMQL	Workpiece - AISI 321 Tool - TiAlN-coated carbide	Sunflower oil outperforms mineral oil in terms of temperature control. Further, nanomaterials dispersed in vegetable oil improve both the process's environmental friendliness and its cooling efficiency.	Pal et al. [77]
6	Rice bran (RB), sunflower (SF) with CuO nanoparticles	Grinding under dry and NMQL	Workpiece - Ni-Cr Alloy Tool - Alumina AA80	Sunflower oil's cooling and lubrication performance under nanofluid conditions are superior to that of pure lubrication and flood lubrication. Triglyceride in vegetable oils produces a viscous layer that improves the lubrication effect.	Virdi et al. [78]

(*Continued*)

Table 5.7 (Continued)

Sr No.	*MWF compositions*	*Machining process*	*Workpiece and tool materials*	*Main findings*	*References*
7	Emulsion (Water + coconut oil) with boric acid nanoparticle	Turning under flood, dry, and NMQL	Workpiece - AISI 1040 steel Tool - CNMG120408NC6110	Nano-cutting fluid gave significant performance enhancement, that is, reduced cutting force (14.28%), surface roughness (24.74%), and tool wear (3.5%) compared to dry machining conditions.	Gugulothu et al. [79]
8	Vegetable oil with MoS_2	Milling under dry, VMQL, and NMQL	Workpiece - Martensitic stainless steel Tool - Uncoated tungsten carbide	MoS_2 nanoparticle-enriched lubricating oil gave the lowest tool wear and surface roughness for milling under MQL.	Uysal et al. [80] Alper u, Furkan
10	Formulated green cutting fluid with hybrid nanoparticles (MoS_2 + CaF_2)	VMQL and NMQL	Workpiece - AISI H-13 steel Tool - Tungsten carbide	Nanoparticles improved thermal conductivity, specific heat, viscosities, and wettability. Excellent improvement in machining properties	Gajrani et al. [81] Kishor Kumar Gajrani, P. S. Suvin
11	ECOCUT SSN 322 neat lubricant oil with SiO_2 dispersed in mineral oil	Milling under MQL and NMQL	Workpiece - Aluminum AA6061-T6 Tool - HSS	Nanofluid reduced COF, cutting forces, and smaller specific energy, and power consumption compared to neat fluid.	Sarhan et al. [82]
12	Sesame oil with three nanoparticles (Fe_3O_4, Carbon, and Al_2O_3)	Broaching operation under NMQL	Workpiece - AISI 1045 steel Tool - Steel HSS-6542	Carbon nanofluid at 160 mg concentration gave superior performance than the other two nanofluids.	Ni J et al. [83] Jing, Zhi
13	Rice bran oil with graphene and TiO_2 nanoparticles	Turning under NMQL	Workpiece - AISI M2 HSS material Tool - Steel	Hybrid nanofluid enhanced tool life, tool wear, and surface roughness and could be used in ecological sustainability.	Anand et al. [84] Rahul A and Ankush

* *MQL*: Minimum quantity lubrication, VMQL: MQL with vegetable oil, NMQL: MQL with nano-fluids

roughness and maximum grinding energy. Ni J et al. prepared different nanofluids by varying concentrations of three nanoparticles (Al_2O_3, Fe_3O_4, and carbon) dispersed in sesame oil for broaching AISI-1045 workpiece material under the MQL lubrication method [83]. They reported good dispersion up to 80 mg for all three nanoparticles. Beyond that, particles tend to agglomerate and sediment, while carbon nanoparticles evenly dispersed up to 160 mg. With the increase in nanoparticle concentration, the load and vibration initially increased and then decreased. The optimum performance was observed for copper nanofluid at 160 mg concentration with mean broaching valley and load peak reductions of 614 N and 725 N, respectively, which were 118.5% and 115.5% greater than the base fluid. They further recommended Carbon nanofluid with sesame-based oil for both economic and environmental benefits for industrial usage. In another study, copper oxide (CuO) outperformed Al_2O_3 and Fe_2O_3 nanoparticles for machining HSLA AISI 4340 steel material under the MQL lubrication method [85]. Nanoparticles' effectiveness as cutting fluid additives depends on multiple factors, such as particle size and other physicochemical properties. Le Gond et al. reported smaller graphene particles (5 μm) outperformed relatively larger particles (15 μm) on every parameter, indicating that nanoparticles' performance is highly influenced by particle size, which further controls physicochemical properties [86]. Anand et al. explored the overall synergetic effect of graphene and titanium dioxide nanoparticles dispersed in biodegradable Rice bran (vegetable oil) for machining M2 steel under MQL conditions [84]. Both single additives and mixed (graphene + TiO_2 + rice bran) nanofluids were explored at 0.5% and 1% weight concentrations. They observed increased viscosity with an increase in nanoparticle concentration for single and hybrid nanofluids. TiO_2 reduced tool wear by 12%, while hybrid nanofluid reduced by 21.7% with respect to the base oil. SEM analysis of inserts indicated different wear mechanisms for both single and hybrid nanofluids, indicating that TiO_2 and GnP improve tribological performance. Finally, they suggested that GnP and TiO_2, applied to rice bran oil, can enhance tool life, tool wear, and surface roughness and could be used for ecological sustainability.

5.5 Conclusions and future scope

MWFs are one of the most important components of any manufacturing system. It can be called a complex compound and can significantly affect the productivity, resource efficiency, and energy of any manufacturing process. There cannot be a single universal MWFs composition that suits all types of applications due to the wide range of performance parameters and dynamic demanding conditions, prompting a multidisciplinary approach to MWF development. Since the discovery of petroleum in the second half of the 19th century, mineral oil has dominated most of the metalworking fluids (MWFs) applications and could be attributed to the inherent negative impact on the environment and operator's health. The ever-increasing ecological awareness has forced lubricant manufacturers and scientists to explore alternative materials and manufacturing methods to achieve progressive sustainability, which could be achieved with the development of ecologically benign MWFs.

Synthetic oils possess several advantages over mineral oil but are economically restrictive (expensive) and suitable for task-specific applications. Moreover, a significant number of synthetic oils are derived from mineral bases hence are not green and lack biodegradability. Vegetable oil checks many desirable brackets for sustainability (biodegradable, green, and environmentally benign) and inherits excellent lubricity. The literature reviewed in this chapter shows that vegetable oil enriched with additives (ionic liquid or nanoparticle) provided better machining than mineral and synthetic-based oil. The drawbacks of vegetable oil, such as poor oxidation stability and relatively higher pour point, could be overcome with chemical modifications (FAME) and specific additives (antioxidants and pour point depressants).

Future studies should involve different lubricant methods for super alloys and nonferrous materials. The novel nanoparticles should be explored in MWF formulation, and the study should explore colloidal stability, machining parameters, and sustainability. Since the synthesis of the first oil-miscible ILs in 2012, several unique ILs with excellent miscibility in polar and non-polar base fluids have been reported in the field of tribology. These ILs should be explored in the field of machining, considering their friction performance and cost. All the published literature on IL research in machining deals with mixing ILs in base fluids, while no report is available on the compatibility of ILs with other additives. Future studies should explore the synergy between ILs and other common MWFs additives, such as antioxidants, corrosion inhibitors, detergents, dispersants, etc., first at the laboratory scale and then at dynamic conditions of the actual industrial scale. Further IL and nanoparticle performance in fully formulated MWFs should also be explored to understand their compatibility with bulk additive packages for actual machining applications.

References

1. Brinksmeier, E., Garbrecht, M., Heinzel, C., Koch, T., Eckebrecht, J.: Current approaches in design and supply of metalworking fluids. Tribology Transactions. 52, 591–601 (2009). https://doi.org/10.1080/10402000902825739
2. Brinksmeier, E., Meyer, D., Huesmann-Cordes, A.G., Herrmann, C.: Metalworking fluids: Mechanisms and performance. CIRP Annals—Manufacturing Technology. 64, 605–628 (2015). https://doi.org/10.1016/j.cirp.2015.05.003
3. Osama, M., Singh, A., Walvekar, R., Khalid, M., Gupta, T.C.S.M., Yin, W.W.: Recent developments and performance review of metal working fluids. Tribology International. 114, 389–401(2017). https://doi.org/10.1016/j.triboint.2017.04.050
4. Gajrani, K.K., Sankar, M.R.: Sustainable machining with self-lubricating coated mechanical micro-textured cutting tools. Reference Module in Materials Science and Materials Engineering, 2018. Elsevier. https://doi.org/10.1016/B978-0-12-813195-4.11325-2
5. Peng, Y., Miao, H., Peng, Z.: Development of TiCN-based cermets: Mechanical properties and wear mechanism. International Journal of Refractory Metals and Hard Materials. 39, 78–89 (2013). https://doi.org/10.1016/j.ijrmhm.2012.07.001
6. Gajrani, K.K., Prasad, A., Kumar, A. (Eds.).: *Advances in Sustainable Machining and Manufacturing Processes*. CRC Press (2022). ISBN: 9781003284574. https://doi.org/10.1201/9781003284574
7. Biksa, A., Yamamoto, K., Dosbaeva, G., Veldhuis, S.C., Fox-Rabinovich, G.S., Elfizy, A., Wagg, T., Shuster, L.S.: Wear behavior of adaptive nano-multilayered AlTiN/MexN PVD coatings during machining of aerospace alloys. Tribology International. 43, 1491–1499 (2010). https://doi.org/10.1016/j.triboint.2010.02.008
8. Zhang, S., Guo, Y.B.: An experimental and analytical analysis on chip morphology, phase transformation, oxidation, and their relationships in finish hard milling. International Journal of Machine Tools and Manufacture. 49, 805–813 (2009). https://doi.org/10.1016/j.ijmachtools.2009.06.006
9. Fox-Rabinovich, G.S., Kovalev, A.I., Aguirre, M.H., Beake, B.D., Yamamoto, K., Veldhuis, S.C., Endrino, J.L., Wainstein, D.L., Rashkovskiy, A.Y.: Design and performance of AlTiN and TiAlCrN PVD coatings for machining of hard to cut materials. Surface and Coatings Technology. 204, 489–496 (2009). https://doi.org/10.1016/j.surfcoat.2009.08.021
10. Rahman Rashid, R.A., Sun, S., Wang, G., Dargusch, M.S.: An investigation of cutting forces and cutting temperatures during laser-assisted machining of the Ti-6Cr-5Mo-5V-4Al beta titanium

alloy. International Journal of Machine Tools and Manufacture. 63, 58–69 (2012). https://doi.org/10.1016/j.ijmachtools.2012.06.004
11. Schubert, A., Nestler, A., Pinternagel, S., Zeidler, H.: Influence of ultrasonic vibration assistance on the surface integrity in turning of the aluminium alloy AA2017. Materwiss Werksttech. 42, 658–665 (2011). https://doi.org/10.1002/mawe.201100834
12. Sun, S., Brandt, M., Dargusch, M.S.: Machining Ti-6Al-4V alloy with cryogenic compressed air cooling. International Journal of Machine Tools and Manufacture. 50, 933–942 (2010). https://doi.org/10.1016/j.ijmachtools.2010.08.003
13. Umbrello, D., Micari, F., Jawahir, I.S.: The effects of cryogenic cooling on surface integrity in hard machining: A comparison with dry machining. CIRP Annals—Manufacturing Technology. 61, 103–106 (2012). https://doi.org/10.1016/j.cirp.2012.03.052
14. Dhar, N.R., Kamruzzaman, M.: Cutting temperature, tool wear, surface roughness and dimensional deviation in turning AISI-4037 steel under cryogenic condition. International Journal of Machine Tools and Manufacture. 47, 754–759 (2007). https://doi.org/10.1016/j.ijmachtools.2006.09.018
15. Shokrani, A., Dhokia, V., Muñoz-Escalona, P., Newman, S.T.: State-of-the-art cryogenic machining and processing. International Journal of Computer Integrated Manufacturing. 26, 616–648 (2013). https://doi.org/10.1080/0951192X.2012.749531
16. Kuzu, A.T., Berenji, K.R., Ekim, B.C., Bakkal, M.: The thermal modeling of deep-hole drilling process under MQL condition. Journal of Manufacturing Processes. 29, 194–203 (2017). https://doi.org/10.1016/j.jmapro.2017.07.020
17. Sultan, A.A., Okafor, A.C.: Effects of geometric parameters of wavy-edge bull-nose helical end-mill on cutting force prediction in end-milling of Inconel 718 under MQL cooling strategy. Journal of Manufacturing Processes. 23, 102–114 (2016). https://doi.org/10.1016/j.jmapro.2016.05.015
18. Weinert, K., Inasaki, I., Sutherland, J.W., Wakabayashi, T.: Dry machining and minimum quantity lubrication. CIRP Annals—Manufacturing Technology. 53, 511–537 (2004). https://doi.org/10.1016/S0007-8506(07)60027-4
19. Emami, M., Sadeghi, M.H., Sarhan, A.A.D.: Investigating the effects of liquid atomization and delivery parameters of minimum quantity lubrication on the grinding process of Al_2O_3 engineering ceramics. Journal of Manufacturing Processes. 15, 374–388 (2013). https://doi.org/10.1016/j.jmapro.2013.02.004
20. Liew, W.Y.H.: Low-speed milling of stainless steel with TiAlN single-layer and TiAlN/AlCrN nano-multilayer coated carbide tools under different lubrication conditions. Wear. 269, 617–631 (2010). https://doi.org/10.1016/j.wear.2010.06.012
21. Maruda, R.W., Krolczyk, G.M., Nieslony, P., Wojciechowski, S., Michalski, M., Legutko, S.: The influence of the cooling conditions on the cutting tool wear and the chip formation mechanism. Journal of Manufacturing Processes. 24, 107–115 (2016). https://doi.org/10.1016/j.jmapro.2016.08.006
22. Obikawa, T., Kamata, Y., Asano, Y., Nakayama, K., Otieno, A.W.: Micro-liter lubrication machining of Inconel 718. International Journal of Machine Tools and Manufacture. 48, 1605–1612 (2008). https://doi.org/10.1016/j.ijmachtools.2008.07.011
23. Suda, S., Yokota, H., Inasaki, I., Wakabayashi, T.: A synthetic ester as an optimal cutting fluid for minimal quantity lubrication machining. CIRP Annals—Manufacturing Technology. 51, 95–98 (2002). https://doi.org/10.1016/S0007-8506(07)61474-7
24. Wang, Z.Y., Rajurkar, K.P.: Cryogenic machining of hard-to-cut materials. Wear. 239, 168–175 (2000). https://doi.org/10.1016/S0043-1648(99)00361-0
25. Aramcharoen, A., Chuan, S.K.: An experimental investigation on cryogenic milling of Inconel 718 and its sustainability assessment. Procedia CIRP. 14, 529–534 (2014). https://doi.org/10.1016/j.procir.2014.03.076
26. Bermingham, M.J., Palanisamy, S., Kent, D., Dargusch, M.S.: A comparison of cryogenic and high pressure emulsion cooling technologies on tool life and chip morphology in Ti-6Al-4V cutting. Journal of Materials Processing Technology. 212, 752–765 (2012). https://doi.org/10.1016/j.jmatprotec.2011.10.027

27. Kenda, J., Pusavec, F., Kopac, J.: Analysis of residual stresses in sustainable cryogenic machining of nickel based alloy: Inconel 718. Journal of Manufacturing Science and Engineering, Transactions of the ASME. 133, 1–7 (2011). https://doi.org/10.1115/1.4004610
28. Zisman, W.A.: Relation of the equilibrium contact angle to liquid and solid constitution. Advances in Chemistry. 43, 1–51 (1964). https://doi.org/10.1021/ba-1964-0043.ch001
29. Evans, C., Bryan, J.B.: Cryogenic diamond turning of stainless steel. CIRP Annals—Manufacturing Technology. 40, 571–575 (1991). https://doi.org/10.1016/S0007-8506(07)62056-3
30. Madanchi, N., Winter, M., Thiede, S., Herrmann, C.: Energy efficient cutting fluid supply: The impact of nozzle design. Procedia CIRP. 61, 564–569 (2017). https://doi.org/10.1016/j.procir.2016.11.192
31. Goindi, G.S., Jayal, A.D., Sarkar, P.: Application of ionic liquids in interrupted minimum quantity lubrication machining of plain medium carbon steel: Effects of ionic liquid properties and cutting conditions. Journal of Manufacturing Processes. 32, 357–371 (2018). https://doi.org/10.1016/j.jmapro.2018.03.007
32. Bambam, A.K., Dhanola, A., Gajrani, K.K.: Machining of titanium alloys using phosphonium-based halogen-free ionic liquid as lubricant additives. Industrial Lubrication and Tribology. 74 (6), 722–728 (2022). https://doi.org/10.1108/ILT-03-2022-0083
33. Spikes, H.: The history and mechanisms of ZDDP. Tribology Letters. 17, 469–489 (2004). https://doi.org/10.1023/B:TRIL.0000044495.26882.b5
34. Spikes, H.: Friction modifier additives. Tribology Letters. 60 (2015). https://doi.org/10.1007/s11249-015-0589-z
35. Gajrani, K.K., Sankar, M.R.: Role of eco-friendly cutting fluids and cooling techniques in machining. Materials Forming, Machining and Post Processing, 159–181 (2020). https://doi.org/10.1007/978-3-030-18854-2_7
36. Geier, J., Lessmann, H., Schnuch, A., Uter, W.: Contact sensitizations in metalworkers with Occupational dermatitis exposed to water-based metalworking fluids: Results of the research project "FaSt." International Archives of Occupational and Environmental Health. 77, 543–551 (2004). https://doi.org/10.1007/s00420-004-0539-9
37. IARC monographs programme on the evaluation of carcinogenic risks to humans. IARC Monographs on the Evaluation of Carcinogenic Risks to Humans. 73, 9–641 (1999).
38. Skerlos, S.J., Hayes, K.F., Clarens, A.F., Zhao, F.: Current advances in sustainable metalworking fluids research. International Journal of Sustainable Manufacturing. 1, 180–202 (2008). https://doi.org/10.1504/IJSM.2008.019233
39. Bart, J.C.J., Gucciardi, E., Cavallaro, S.: *Biolubricants: Science and Technology*. Woodhead Publishing Series in Energy. Elsevier (2012). ISBN-085709632X, 9780857096326
40. Syahir, A.Z., Zulkifli, N.W.M., Masjuki, H.H., Kalam, M.A., Alabdulkarem, A., Gulzar, M., Khuong, L.S., Harith, M.H.: A review on bio-based lubricants and their applications. Journal of Cleaner Production. 168, 997–1016 (2017). https://doi.org/10.1016/j.jclepro.2017.09.106
41. Spikes, H.A.: Additive-additive and additive-surface interactions in lubrication. Lubrication Science. 2, 3–23 (1989). https://doi.org/10.1002/ls.3010020102
42. Gajrani, K.K., Sankar, M.R. Past and current status of eco-friendly vegetable oil based metal cutting fluids. Materials Today: Proceedings. 4 (2), 3786–3795 (2017).
43. Gajrani, K.K., Suvin, P.S., Kailas, S.V., Rajurkar, K.P., Sankar, M.R.: Machining of hard materials using textured tool with minimum quantity nano-green cutting fluid. CIRP Journal of Manufacturing Science and Technology. 35, 410–421 (2021).
44. Belluco, W., de Chiffre, L.: Performance evaluation of vegetable-based oils in drilling austenitic stainless steel. Journal of Materials Processing Technology. 148, 171–176 (2004). https://doi.org/10.1016/S0924-0136(03)00679-4
45. Ojolo, S.J.I., Amuda, M.O.H., Ogunmola, O.Y.I., Ononiwu, C.U.I.: Experimental determination of the effect of some straight biological oils on cutting force during cylindrical turning. Matérial. 13, 650–663 (2008). http://dx.doi.org/10.1590/s1517-70762008000400011

46. Khan, M.M.A., Mithu, M.A.H., Dhar, N.R.: Effects of minimum quantity lubrication on turning AISI 9310 alloy steel using vegetable oil-based cutting fluid. Journal of Materials Processing Technology. 209, 5573–5583 (2009). https://doi.org/10.1016/J.JMATPROTEC.2009.05.014
47. Gupta, M.K., Song, Q., Liu, Z., Sarikaya, M., Jamil, M., Mia, M., Singla, A.K., Khan, A.M., Khanna, N., Pimenov, D.Y.: Environment and economic burden of sustainable cooling/lubrication methods in machining of Inconel-800. Journal of Cleaner Production. 287 (2021). https://doi.org/10.1016/j.jclepro.2020.125074
48. Singh, G., Aggarwal, V., Singh, S.: Experimental investigations into machining performance of Hastelloy C-276 in different cooling environments. Materials and Manufacturing Processes. 36, 1789–1799 (2021). https://doi.org/10.1080/10426914.2021.1945099
49. Makhesana, M.A., Baravaliya, J.A., Parmar, R.J., Mawandiya, B.K., Patel, K.M.: Machinability improvement and sustainability assessment during machining of AISI 4140 using vegetable oil-based MQL. Journal of the Brazilian Society of Mechanical Sciences and Engineering. 43 (2021). https://doi.org/10.1007/s40430-021-03256-2
50. Shukla, A., Kotwani, A., Unune, D.R.: Performance comparison of dry, flood and vegetable oil based minimum quantity lubrication environments during CNC milling of Aluminium 6061. In: Materials Today: Proceedings. Pp. 1483–1488. Elsevier Ltd (2020). https://doi.org/10.1016/j.matpr.2019.11.060
51. Agrawal, S.M., Patil, N.G.: Experimental study of non edible vegetable oil as a cutting fluid in machining of M2 Steel using MQL. In: Procedia Manufacturing. Pp. 207–212. Elsevier B.V. (2018). https://doi.org/10.1016/j.promfg.2018.02.030
52. Majak, D., Olugu, E.U., Lawal, S.A.: Analysis of the effect of sustainable lubricants in the turning of AISI 304 stainless steel. In: Procedia Manufacturing. Pp. 495–502. Elsevier B.V. (2020). https://doi.org/10.1016/j.promfg.2020.02.183
53. Gaurav, G., Sharma, A., Dangayach, G.S., Meena, M.L.: Assessment of jojoba as a pure and nanofluid base oil in minimum quantity lubrication (MQL) hard-turning of Ti-6Al-4V: A step towards sustainable machining. Journal of Cleaner Production. 272 (2020). https://doi.org/10.1016/j.jclepro.2020.122553
54. Eltaggaz, A., Hegab, H., Deiab, I., Kishawy, H.A.: Hybrid nano-fluid-minimum quantity lubrication strategy for machining austempered ductile iron (ADI). International Journal on Interactive Design and Manufacturing. 12, 1273–1281 (2018). https://doi.org/10.1007/s12008-018-0491-7
55. Ajay Vardhaman, B.S., Amarnath, M., Jhodkar, D., Ramkumar, J., Chelladurai, H., Roy, M.K.: Influence of coconut oil on tribological behavior of carbide cutting tool insert during turning operation. Journal of the Brazilian Society of Mechanical Sciences and Engineering. 40 (2018). https://doi.org/10.1007/s40430-018-1379-y
56. Rahim, E.A., Sasahara, H.: A study of the effect of palm oil as MQL lubricant on high speed drilling of titanium alloys. Tribology International. 44, 309–317 (2011). https://doi.org/10.1016/j.triboint.2010.10.032
57. Upendra, M., Vasu, V.: Synergistic effect between phosphonium-based ionic liquid and three oxide nanoparticles as hybrid lubricant additives. Journal of Tribology. 142 (5), 052101 (2020). https://doi.org/10.1115/1.4045769
58. Maurya, U., Vasu, V., Kashinath, D.: Ionic liquid-nanoparticle-based hybrid-nanolubricant additives for potential enhancement of tribological properties of lubricants and their comparative study with ZDDP. Tribology Letters. 70, 11 (2022). https://doi.org/10.1007/s11249-021-01551-6
59. Maurya, U., Vasu, V., Kashinath, D.: Three-way compatibility study among Nanoparticles, Ionic Liquid, and Dispersant for potential in lubricant formulation. Materials Today: Proceedings. 59 (3), 1651–1658 (2022). https://doi.org/10.1016/J.MATPR.2022.03.329
60. Upendra, M., Vasu, V.: Ionic liquids and its potential in lubricants: A review. Proceedings of TRIBOINDIA-2018 An International Conference on Tribology. (December 13, 2018). Available at SSRN: https://ssrn.com/abstract=3320367 or http://dx.doi.org/10.2139/ssrn.3320367

61. Zhou, Y., Qu, J.: Ionic liquids as lubricant additives: A review. ACS Applied Materials and Interfaces. 9, 3209–3222 (2017). https://doi.org/10.1021/acsami.6b12489
62. Pham, M.Q., Yoon, H.S., Khare, V., Ahn, S.H.: Evaluation of ionic liquids as lubricants in micro milling: Process capability and sustainability. Journal of Cleaner Production. 76, 167–173 (2014). https://doi.org/10.1016/j.jclepro.2014.04.055
63. Davis, B., Schueller, J.K., Huang, Y.: Study of ionic liquid as effective additive for minimum quantity lubrication during titanium machining. Manufacturing Letters. 5, 1–6 (2015). https://doi.org/10.1016/j.mfglet.2015.04.001
64. Karthikey, V.V.: Investigations on the effect of Ionic Liquid (BMIMBF4) based metal working fluids applied using minimum quantity lubrication on surface roughness. International Journal of Pure and Applied Mathematics. 118, 2283–2293 (2018).
65. Abdul Sani, A.S., Rahim, E.A., Sharif, S., Sasahara, H.: Machining performance of vegetable oil with phosphonium- and ammonium-based ionic liquids via MQL technique. Journal of Cleaner Production. 209, 947–964 (2019). https://doi.org/10.1016/j.jclepro.2018.10.317
66. Abdul Sani, A.S., Rahim, E.A., Sharif, S., Sasahara, H.: The influence of modified vegetable oils on tool failure mode and wear mechanisms when turning AISI 1045. Tribology International. 129, 347–362 (2019). https://doi.org/10.1016/j.triboint.2018.08.038
67. Pandey, A., Kumar, R., Sahoo, A.K., Paul, A., Panda, A.: Performance analysis of trihexyltetradecylphosphonium chloride ionic fluid under MQL condition in hard turning. International Journal of Automotive and Mechanical Engineering. 17, 7629–7647 (2020). https://doi.org/10.15282/IJAME.17.1.2020.12.0567
68. Naresh Babu, M., Anandan, V., Dinesh Babu, M.: Performance of ionic liquid as a lubricant in turning Inconel 825 via minimum quantity lubrication method. Journal of Manufacturing Processes. 64, 793–804 (2021). https://doi.org/10.1016/j.jmapro.2021.02.011
69. Abdul Sani, A.S., Rahim, E.A., Sharif, S., Sasahara, H.: Machining performance of vegetable oil with phosphonium- and ammonium-based ionic liquids via MQL technique. Journal of Cleaner Production. 209, 947–964 (2019). https://doi.org/10.1016/j.jclepro.2018.10.317
70. Abdul Sani, A.S., Rahim, E.A., Sharif, S., Sasahara, H.: Machining performance of vegetable oil with phosphonium- and ammonium-based ionic liquids via MQL technique. Journal of Cleaner Production. 209, 947–964 (2019). https://doi.org/10.1016/j.jclepro.2018.10.317
71. Cai, M., Yu, Q., Liu, W., Zhou, F.: Ionic liquid lubricants: When chemistry meets tribology. Chemical Society Reviews. 49, 7753–7818 (2020). https://doi.org/10.1039/d0cs00126k
72. Maurya, U., Vasu, V.: Boehmite nanoparticles for potential enhancement of tribological performance of lubricants. Wear. 498–499, 204311 (2022). https://doi.org/10.1016/J.WEAR.2022.204311
73. Jamil, M., Khan, A.M., Hegab, H., Gong, L., Mia, M., Gupta, M.K., He, N.: Effects of hybrid Al_2O_3-CNT nanofluids and cryogenic cooling on machining of Ti-6Al-4V. International Journal of Advanced Manufacturing Technology. 102, 3895–3909 (2019). https://doi.org/10.1007/s00170-019-03485-9
74. Shen, B., Shih, A.J., Tung, S.C.: Application of nanofluids in minimum quantity lubrication grinding. Tribology Transactions. 51, 730–737 (2008). https://doi.org/10.1080/10402000802071277
75. Naresh Babu, M., Anandan, V., Muthukrishnan, N., Arivalagar, A.A., Dinesh Babu, M.: Evaluation of graphene based nano fluids with minimum quantity lubrication in turning of AISI D3 steel. SN Applied Sciences. 1 (2019). https://doi.org/10.1007/s42452-019-1182-0
76. Duc, T.M., Long, T.T., van Thanh, D.: Evaluation of minimum quantity lubrication and minimum quantity cooling lubrication performance in hard drilling of Hardox 500 steel using Al_2O_3 nanofluid. Advances in Mechanical Engineering. 12 (2020). https://doi.org/10.1177/1687814019888404
77. Pal, A., Chatha, S.S., Sidhu, H.S.: Tribological characteristics and drilling performance of nano-MoS2-enhanced vegetable oil-based cutting fluid using eco-friendly MQL technique in drilling of AISI 321 stainless steel. Journal of the Brazilian Society of Mechanical Sciences and Engineering. 43 (2021). https://doi.org/10.1007/s40430-021-02899-5
78. Virdi, R.L., Chatha, S.S., Singh, H.: Processing characteristics of different vegetable oil-based nanofluid MQL for grinding of Ni-Cr alloy. Advances in Materials and Processing Technologies. (2020). https://doi.org/10.1080/2374068X.2020.1800312

79. Gugulothu, S., Pasam, V.K., Revuru, R.S.: Machining performance and sustainability of vegetable oil based nano cutting fluids in turning. Proceedings of 10th International Conference on Precision, Meso, Micro and Nano Engineering (COPEN 10), 871–875. (2017). ISBN: 978-93-80689-28-9
80. Uysal, A., Demiren, F., Altan, E.: Applying minimum quantity lubrication (MQL) method on milling of martensitic stainless steel by using nano MoS_2 reinforced vegetable cutting fluid. Procedia—Social and Behavioral Sciences. 195, 2742–2747 (2015). https://doi.org/10.1016/j.sbspro.2015.06.384
81. Gajrani, K.K., Suvin, P.S., Kailas, S.V., Mamilla, R.S.: Thermal, rheological, wettability and hard machining performance of MoS_2 and CaF_2 based minimum quantity hybrid nano-green cutting fluids. Journal of Materials Processing Technology. 266, 125–139 (2019). https://doi.org/10.1016/j.jmatprotec.2018.10.036
82. Sarhan, A.A.D., Sayuti, M., Hamdi, M.: Reduction of power and lubricant oil consumption in milling process using a new SiO_2 nanolubrication system. International Journal of Advanced Manufacturing Technology. 63, 505–512 (2012). https://doi.org/10.1007/s00170-012-3940-7
83. Ni, J., Cui, Z., Wu, C., Sun, J., Zhou, J.: Evaluation of MQL broaching AISI 1045 steel with sesame oil containing nano-particles under best concentration. Journal of Cleaner Production. 320 (2021). https://doi.org/10.1016/j.jclepro.2021.128888
84. Anand, R., Raina, A., Irfan Ul Haq, M., Mir, M.J., Gulzar, O., Wani, M.F.: Synergism of TiO_2 and graphene as nano-additives in bio-based cutting fluid: An experimental investigation. Tribology Transactions. 64, 350–366 (2021). https://doi.org/10.1080/10402004.2020.1842953
85. Das, A., Patel, S.K., Das, S.R.: Performance comparison of vegetable oil based nanofluids towards machinability improvement in hard turning of HSLA steel using minimum quantity lubrication. Mechanics and Industry. 20 (2019). https://doi.org/10.1051/meca/2019036
86. Gong, L., Bertolini, R., Ghiotti, A., He, N., Bruschi, S.: Sustainable turning of Inconel 718 nickel alloy using MQL strategy based on graphene nanofluids. The International Journal of Advanced Manufacturing Technology. 108, 3159–3174 (2020). https://doi.org/10.1007/s00170-020-05626-x/Published

Chapter 6

Laser micro-texturing of silicon for reduced reflectivity

S. Purushothaman, M. S. Srinivas, R. R. Behera, N. Venkaiah, and M. R. Sankar

Contents

Abbreviations

A: Adsorbed radiation
ARC: Anti-reflective coating
c-Si: Single crystalline Si
d: Diameter of micro-hole
D: Focusing depth
F: Pulse repetition frequency
IP: Inverted pyramid
LIPSS: Laser-induced periodic surface structures
mc-Si: Multi-crystalline Si
OF: Overlapping factor
R: Reflected radiation
Ra: Surface roughness
SS: Scanning speed

DOI: 10.1201/9781003291961-7

T: Transmitted radiation
Λ: Wavelength

6.1 Introduction

Solar energy is one of the most commonly available renewable energy, and a lot of research is being done for the efficient conversion of solar to electrical energy. The unique electronic configuration and ease of availability make silicon an ideal material to perform this conversion [1, 2]. But dependence on sunlight for energy has some disadvantages [1], that is, the incoming solar energy depends upon time, climate, location, season, and atmospheric pollution, and a large surface area is required to produce an appreciable amount of electricity. For the proper performance of Si cells, the factors mentioned above should be considered. When sunlight, which is a spectrum of different electromagnetic radiations, falls on the Si cell, only those radiations which are adsorbed (A) into the cell will be converted into electricity. The rest of the radiations will simply pass through as transmitted radiations (T) or reflected (R) away, as shown in Eq. (6.1).

$$A = 1 - T - R \tag{6.1}$$

Thus, the ideal way to increase the electricity output from a solar cell is to reduce the transmitted and reflected radiation. The reduction in transmitted radiations is possible with the texturing of Si cells or by increasing the thickness of Si cells.

6.2 Texturing of silicon

The reflectance value (R) of a typical Si cell varies between 30% and 60% [3] because of the high difference in refractive index between air and Si [4]. Hence, reducing the reflection of incoming radiations is one of the most effective ways to improve the efficiency of the Si cell. This is possible by a combination of surface texturing and anti-reflective coating (ARC). Surface texturing was found to enhance the light absorption of Si in the following ways:

- The incoming radiation, which got reflected away from one surface feature (e.g., a cone), may strike another surface feature, increasing the chances of being adsorbed [5].
- The radiation entering Si will be entering at an angle (due to refraction). Hence, the light rays have to travel more distance than the thickness of the Si cell, thereby reducing the transmission losses [6].
- Additionally, the long-wavelength photons reflected from other surface features may encounter a tilted Si surface, improving the chances of being internally reflected, thereby increasing the distance traveled within the Si and, consequently, the chances of being adsorbed [7, 8].
- The surface micro texturing was found to introduce hydrophobicity, which repels the dust particles that fall over the Si cell, thereby keeping the surface clean [9, 10]. Furthermore, extra-fine particles can settle between the textures, reducing the effectiveness of the texturing. Hence, various techniques, such as ultrasonic cleaning [11] and surface coating [12], are used in addition to laser texturing to ensure the prevention or easy removal of the settled dust particles.

To bring about these benefits, multiple anti-reflective textures, such as grooves [13–16], honeycomb structures [17], inverted pyramids [18, 19], and cones [20], have been developed. The influence of these textures will be discussed in upcoming units.

6.2.1 Laser texturing of silicon

Texturing of Si can be performed using multiple techniques, including wet chemical etching [21], reactive ion etching [22], mechanical grooving [23], electrochemical etching [24], masked laser texturing [25], and direct laser texturing [26].

Wet chemical etching and reactive ion etching involve the usage of liquid chemicals to remove material. They have a high material removal rate with a surface finish, but the disposal of chemicals is challenging. Mechanical grooving involves the usage of cutting tools to create textures over Si. It is the cheapest mode of texturing, but the major limitation of mechanical grooving is its difficulty in performing complex structures. During electrochemical etching, the material removal occurs due to an anodic dissolution process in the presence of electric energy and dielectric fluid. It is an environment-friendly texturing technique with a high material removal rate but is applicable only for electrically conductive materials.

Laser texturing utilizes the thermal energy from lasers to perform texturing. Using lasers, complicated textures can be made for any material with no environmental impact. Hence, compared to the above-mentioned techniques, laser texturing is one of the most suitable processes for Si texturing. Over the last few decades, laser texturing has been performed successfully on a wide range of materials, including metals [27–29], semiconductors [30], dielectrics [31, 32], and ceramics [33, 34]. In 1989, Zolper et al. [35] performed the laser texturing of Si to reduce reflectivity. Since then, a lot of research has been done to perform laser texturing on Si for multiple reasons, including increasing the absorption of light [36], dielectric layer ablation [18], selective emitter formation [37], hydrophobic texturing [38], etc. Also, the laser machining process is used in drilling [39], scribing [40], joining [41], doping [42], etc., Si for the production of solar cells. Hence, laser machining is a highly versatile machining process.

Additionally, various chemical and electrochemical processes specified earlier are not entirely effective in machining multi-crystalline silicon due to the random orientation of crystallographic grain and high etching sensitivity towards specific directions [43]. Similarly, mechanical processes are not satisfactory due to silicon's high fragility and breakability and the low thickness of Si wafers. Thus, laser machining is the only way to overcome these difficulties [44]. The experimental setup and schematic of a typical laser texturing system are depicted in Figure 6.1 (a), and the insert in Figure 6.1 (b) indicates a laser-textured surface.

A typical Si texturing by laser consists of three steps: masking, ablation, and etching. Masking is the process wherein a silicon nitride (SiN) layer is deposited over the Si surface by chemical vapor deposition. The primary purpose of the mask is to protect the untextured Si surface from the etching process. After masking, laser texturing is done, which creates the preferred texture over the Si wafer. It is followed by the etching process, where chemicals are used to remove the ablation slag formed due to laser ablation.

Laser machining is a complicated thermal material removal process comprising various physical phenomena, such as heat transfer, melting, vaporization, and solidification [45]. Hence, proper control of machining parameters is unavoidable. In the next section, the influence of various machining parameters on the final texture is presented in detail.

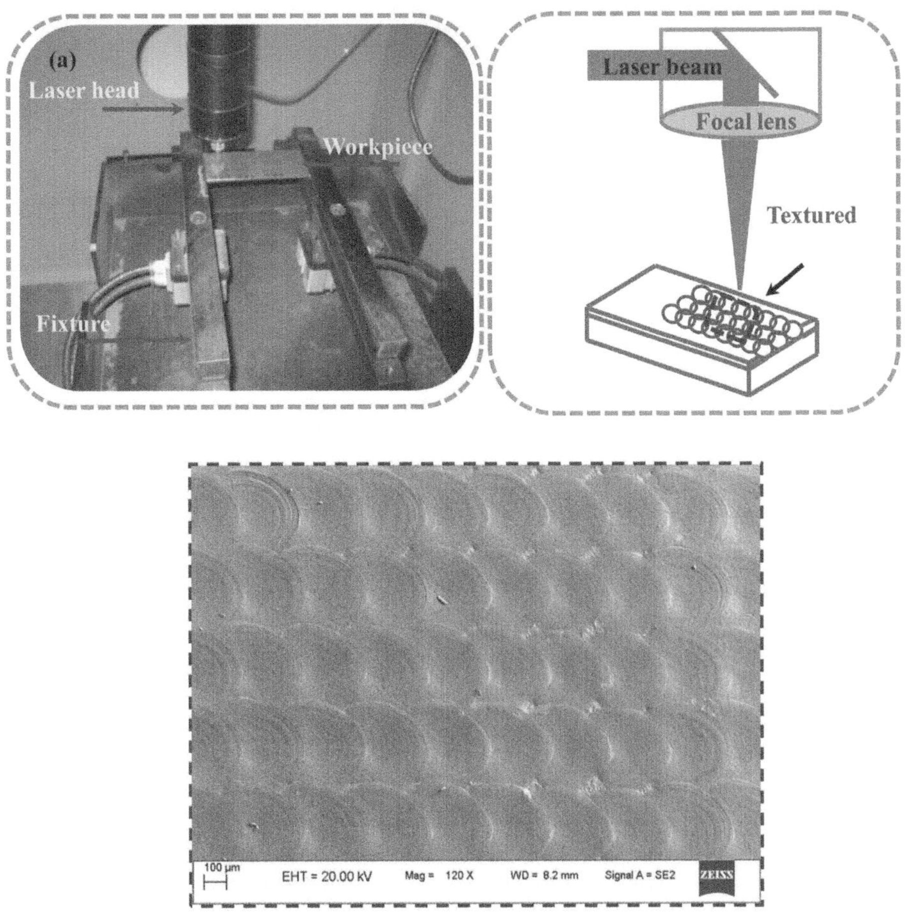

Figure 6.1 (a) Experimental setup and (b) schematics of a typical laser texturing system, and (c) a typical laser textured surface [27].

6.3 Factors influencing the surface reflectivity

The surface reflectivity of the Si wafer is a function of the surface topology, which depends upon various laser parameters used to perform laser texturing. It is a well-known fact that any surface with a mirror-like finish has a high reflectivity [46]. On the other hand, it can be said that any surface with higher roughness exhibits less surface reflectivity. During laser texturing, the surface roughness is affected, which depends upon the number of laser processing parameters. The influence of these parameters is discussed in this section.

6.3.1 Ablation pit depth

With the increase in the number of pulses per texture pit, the pit depth increases. This variation in pit depth was found to create variations in the surface reflections on Si [14, 24], as shown in Figure 6.2. Initially, with the increase in pit depth, the reflection decreased

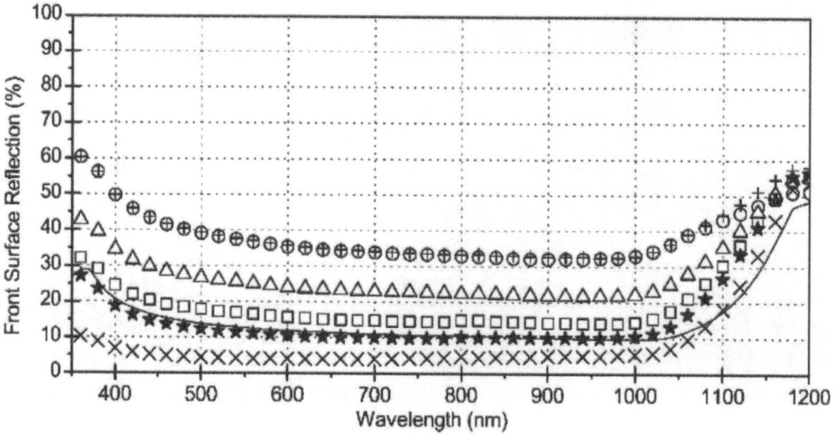

Figure 6.2 Reflection of the front surface of laser textured samples for different ablation pit depths of (○) 10 mm, (Δ) 20 mm, (□) 30 mm, (*) 40 mm, and (x) 50 mm with residual slag, (+) planar silicon, and (line) random pyramid textured silicon [24].

proportionally until a specific limit. During etching, the removal of Si occurs such that more material removal occurs at the top rather than the bottom. Hence, with the increase in etching time, the pits start to open up, decreasing the reflection due to the double bounce of the incident light [14].

After a specific limit, it is impossible to decrease the reflectivity, and a further increase in pit depth leads to an increase in reflectivity. This happens because, as the pits become wider, a limit is reached where the pits join. Any further etching will cause an increase in reflection due to the flattening of the top surfaces and inefficient slag removal of the bottom surface of the pits (which is detrimental to electrical performance). Thus, the pit depth, which results in complete slag removal and causes the pits to join each other, is the ideal pit depth. As shown in Figure 6.2, a 40-mm pit depth can be considered the ideal pit depth [14].

6.3.2 Etching time

Spatter is the molten metal ejected during laser machining due to intense pressure generated in the laser-irradiated region. This molten metal (remelt) then solidifies over the surface, thereby declining the textured surface quality. Hence, the removal of spatter is necessary. Etching is one of the techniques utilized to remove the surface remelt. The amount of material removed by etching is a function of etching time.

An increase in etching time causes the reflection to reduce up to a level, then starts increasing [16]. Initially, with an increase in etching time, the dimensions (aspect ratio) of the texture increase, and along with it, the reflection of the silicon wafer reduces (as in Figure 6.3 (a) and (b)). This is due to the removal of the spatter from the surface. This

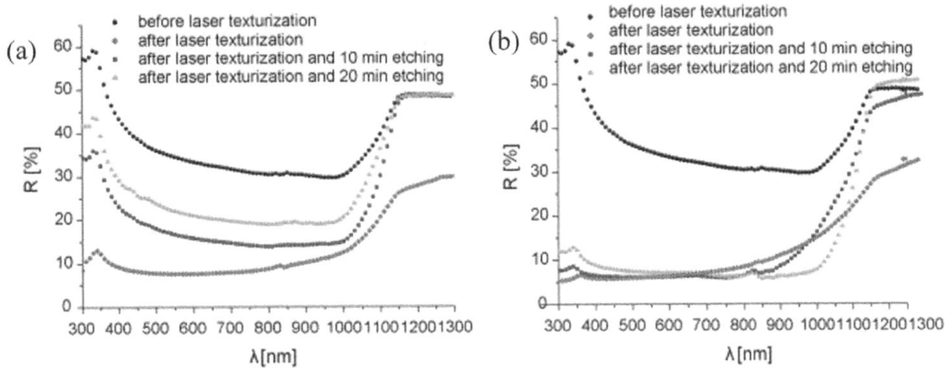

Figure 6.3 Reflection vs. wavelength for varying etching time for (a) parallel groove and (b) perpendicular groove [47].

reflection reduction continues up to a limit of texture dimensions, above which a steep rise in reflection can be observed. This is because of the overlapping of the textures, which reduces the overall aspect ratio and thus leads to a sudden surge in reflectivity. The reason for the increase in reflection due to overlapping is explained in Section 6.4.1.

6.3.3 *Pulse energy/focusing depth*

The pulse energy and focusing depth influence the spike height of the texture [18, 19]. The spike height refers to the height of the microstructure from the base of Si. From the previous subtopics, it is observed that the increase in the height of the spike can lower the reflectivity of Si. This parameter applies to the cone, groove, and honeycomb textures. The increase in spike height increases the adsorption [18, 44] and reduces the reflection [19] of the incident light energy.

The relation between the average spike height and laser focusing depth is explained in Figure 6.4 (a). Initially, while focusing the laser beam on the Si surface, that is, when the focusing distance is 0 mm, the energy density on the surface is too high, and less energy is transferred to the material. Hence, the Si gets converted to liquid, and spikes cannot be formed. With the increase in focal depth, the energy density at the surface reduces, and hence more energy is distributed to the material. Thus, surface defects and ripples of the order of the laser wavelength will be formed, thereby facilitating spike formation. This explanation is proved in Figure 6.4 (b), where a reduction in reflectivity is observed when the focusing depth rises from 0 mm to 20 mm. But the reflectivity increases with a further rise in focusing depth above 20 mm, as in Figure 6.4 (b). This is because the energy density at the surface becomes too low to create the spike at a high focusing depth [18, 25]. Additionally, the variation of pulse energy has the same outcome as the variation of focusing depth. This is evident in Figure 6.5 (a), where the average spike height vs. variation in a single pulse energy graph looks similar to Figure 6.4 (a).

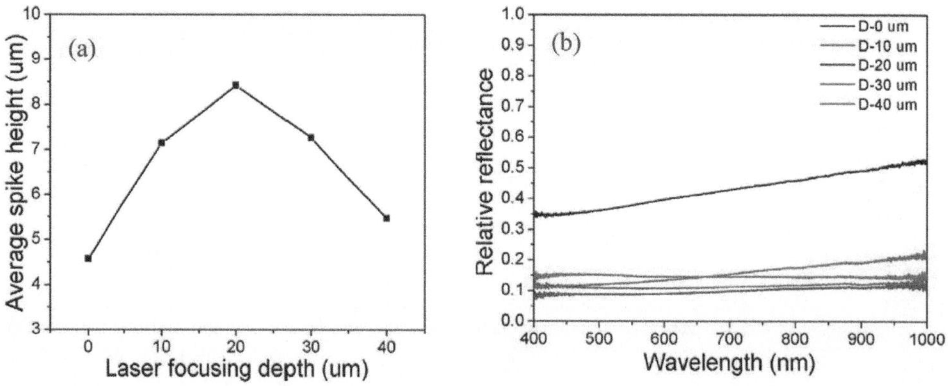

Figure 6.4 (a) The average height of spike vs. laser focusing depth. (b) Relative reflectance vs. wavelength for monocrystalline Si for varying laser focusing depth (D) [19].

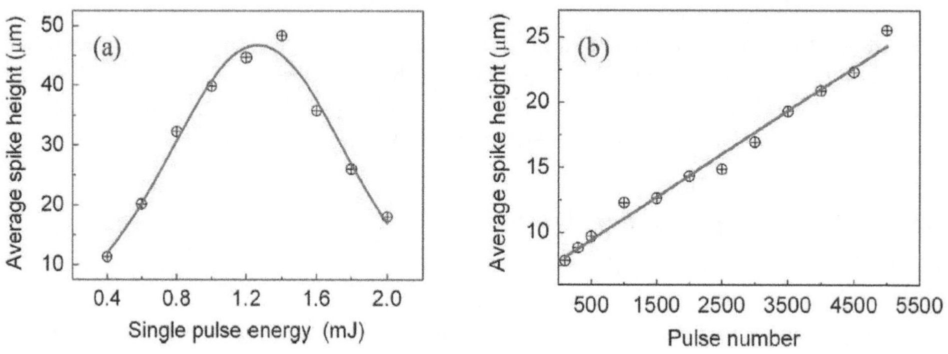

Figure 6.5 (a) Average spike height vs. single pulse energy; □ – experimental data and solid line-fitted curve (gauss function). (b) Average spike height vs. pulse number; – experimental data and solid line-fitted curve (straight line) [18].

6.3.4 Pulse number

The increase in the number of pulses means an increase in the spike height, as observed in Figure 6.5 (b). This is because more energy is available with the increase in the number of pulses. Since the energy is in the form of pulses, the energy is absorbed into the bulk of the material rather than being conducted away, as in the case of a continuous laser beam. Thus, the rise in spike height indicates a reduction in the reflectance of the Si surface. Therefore, the supplied energy is directly utilized to increase spike height [14, 18].

6.3.5 Scanning speed

The increased scanning speed means fewer incident pulses at a spot, that is, lower incident energy. At high scanning speeds, only minor surface modification can be observed on the Si surface. Also, the distance between these microstructures is relatively small, acting as a flat surface.

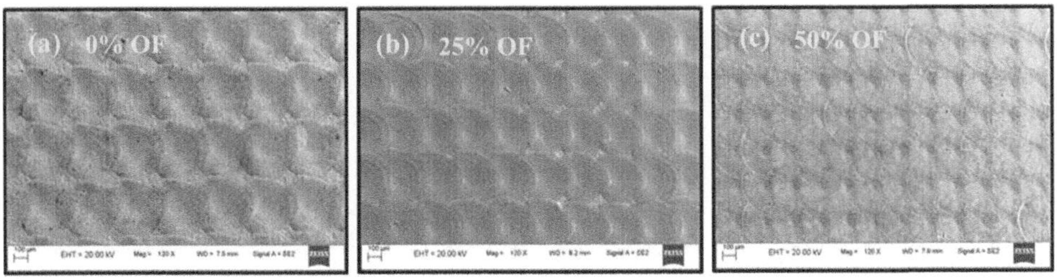

Figure 6.6 Laser textured surface having 0% OF, 25% OF, and 50% OF between two consecutive spots and lines [49].

Hence the reflection is high at high scanning speeds. With the decrease in scanning speeds, the inter-microstructure distance also increases, leading to better anti-reflection properties [25].

By keeping different laser parameters, like input current, pulse duration, pulse repetition frequency, stand-off distance, and focus spot diameter constant, a micro-hole having a constant diameter and depth can be obtained [48]. At this condition, if the scanning speed is varied, a difference in overlapping factor between the microholes can be observed. [49]. When the scanning speed increases, the distance between the micro-hole is increased. However, when the scanning speed decreases, the overlapping between the micro-hole increases. Hence, in order to obtain a particular overlapping factor (OF) between the micro-holes, the scanning speed can be changed according to the following equation [50].

$$SS = f \times d \times (1 - OF/100) \tag{6.2}$$

where SS = scanning speed, f = pulse repetition frequency, and d = diameter of micro-hole. Figure 6.6 shows the micro-textured surface having 0%, 25%, and 50% OF between the two consecutive laser spots as well as laser lines [49]. With the increasing OF, the surface roughness (R_a) increases. This is because the frequency of laser interaction increases with the increase in OF. This creates pits and waves on the surface of molten material after solidification. Hence, the R_a increases, which may reduce the reflectivity of a surface.

6.4 Textures

Texture is the most crucial factor that decides the level of reflectivity of the Si component. In this section, some of the textures used and the literature associated with those textures are discussed.

6.4.1 Grooves

Groove is one of the most studied textures, consisting of long and narrow lines ablated over the surface of Si. The lines can be parallel or perpendicular, giving rise to different surfaces, as shown in Figures 6.7 (a) and (b). To produce groove textures, the grooves were scribed over the Si wafer at a constant spacing. During laser texturing, an intense laser pulse causes silicon to melt, out of which some amount can flow out to form the rim, and the rest can fill

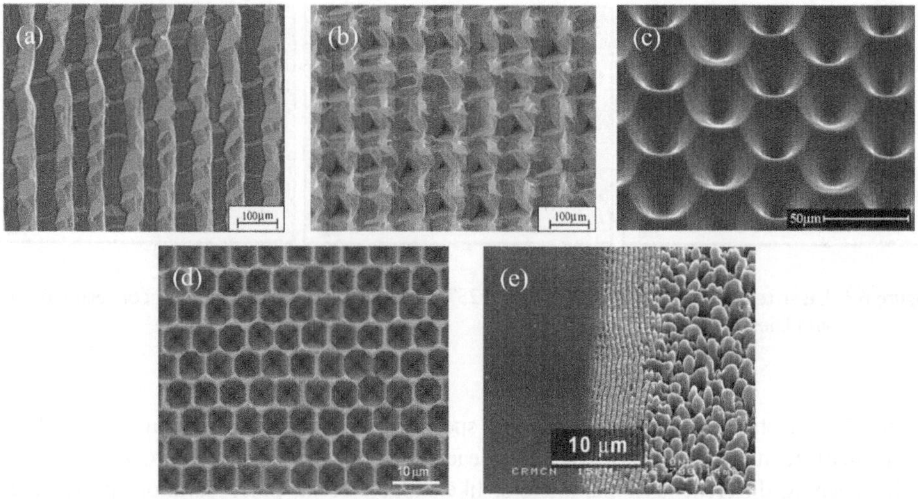

Figure 6.7 Different types of textures over Si surface: (a) parallel grooves [47]; (b) perpendicular grooves [47]; (c) honeycomb [24]; (d) inverted pyramid [53]; and (e) laser-induced periodic surface structures [54].

the existing grooves [13]. Hence, it is followed by a wet etching treatment to remove the laser-damaged layer. Figures 6.7 (a) and (b) illustrate a typical parallel and perpendicular groove configuration on Si. One of the variants of groove texturing is black Si, where micro-grooves are made close to each other [44].

Engelhart et al. [45] successfully performed parallel grove texturing on single crystalline Si (c-Si) and compared the effectiveness of different lasers in performing laser texturing of Si. Dobrzański et al. [37] performed perpendicular groove texturing on Si and observed the existence of a laser-damaged layer, which hindered the efficiency of the solar cells. Thus, they removed this layer by etching with a potassium alkali, and the overall efficiency went up by 10.2%. In 2008, Dobrzański et al. [47] performed groove texturing over multi-crystalline Si (mc-Si) to reduce the reflectivity of Si. In the following year, they [43] proved that it's possible to perform surface texturing of mc-Si, independent of the crystallographic orientation of grains using groove texturing.

Dobrzański et al. [13] compared parallel and perpendicular grooves, and a significant reduction in effective reflectance was observed for perpendicular grooves compared to the data from his paper in 2009 [15]. Vorobyev et al. [44] was able to bring down the total reflectivity to less than 5% by creating black Si. Later, Yang et al. [51] fabricated Si nanowires over the Si surface, making it possible to reduce the light reflectivity to less than 1% (wavelength between 300 nm and 1000 nm) by light trapping.

6.4.2 Honeycomb

Honeycomb texture comprises thin hollow structures of a specific geometry separated by thin walls (Figure 6.7 (c)). The most common geometry is columnar or hexagonal. Using honeycomb texturing, a minimum reflectance of 10% can be achieved. The first honeycomb texturing over Si was performed by Abbott and Cotter [24], wherein pits were ablated over an Si surface in an interlocking pattern, as shown in Figure 6.7 (c). It is followed by a two-stage chemical etching, which removes the slag, smoothens the resulting surface, and removes residual laser damage.

Grischke et al. [14], Niinobe et al. [15], and Volk et al. [16] did honeycomb texturing over Si by masking method, followed by laser texturing and etching. During masking, a uniform layer of silicon nitride is deposited over Si using the chemical vapor deposition technique. The mask is partially removed using laser texturing, which is followed by chemical etching, which results in the generation of a honeycomb texture. Yang et al. [52] used a wave interference technique using three laser beams to create a honeycomb texture over Si. When three laser beams were made to intersect over a point, laser irradiation of the material occurs along the generated interference pattern, resulting in the formation of concave holes. Here, masking is not involved, which simplifies the fabrication process.

Niinobe et al. [15] performed honeycomb texturing over mc-Si, and the resulting cell observed an overall conversion efficiency of 19.1%. Volk et al. [16] studied the reflection variation for different etching times and found that it is possible to machine a homogenous honeycomb texture in 80 seconds. In the same year, Yang and Lee [52] created a honeycomb mesh using laser interferometry. They were able to manipulate the shape of the resulting texture by changing the chemicals used in etching. The following year, they [4] studied the importance of the in-plane orientation relationship between interfering laser beams.

6.4.3 *Inverted pyramid*

The inverted pyramid (IP) is one of the most efficient patterns, owing to its excellent anti-reflection properties (5–10%), light trapping properties, and minimal material removal [3]. There are multiple ways to perform an inverted pyramid texturing over Si. Kumar et al. [3] performed IP texturing over the Si surface using masking technique. Figure 6.7 (d) shows the resultant IP textured Si surface. They studied the influence of the size of the inverted pyramid over the reflectance and observed that 1.3 μm sized textures resulted in the least reflectance. Yang et al. [4] used the laser interference technique to produce IP texture. Here, the laser beam was split into three beams using a trigonal pyramid-shaped prism and was made to overlap on the wafer symmetrically. Using this method, a surface reflectance lesser than 20% was attained at low costs. Ji et al. [17] created micro-pores over Si surface using laser ablation followed by etching. The resulting IP texture reduced the reflectivity to 6% in the wavelength range of 400–1000 nm.

6.4.4 *Self-assembled texture/laser-induced periodic surface structures*

Self-assembly is a process wherein a pre-existing disordered surface is converted into a pattern due to specific, local interactions between the material components without external direction. Laser-induced periodic surface structures (LIPSS) can be generated using a self-assembly technique. According to Gurevich [55], ablative Rayleigh–Taylor instability is the basic mechanism behind the formation of LIPSS. Accordingly, when material ablation takes place, a recoil force is generated. This recoil force decides the final surface topology of the LIPSS. This texturing technique has advantages over other laser textures owing to its high reproducibility, uniformity of micro/nanostructures, and applicability to c-Si, mc-Si, and thin films [56]. Additionally, it is simple, quick, and cost-effective. Hence, a lot of research has been done on LIPSS texturing of Si wafers [25].

For the first time, a conical self-assembled texture was attained over Si in 1998 by Her et al. [57] by repeatedly irradiating the c-Si surface with femtosecond laser pulses in an atmosphere of SF_6 or Cl_2. Pedraza and Fowlkes [58] did self-assembled texturing over Si for different orientations of c-Si and studied the influence of the gas environment over the dimensions of conical protrusion over Si. In 2008, Halbwax et al. [54] did a detailed study on the variation of laser-induced periodic surface structures (LIPSS) by varying the machining conditions and observed that the spike was the best structure in terms of light absorption. Figure 6.7 (e) shows a typical LIPSS texture on Si by Halbwax et al. They produced capillary waves with a periodicity of 800 nm in the center and beads with a periodicity of about 2 µm towards the right. Lee et al. [26] textured 20 µm and 400 µm thick Si wafers and found that the final absorption is 80% and 90%, respectively, for the wavelength range of 400–1000 nm.

Tavera et al. [59] created LIPSS over Si using a nanosecond laser using the laser interference technique. For this purpose, they used a four-beam configuration and experimented with the modification of surface morphology for different laser fluences. They observed the formation of micro and nano ripples, wherein micro ripples were observed for low fluences. With the increase in laser fluence, micro ripples diminished due to the formation of nano ripples.

Binetti et al. [60] performed LIPSS texturing over mc-Si for different wavelengths and observed that the lowest reflectivity of 8% is achieved using IR laser, followed by 13% and 11% for green and UV lasers, respectively. Additionally, UV laser induced the lowest subsurface damage to Si wafers. Sarbada et al. [61] performed a simulation of LIPSS texturing for different types of profiles and observed that the elliptical profiles resulted in the lowest average reflectance.

Parmar and Shin [25] observed that LIPSS texturing with high energy, that is, low scan speed and high fluence, results in a sparsely packed nano/microstructure due to material ablation, thereby resulting in efficient trapping of incident light. But texturing with low energy will result in densely packed nano/microstructure, as there is no material ablation, thereby leading to a high reflectance. Additionally, they observed that globular nano/microstructures are formed when laser texturing is performed in air, while the same process results in elongated columnar microstructure when performed in water. Singh and Das [62] were able to generate sub-wavelength LIPSS over Si using a low-cost N_2 laser over a large surface area and successfully reduced the reflection to 10%.

6.5 Conclusion

Ablation pit depth, etching time, focusing time, pulse energy, pulse number, and scanning speed are the factors to be considered while laser texturing Si wafer and among them. Pulse energy is the most influential parameter, as the pulse energy is directly proportional to the maximum temperature attained on the material surface. Thus, an increase in pulse energy results in a significant rise in the ablation rate, thereby increasing the height of the textures developed. This leads to a substantial reduction in reflectivity. Also, a brief discussion is devoted to some of the most used textures over Si wafers, such as grooves, honeycombs, and inverted pyramids. Different researchers devised multiple methods to realize these textures on the Si wafer using a combination of laser texturing with/without chemical etching.

Acknowledgments

Authors are grateful to "Elsevier (License numbers: 5315901452502, 4916891365269, 4916910898794, 4916931097712, 4916941207887, 4917110818314, 4916950124830" for granting the copyright permission to various figures in this chapter.

References

1. A. Goetzberger, V. U. Hoffmann.: Photovoltaic solar energy generation. Springer Science & Business Media, 112 (2005).
2. Z. M. Jarzębski.: Solar energy: Photovoltaic conversion. PWN, Warsaw (1990).
3. K. Kumar, K. C. Lee, J. Nogami, P. R. Herman, N. P. Kherani.: Ultrafast laser direct hard-mask writing for high performance inverted-pyramidal texturing of silicon In 2012 38th IEEE Photovoltaic Specialists Conference, 002182–002185 (2012).
4. B. Yang, M. Lee.: Laser interference-driven fabrication of regular inverted-pyramid texture on monocrystalline Si. Microelectronic Engineering 130, 52–56 (2014).
5. M. Lipiński, P. Zięba, A. Kamiński.: Crystalline silicon solar cells, in foundation of materials design. Research Signpost, 285–308 (2006).
6. M. A. Green, J. Zhao, A. Wang, A. S. R. Wenham.: Progress and outlook for high-efficiency crystalline silicon solar cells. Solar Energy Materials and Solar Cells 65(1–4), 9–16 (2001).
7. J. F. Nijs, J. Szlufcik, J. Poortmans, S. Sivoththaman, R. P. Mertens.: Advanced cost-effective crystalline silicon solar cell technologies. Solar Energy Materials and Solar Cells 65(1–4), 249–259 (2001).
8. D. H. Macdonald, A. Cuevas, M. J. Kerr, C. Samundsett, D. Ruby, S. Winderbaum, A. Leo.: Texturing industrial multicrystalline silicon solar cells. Solar Energy 76(1–3), 277–283 (2004).
9. B. S. Yilbas, H. Al-Qahtani, A. Al-Sharafi, S. Bahattab, G. Hassan, N. Al-Aqeeli, M. Kassas.: Environmental dust particles repelling from a hydrophobic surface under electrostatic influence. Scientific Reports 9(1), 1–18 (2019).
10. Y. Song, R. P. Nair, M. Zou, Y. Wang.: Superhydrophobic surfaces produced by applying a self-assembled monolayer to silicon micro/nano-textured surfaces. Nano Research 2(2), 143–150 (2009).
11. A. A. Abubakar, B. S. Yilbas, H. Al-Qahtani, A. Alzaydi.: Environmental dust repelling from hydrophilic/hydrophobic surfaces under sonic excitations. Scientific Reports, 10(1), 1–22 (2020)
12. S. Maharjan, K. S. Liao, A. J. Wang, K. Barton, A. Haldar, N. J. Alley, H. J. Byrne, S. A. Curran.: Self-cleaning hydrophobic nanocoating on glass: A scalable manufacturing process. Materials Chemistry and Physics, 239, 122000 (2020).
13. L. A. Dobrzański, A. Drygała.: Influence of laser processing on polycrystalline silicon surface. In Materials Science Forum. Trans Tech Publications Ltd. 706, 829–834. (2012).
14. R. Grischke, B. Terheiden, S. Mau, N. P. Harder, A. Schoonderbeek, R. Kling, A. Ostendorf, B. Denkena, R. Brendel.: Laser surface texturing for reducing reflection losses in multicrystalline silicon solar cells. International Congress on Applications of Lasers & Electro-Optics, Laser Institute of America 2007(1), 305 (2007).
15. D. Niinobe, H. Morikawa, S. Hiza, T. Sato, S. Matsuno, H. Fujioka, T. Katsura, T. Okamoto, S. Hamamoto, T. Ishihara, S. Arimoto.: Large-size multi-crystalline silicon solar cells with honeycomb textured surface and point-contacted rear toward industrial production. Solar Energy Materials and Solar Cells 95(1), 49–52 (2011).
16. A. K. Volk, S. Gutscher, A. Brand, W. Wolke, M. Zimmer, H. Reinecke.: Laser assisted honeycomb-texturing on multicrystalline silicon for industrial applications. In 28th European PV Solar Energy Conference and Exhibition, 1024–1028 (2013).
17. L. Ji, X. Lv, Y. Wu, Z. Lin, Y. Jiang.: Hydrophobic light-trapping structures fabricated on silicon surfaces by picosecond laser texturing and chemical etching. Journal of Photonics for Energy 5(1), 053094 (2015).

18. Y. Peng, Y. Zhou, X. Chen, Y. Zhu.: The fabrication and characteristic investigation of microstructured silicon with different spike heights. Optics Communications 334, 122–128 (2015).
19. Q. Wang, W. Zhou.: Direct fabrication of cone array microstructure on monocrystalline silicon surface by femtosecond laser texturing. Optical Materials 72, 508–512 (2017).
20. S. Koynov, M. S. Brandt, M. Stutzmann.: Black nonreflecting silicon surfaces for solar cells. Applied Physics Letters 88(20), 203107 (2006).
21. Y. J. Hung, S. L. Lee, K. C. Wu, Y. Tai, Y. T. Pan.: Anti-reflective silicon surface with vertical-aligned silicon nanowires realized by simple wet chemical etching processes. Optics Express 19(17), 15792–15802 (2011).
22. L. L. Ma, Y. C. Zhou, N. Jiang, X. Lu, J. Shao, W. Lu, J. Ge, X. M. Ding, X. Y. Hou.: Wide-band "black silicon" based on porous silicon. Applied Physics Letters 88(17), 171907 (2006).
23. M. Ben Rabha, S. B. Mohamed, W. Dimassi, M. Gaidi, H. Ezzaouia, B. Bessais.: Reduction of absorption loss in multicrystalline silicon via combination of mechanical grooving and porous silicon. Physica Status Solidic 8(3), 883–886 (2011).
24. M. Abbott, J. Cotter.: Optical and electrical properties of laser texturing for high-efficiency solar cells. Progress in Photovoltaics: Research and Applications 14(3), 225–235 (2006).
25. V. Parmar, Y. C., Shin.: Wideband anti-reflective silicon surface structures fabricated by femtosecond laser texturing. Applied Surface Science 459, 86–91 (2018).
26. B. G. Lee, Y. T. Lin, M. J. Sher, E. Mazur, H. M. Branz.: Light trapping for thin silicon solar cells by femtosecond laser texturing In 2012 38th IEEE Photovoltaic Specialists Conference, 001606–001608 (2012).
27. R. R. Behera, A. Das, A. Hasan, D. Pamu, L. M. Pandey, M. R. Sankar: Deposition of biphasic calcium phosphate film on laser surface textured Ti-6Al-4V and its effect on different biological properties for orthopedic applications. Journal of Alloys and Compounds. 842, 155683 (2020).
28. R. R. Behera, P. M. Babu, K. K. Gajrani, M. R. Sankar: Fabrication of micro-features on 304 stainless steel (SS-304) using Nd: YAG laser beam micro-machining. International Journal of Additive and Subtractive Materials Manufacturing 1(3–4), 338–359 (2017). Doi: 10.1504/IJASMM.2017.10010934.
29. K. K. Gajrani, M. R. Sankar. State of the art on micro to nano textured cutting tools. Materials Today: Proceedings 4(2), 3776–3785 (2017).
30. C. Y. Chen, C. J. Chung, B. H. Wu, W. L. Li, C. W. Chien, P. H. Wu, C. W. Cheng.: Microstructure and lubricating property of ultra-fast laser pulse textured silicon carbide seals. Applied Physics A 107(2), 345–350 (2012).
31. S. Lei, Y. C. Shin, F. P. Incropera.: Deformation mechanisms and constitutive modeling for silicon nitride undergoing laser-assisted machining. International Journal of Machine tools and manufacture 40(15), 2213–2233 (2000).
32. H. Yamakiri, S. Sasaki, T. Kurita, N. Kasashima.: Effects of laser surface texturing on friction behavior of silicon nitride under lubrication with water. Tribology International 44(5), 579–584 (2011).
33. T. Akova, O. Yoldas, M. S. Toroglu, H. Uysal.: Porcelain surface treatment by laser for bracket-porcelain bonding. American Journal of Orthodontics and Dentofacial Orthopedics 128(5), 630–637 (2005).
34. B. S. Yilbas, M. Khaled, N. Abu-Dheir, N. Aqeeli, S. Z. Furquan.: Laser texturing of alumina surface for improved hydrophobicity. Applied Surface Science 286, 161–170 (2013).
35. S. A. G. D. Correia, J. Lossen, M Wald, K. Neckermann, M. Bähr.: Selective laser ablation of dielectric layers. In Proce[ed]ings of 22[nd] European Photovoltaic Solar Energy Conference, Milan (2007).
36. R. S. Davidsen, H. Li, A. To, X. Wang, A. Han, J. An, J. Colwell, C. Chan, A. Wenham, M. S. Schmidt, A. Boisen.: Black silicon laser-doped selective emitter solar cell with 18.1% efficiency. Solar Energy Materials and Solar Cells, 144, 740–747 (2016).
37. L. A. Dobrzański and A. Drygała.: Laser processing of multicrystalline silicon for texturization of solar cells. Journal of Materials Processing Technology 191(1–3), 228–231 (2007).

38. B. Nunes, A. P. Serro, V. Oliveira, M. F. Montemor, E. Alves, B. Saramago, R. Colaco.: Ageing effects on the wettability behavior of laser textured silicon. Applied Surface Science 257(7), 2604–2609 (2011).
39. S. Ahn, D. J. Hwang, H. K. Park, C. P. Grigoropoulos.: Femtosecond laser drilling of crystalline and multicrystalline silicon for advanced solar cell fabrication. Applied Physics A 108(1), 113–120 (2012).
40. P. Gecys and G. Raciukaitis.: Scribing of a-Si thin-film solar cells with picosecond laser. European Physical Journal- Applied Physics 51(3), (2010).
41. I. H. W. Nordin, Y. Okamoto, A. Okada, H. Jiang, T. Sakagawa.: Effect of wavelength and pulse duration on laser micro-welding of monocrystalline silicon and glass. Applied Physics A 122(4), 400 (2016).
42. A. Slaoui, F. Foulon, P. Siffert.: Excimer laser induced doping of phosphorus into silicon. Journal of Applied Physics 67(10), 6197–6201 (1990).
43. L. A. Dobrzañski, A. Drygaa, P. Panek, M. Lipiñski, P. Ziêba.: Development of the laser method of multicrystalline silicon surface texturization. Archives of Materials Science 6, 6 (2009).
44. A. Y. Vorobyev, C. Guo.: Direct creation of black silicon using femtosecond laser pulses. Applied Surface Science 257(16), 7291–7294 (2011).
45. P. Engelhart, R. Grischke, S. Eidelloth, R. Meyer, A. Schoonderbeek, U. Stute, A. Ostendorf, R. Brendel.: Laser processing for back-contacted silicon solar cells. International Congress on Applications of Lasers & Electro-Optics, Laser Institute of America 2006(1), M703 (2006).
46. S. Singh, M. R. Sankar, V. K. Jain.: Simulation and experimental investigations into abrasive flow nanofinishing of surgical stainless steel tubes. Machining Science and Technology 22(3), 454–475 (2018).
47. L. A. Dobrzañski, A. Drygaa, K. Gołombek, P. Panek, E. Bielańska, P. Zięba.: Laser surface treatment of multicrystalline silicon for enhancing optical properties. Journal of Materials Processing Technology 201(1–3), 291–296 (2008).
48. R. R. Behera, P. M. Babu, K. K. Gajrani, M. R. Sankar.: Fabrication of micro-features on 304 stainless steel (SS-304) using Nd: YAG laser beam micro-machining. International Journal of Additive and Subtractive Materials Manufacturing 1(3–4), 338–359 (2017).
49. R. R. Behera, A. Das, A. Hasan, D. Pamu, L. M. Pandey, M. R. Sankar.: Deposition of biphasic calcium phosphate film on laser surface textured Ti-6Al-4V and its effect on different biological properties for orthopedic applications. Journal of Alloys and Compounds 842, 155683 (2020).
50. C. H. Sahoo, M. Masanta.: Effect of pulse laser parameters on TiC reinforced AISI 304 stainless steel composite coating by laser surface engineering process. Optics and Lasers in Engineering 67, 36–48 (2015).
51. J. Yang, F. Luo, T. S. Kao, X. Li, G. W. Ho, J. Teng, X. Luo, M. Hong.: Design and fabrication of broadband ultralow reflectivity black Si surfaces by laser micro/nanoprocessing. Light: Science & Applications 3(7), e185–e185 (2014).
52. B. Yang, M. Lee.: Fabrication of honeycomb texture on poly-Si by laser interference and chemical etching. Applied Surface Science 284, 565–568 (2013).
53. B. Yang, M. Lee.: Mask-free fabrication of inverted-pyramid texture on single-crystalline Si wafer. Optics & Laser Technology 63, 120–124 (2014).
54. M. Halbwax, T. Sarnet, P. Delaporte, M. Sentis, H. Etienne, F. Torregrosa, V. Vervisch, I. Perichaud, S. Martinuzzi.: Micro and nano-structuration of silicon by femtosecond laser: Application to silicon photovoltaic cells fabrication. Thin solid films 516(20), 6791–6795 (2008).
55. E. L. Gurevich.: Mechanisms of femtosecond LIPSS formation induced by periodic surface temperature modulation. Applied Surface Science 374, 56–60 (2016).
56. B. K. Nayak, V. V. Iyengar, M. C. Gupta.: Efficient light trapping in silicon solar cells by ultrafast-laser-induced self-assembled micro/nano structures. Progress in Photovoltaics: Research and Applications 19(6), 631–639 (2011).
57. T. H. Her, R. J. Finlay, C. Wu, S. Deliwala, E. Mazur.: Microstructuring of silicon with femtosecond laser pulses. Applied Physics Letters 73(12), 1673–1675 (1998).

58. A. J. Pedraza, J. D. Fowlkes, D. H. Lowndes.: Silicon microcolumn arrays grown by nanosecond pulsed-excimer laser irradiation. Applied Physics Letters 74(16), 2322–2324 (1999).
59. T. Tavera, N. Pérez, A. Rodriguez, P. Yurrita, S. M. Olaizola, E. Castano.: Periodic patterning of silicon by direct nanosecond laser interference ablation. Applied Surface Science 258(3), 1175–1180 (2011).
60. S. Binetti, A. L. Donne, A. Rolfi, B. Jaggi, B. Neuenschwander, C. Busto, C. Frigeri, D. Scorticati, L. Longoni, S. Pellegrino.: Picosecond laser texturization of mc-silicon for photovoltaics: A comparison between 1064 nm, 532 nm and 355 nm radiation wavelengths. Applied Surface Science 371, 196–202 (2016).
61. S. Sarbada, Z. Huang, Y. C. Shin, X. Ruan.: Low-reflectance laser-induced surface nanostructures created with a picosecond laser. Applied Physics A 122(4), 1–10 (2016).
62. P. C. Singh, S. K. Das.: Generation of microstructures and extreme sub-wavelength laser-induced periodic structures on the Si surface using N2 nanosecond pulsed laser for the reduction of reflectance. Pramana 95(1), 1–7 (2021).

Chapter 7

Application of ML & AI for energy efficiency in manufacturing

Anuj Kumar and V. Vasu

Contents

Abbreviations

AI: Artificial Intelligence
ML: Machine Learning
ANN: Artificial Neural Network
IoT: Internet of Things
IIoT: Industrial Internet of Things
MQTT: Message Queuing Telemetry Transport
GPIO: General Purpose Input/Output

7.1 Introduction

The industry 4.0 revolution involves various advanced technologies such as artificial intelligence (AI), big data, data analytics, cloud computing, machine-to-machine communication, the Internet of things (IoT), 3D printing, augmented reality, etc. It is primarily based on the digitalization or automation of industries through the use of IoT. These technological revolutions required a continuous supply of energy. The industrial sector is the biggest energy consumer. As per Energy Agency (IEA) study, manufacturing systems' energy consumption increased rapidly from 2000 [2]. The data from United National Industrial Development showed that the energy consumption growth rate of the manufacturing industry increases

DOI: 10.1201/9781003291961-8

by 1.8–3.1% per year [3]. This condition, combined with depleting natural resources, has forced the manufacturing industry to seek long-term development. Therefore, managing the energy demand and improving the manufacturing industry's efficiency is very crucial for international reductions in carbon emissions and decreasing manufacturing energy uses. Technology can help reduce energy consumption, and enhance energy efficiency [4],[5]. However, the current energy efficiency of manufacturing is significantly below the technically possible levels. With the emergence of the technological revolution, AI and ML has great potential to contribute significantly to the advancement of the manufacturing industry. These technologies can help manufacturing industries to predict energy consumption and reduce energy uses, and these predictions are based on some key capabilities within the domains of Al and ML.

7.2 Fundamentals of AI and ML

The foundations of modern AI are introduced in Fundamentals of Artificial Intelligence, which also covers recent developments in AI, like, adversarial search and game theory, automated planning, intelligent agents, statistical learning theory, information retrieval, constraint satisfaction problems, and natural language processing.

7.2.1 *Artificial intelligence (AI)*

It has the ability of a virtual machine to perform a task that normally requires human knowledge and judgment. However, no AI can accomplish the wide variety of tasks that a human can, some AI can compete with humans in specific fields.

7.2.2 *Machine learning (ML)*

ML (stream of AI) focuses on using the data and algorithm to mimic the way humans learn and continuously improve its efficiency. It has many popular algorithms that can be used in many application areas, such as image processing and voice recognition, automatic translations, and disease diagnosis. In the past few years, ML has appeared as a strong tool.

7.2.3 *AI vs. ML*

The phrases "artificial intelligence" and "machine learning" are frequently mixed up. Both terminologies are sometimes used interchangeably, but they are distinct. ML is a branch of AI that aids in the advancement of AI. Figure 7.1 shows the relationship between AL, ML, and deep learning.

AI is a technology that allows a computer to mimic human behavior. ML is the subset of AI that allows a machine to learn from past data without programming explicitly.

7.2.4 *AI and ML learning approaches*

AI and ML comprise many learning approaches, due to the rapid development of hardware such as CPU and GPU, researchers are working on developing new algorithms that are more precise and accurate (see Table 7.1).

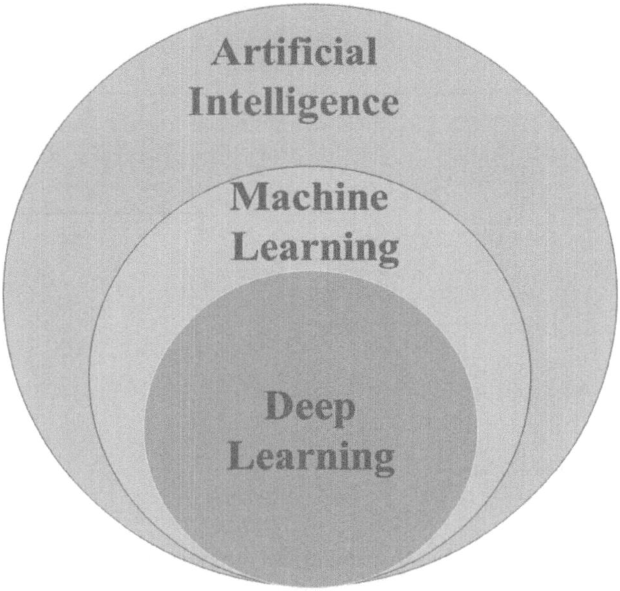

Figure 7.1 Deep learning vs. ML vs. AI.

Table 7.1 Classification of AI and ML Algorithms

Types of learning	*Learning task*	*Most widely ML algorithms*
Unsupervised learning	• Clustering	• K-means clustering • Means shift clustering • DBSCAN clustering • Agglomerative hierarchical clustering • Gaussian mixture
	• Dimensionality reduction	• K- means clustering • PCA (Principal Component Analysis) • Deep learning model • Genetic algorithm • Association rule • t-SNE (t-Distributed Stochastic Neighbor) • Artificial neural network • Instant-based learning models
Reinforcement learning	• Model free • Model-based	• Monte-Carlo tree search • Deep Q-learning • Q-learning (SARSA max) • Temporal difference • Dataset aggregation • Asynchronous Actor-Critic Agent (A3C)
Semi-supervised learning	• Clustering	• Linear regression • Logistic regression

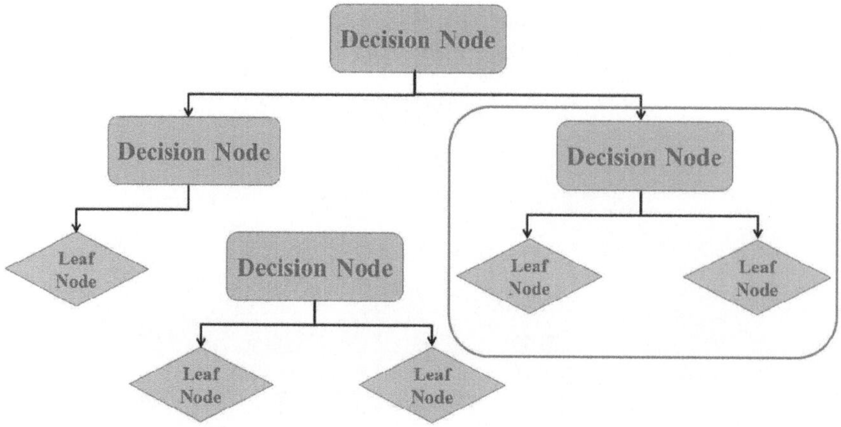

Figure 7.2 Decision tree flow chart.

In ML, there are many approaches to learning, but the most common types of approaches are supervised and unsupervised learning. The supervised learning approach is based on two datasets, the test dataset and the training dataset, a model trained from the output in the training dataset. That is, the previous values of the X attribute are used to know what the results of X are for the never-before-seen examples. Hence, the learning process is based on past data. In this type of learning, many different tasks are there, such as regression and classification, are applied to the existing dataset and the information proposed to be acquired from the output. For example, It is referred to as classification when the label associated with X has a category value. On the other hand, it is referred to as regression when the label associated with X has a numerical value.

In supervised learning, the dataset has no labels. These types of learning models discover the hidden pattern of the dataset without human intervention. Clustering is the most common activity in this form of learning.

Currently, there are many supervised machine learning methods available, such as artificial neural network (ANN), support vector machine (SVM), decision tree (DT), and random forest. In the above ML algorithms, the DT algorithm is widely used and very popular. In classification and regression methods, DTs are used because they are non-parametric supervised learning algorithms. The primary objective of this learning is to create a model that predicts the value of a target variable by using straightforward decision-based rules derived from data properties.

Figure 7.2 shows the structure of the decision tree algorithm. The initial element of the tree is called the root. The elements below it are the root branches, that harbor various decision nodes depending on the best attribute that distinguishes the data and, subsequently, the end node (or leaf).

7.3 Potential of AI and ML in manufacturing

The implementation of AI and ML in the manufacturing industry is getting popular among manufacturers. According to Capgemini's research, in Europe, more than half of the manufacturers (51%) are already using AI technology, followed by Japan (31%) and the US (28%). According to the same report, the most common applications in manufacturing are improving energy efficiency and optimizing process planning. In manufacturing, the use of AI in maintenance is around 29%, and in quality, it is around 27%.

The application of AI has been extensively explored at serval phases of the product life cycle, including operation, visual inspection, evolution, conceptual design, and maintenance [6]. AI is also used in product quality inspection and surface inspection [7]. In the production phase, AI enhances the efficiency and accuracy of production [8]. AI technologies can improve production performance, customer satisfaction, customer experience, and trust from the consumer's perspective. There are numerous uses of AI in industrial design, equipment maintenance, and manufacturing. However, there are only a few applications in various product life cycle stages and relatively little research on AI-enabled production system strategies. Additionally, the integration of AI and IIoT can improve production operations to generate more SMSs, like smart equipment-based monitoring and maintenance [9], and exceptional increases in product agility and accuracy, as well as encourage and collaboratively produce. Figure 7.3 shows how an AI platform-based strategy may be used to integrate

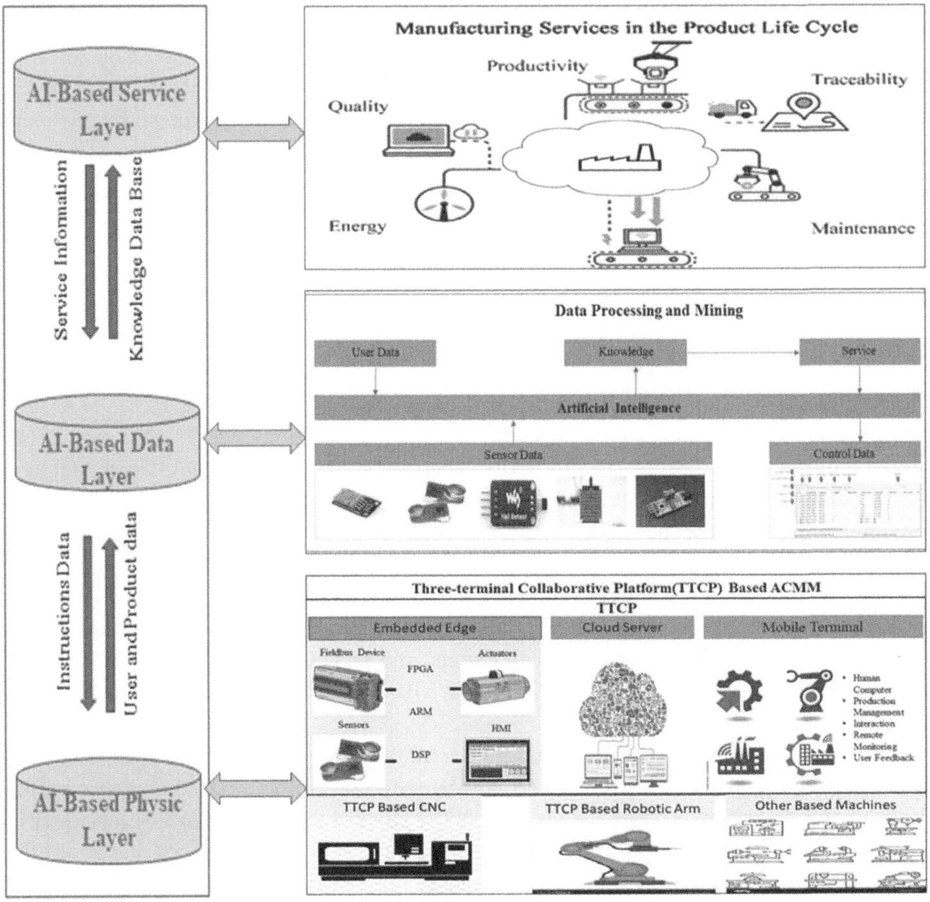

Figure 7.3 The overall framework of a smart manufacturing system.

several technologies into smart manufacturing systems [10]. Fábio Lima proposed a retrofitting model for small- and medium-sized industries. That model was based on digital manufacturing. An industrial energy sensor was used to collect the data from the CNC machine and send it to the cloud with the help of an IoT gateway.

AI and deep learning algorithms are also used to predict machine maintenance, improve product quality, and monitor the real-time machining process. ANN algorithm is used to monitor the tool wear of retrofitted CNC machines in the context of industry 4.0. ANN model has been compared with two other models (KNN and SVM). The precision score of the ANN model was found to be greater than the other two models [11].

In the cutting process, one of the important parameters for quality control is surface roughness. Many studies have been published to predict surface roughness using AI techniques [12].

Figure 7.3 shows the systematic framework of AI and IIoT-based smart manufacturing systems. The data can collect from various machines through the use of different communication protocols, such as MQTT, HTTP, etc. This framework is based on three AI-based layers. The physical layer is the first layer, which contains different TTCP-design based autonomous manufacturing machines. The second layer is the AI-based data layer, which collects heterogeneous data, performs data cleaning and, deep processing, and handless raw data storage and knowledge mining, all with the support of AI technology. And the third layer is the AI-based service layer. During manufacturing, huge number of multi-source data is generated. In a smart manufacturing system, essential information can be mined using various AI-based algorithms. The service layer can facilitate the manufacturing services in all phases of the product life cycle [13].

7.4 AI and ML in energy efficiency

With the continuously increased energy consumption and increasingly prominent energy problem, researchers are focusing their efforts on identifying the elements that influence energy consumption. In literature, two types of significant studies on the influence of technological advancement on industrial energy efficiency. According to one view, technological advancement can help corporate to improve their energy efficiency [14]. Through the empirical study in Germany, according to Welsch and Ochsen, technological advancement can enhance corporate energy efficiency. However, factor substitutions and imbalanced technical advancement are major factors for variation in energy efficiency.

Many researchers predict that advancements in artificial intelligence and information and communication technology can decrease energy costs. In addition, AI can optimize energy efficiency, reduce the gaps between organizations, and increase energy efficiency by boosting technological performance. Table 7.2 lists AI, ML, and energy efficiency-related research and findings.

AI has already been successfully used in different industries to optimize energy efficiency [21]. In the construction industry, AI and big data are used to optimize the energy performance of buildings. Smart meter data is used in the power supply sector to forecast electricity and natural gas use so that the energy supply system can be better planned and controlled [22]. According to Huang and Koroteev, AI technologies such as machine learning and neural network are more effective in waste and energy management. That can be applied in the future to optimize the use of electrical, heat, and gas [23].

Table 7.2 Artificial Intelligence-, Machine Learning-, and Energy Efficiency-Related Research and Findings

Sr. No	*Author*	*Paper title*	*Application area*	*AI methods*	*Findings*
1	Gabrieli D. Silva et al. [15]	Performance and energy efficiency analysis of machine learning algorithms towards green AI: A case study of decision tree algorithms	CPU energy consumption	Decision Tree	This research looked into energy efficiency and computing performance of machine learning algorithms and the predictive evaluation and factors affecting energy efficiency (EDP and CO2e).
2	Yan He et al. [2]	A generic energy prediction model of machine tools using deep learning algorithms	Milling machine And grinding machine	Deep learning (Unsupervised learning)	The energy prediction performance improved 19.14 -74.13% for grinding machines and 64.89–85.61% for milling machine.
3	Dominik Flick et al. [16]	Energy efficiency evaluation of manufacturing systems by considering relevant influencing factors	Automobile industry (Shop floor)	Multiple linear regression models	The model performance is based on a real model. In this study, influencing factors are used in the common regression model for better performance.
4	Rishi Kumar et al. [17]	Development of machine learning algorithm for characterization and estimation of energy consumption of various stages during 3D printing	3D printing	Long Short-Term Memory (LSTM)	The proposed model can provide decision support to a researcher in tuning parameters to energy and time waste in the 3D printing process.
5	Andrea Maria N. C. Ribeiro et al. [18]	Short-term firm-level energy-consumption forecasting for energy-intensive manufacturing: A comparison of machine learning and deep learning models	Manufacturing industry	DEEP LEARNING MODELS: RNN, LSTM, GRU Machine Learning: SVR and Randon Forest	In this study, the GRU model was the best performance model, and RNN was the worst performance model.
6	Chih-Min Yu et al. [19]	Exploit the value of production data to discover opportunities for saving power consumption of production tools	Production tools	Neural network	This study proposed the Neural Network based data mining framework to estimate the kwh/move of specific toolsets. And also analyze the relationship between specific input factors and kwh/move.
7	Arnim Reger et al. [20]	Pattern Recognition in Load Profiles of Electric Drives in Manufacturing Plants	Manufacturing Plant	Pattern Recognition	Discrete Wavelet Transform, Continuous Wavelet Transform, and Short-Time Fourier Transform were used to estimate the current profile in the frequency domain (STFT).

7.5 AI and ML approaches to legacy machine energy efficiency

The production system, which consists of machine tools, fixtures, and cutting tools, is the primary energy consumer in the manufacturing sector. To optimize the energy consumption of the production industry, researchers are doing research from many perspectives, such as production scheduling, production process planning for CNC machines [24] [25], and state control of a machine. Due to the dynamic and complex nature of manufacturing processes, The energy consumption data of the production system contains the elements of continuity and dynamic variation. Rapid advancements in AI, communication, and information technologies have led to reduced energy costs [25].

In manufacturing industries, most of the industries are small and medium, these industries are still using the old legacy machines, and these machines are not able to exchange data. So, the first challenge is to ready the legacy machines for data exchanges. The data exchanges can be possible with the retrofitting of external sensors. These machine data are used to measure the status of the machine, such as spindle speed, power consumption, machine vibration, etc.

7.6 Architecture for retrofitting of legacy machine

This architecture gives the general view of the Industrial Internet of Things (IIoT), which can be used for any legacy machine for retrofitting. This architecture has several layers, including the physical layer, IIoT layer, and cyber layer. The physical layer consists of physical devices, such as sensors, and actuators, which are used to send or receive the data to other devices using the different IoT communication protocols, such as zig bee, MQTT, Bluetooth, wifi, LoRaWan, etc. [26]. The second layer is the IoT gateway. Inside this layer are two more layers – one is the IIoT gateway and the other is IIoT middleware. The main aim of this layer is to provide proper communication between the physical layer and the cloud layer.

The third layer is the cyber layer/cloud layer. The main aim of this layer is to store the machine data coming from the different sensors [27]. From the energy point of view, we can use different sensors to collect the AC data from any machine, such as the non-invasive SCT-013 sensor, which can measure the current load up to 30A. Raspberry-pi or NodeMcu can be used as a controller, and sensors are connected to GPIO pins of the controller (Raspberry-pi or Node Mcu) with wires. Internet connectivity is a very important feature for sending and collecting data from sensors to other software tools step by step.

There are many free gateways available, such as MQTT, Node-Red, etc., that provide the connection between the physical layer and the cyber layer. After storing the data in the cloud, use these data for AI and ML algorithms for training and testing and predict the energy consumption of any legacy machine. Figure 7.4 shows a general architecture for retrofitting.

7.7 Conclusion and outlook

The main aim of this chapter was to study the application of AI and ML to improve the energy efficiency of the manufacturing industry. Many studies show that manufacturing businesses can benefit from AI and ML to increase their energy efficiency. However, we observed that AI-derived benefits to the manufacturing industry is primarily achieved by increasing the R&D, improving the knowledge learning and talent investment of manufacturing enterprises,

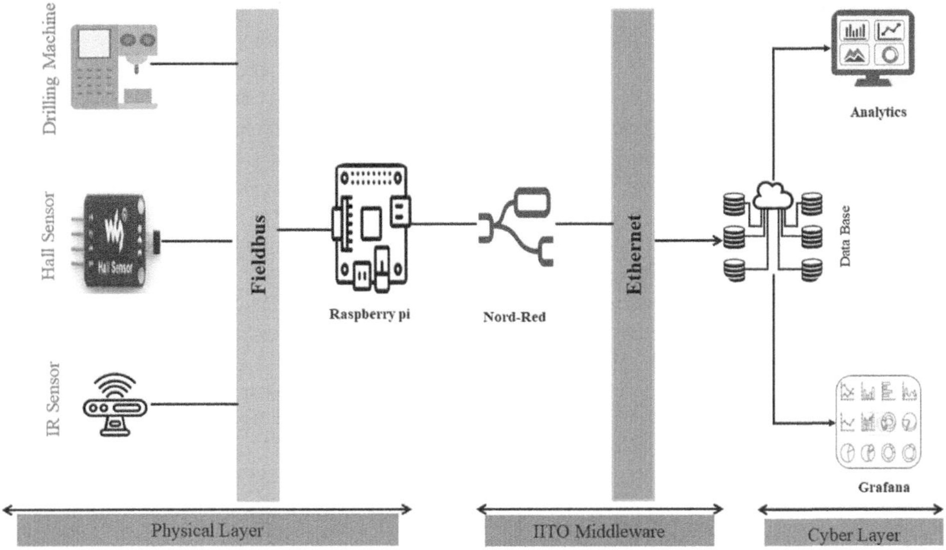

Figure 7.4 A general architecture for retrofitting.

and supporting the manufacturing industry to adopt the new technologies. The study found that the manufacturing industry inhibits the implementation of AI and ML, because of high initial investments. The use of AI and ML increases the energy efficiency performance of the manufacturing industry, but this improvement is visible only when the industry performance is good.

The following policy suggestions are proposed based on the above conclusions:

To accelerate the implementation of artificial intelligence in industries and provide the positive impact of new technology advancements on energy efficiency. Strengthen the manufacturing industries to keep increasing their investment in AI research and development.

Maximize the use of AI and ML to provide the technical support for energy utilization and provide the resources more appropriately in production, which can save energy as well as reduce energy waste, and improve energy efficiency using AI and ML.

In manufacturing industries, a large number of industries are medium and small types, and use very old machines. To adopt new technologies for these industries is very difficult. Till now, very few research article has been published on the use of AI and ML, which is very less as compared to the modern industry. To minimize the energy use of the manufacturing industry, researchers should focus on the innovation of cost-effective technologies, so small and medium enterprises can adopt the technologies, to improve energy optimization.

References

[1] J. Walther and M. Weigold, "A systematic review on predicting and forecasting the electrical energy consumption in the manufacturing industry," *Energies*, vol. 14, no. 4. MDPI AG, Feb. 02, 2021. doi: 10.3390/en14040968

[2] Y. He, P. Wu, Y. Li, Y. Wang, F. Tao, and Y. Wang, "A generic energy prediction model of machine tools using deep learning algorithms," *Applied Energy*, vol. 275, Oct. 2020. doi: 10.1016/j.apenergy.2020.115402
[3] J. Liu, Y. Qian, Y. Yang, and Z. Yang, "Can artificial intelligence improve the energy efficiency of manufacturing companies? Evidence from China," *International Journal of Environmental Research and Public Health*, vol. 19, no. 4, Feb. 2022. doi: 10.3390/ijerph19042091
[4] S. Shan, S. Y. Genç, H. W. Kamran, and G. Dinca, "Role of green technology innovation and renewable energy in carbon neutrality: A sustainable investigation from Turkey," *Journal of Environmental Management*, vol. 294, Sep. 2021. doi: 10.1016/j.jenvman.2021.113004
[5] Y. He, F. Fu, and N. Liao, "Exploring the path of carbon emissions reduction in China's industrial sector through energy efficiency enhancement induced by R&D investment," *Energy*, vol. 225, Jun. 2021. doi: 10.1016/j.energy.2021.120208
[6] M. Qu, S. Yu, D. Chen, J. Chu, and B. Tian, "State-of-the-art of design, evaluation, and operation methodologies in product service systems," *Computers in Industry*, vol. 77. Elsevier B.V., pp. 1–14, Apr. 01, 2016. doi: 10.1016/j.compind.2015.12.004
[7] J. Schmitt, J. Bönig, T. Borggräfe, G. Beitinger, and J. Deuse, "Predictive model-based quality inspection using machine learning and edge cloud computing," *Advanced Engineering Informatics*, vol. 45, Aug. 2020. doi: 10.1016/j.aei.2020.101101.
[8] C. Prentice and M. Nguyen, "Engaging and retaining customers with AI and employee service," *Journal of Retailing and Consumer Services*, vol. 56, Sep. 2020. doi: 10.1016/j.jretconser.2020.102186
[9] M. Cakir, M. A. Guvenc, and S. Mistikoglu, "The experimental application of popular machine learning algorithms on predictive maintenance and the design of IIoT based condition monitoring system," *Computers and Industrial Engineering*, vol. 151, Jan. 2021. doi: 10.1016/j.cie.2020.106948
[10] F. Lima, A. A. Massote, and R. F. Maia, "IoT energy retrofit and the connection of legacy machines inside the industry 4.0 concept," *IECON 2019 – 45th Annual Conference of the IEEE Industrial Electronics Society*, pp. 5499–5504, 2019. doi: 10.1109/IECON.2019.8927799.
[11] D. F. Hesser, and B. Markert. "Tool wear monitoring of a retrofitted CNC milling machine using artificial neural networks," *Manufacturing Letters*, vol. 19, pp. 1-4 2019.
[12] B. P. Huang, J. C. Chen, and Y. Li, "Artificial-neural-networks-based surface roughness Pokayoke system for end-milling operations," *Neurocomputing*, vol. 71, no. 4–6, pp. 544–549, Jan. 2008. doi: 10.1016/j.neucom.2007.07.029
[13] L. Bu, Y. Zhang, H. Liu, X. Yuan, J. Guo, and S. Han, "An IIoT-driven and AI-enabled framework for smart manufacturing system based on three-terminal collaborative platform," *Advanced Engineering Informatics*, vol. 50, Oct. 2021. doi: 10.1016/j.aei.2021.101370
[14] B. Lin and M. Moubarak, "Renewable energy consumption—Economic growth nexus for China," *Renewable and Sustainable Energy Reviews*, vol. 40. Elsevier Ltd, pp. 111–117, 2014. doi: 10.1016/j.rser.2014.07.128
[15] G. Silva, B. Schulze, and M. Ferro, *Performance and energy efficiency analysis of machine learning algorithms towards green AI: A case study of decision tree algorithms* (Doctoral dissertation, Master's thesis, National Lab. for Scientific Computing), 2021. doi: 10.13140/RG.2.2.27740.31363
[16] D. Flick, L. Ji, P. Dehning, S. Thiede, and C. Herrmann, "Energy efficiency evaluation of manufacturing systems by considering relevant influencing factors," *Procedia CIRP*, vol. 63, pp. 586–591, 2017. doi: 10.1016/j.procir.2017.03.097
[17] R. Kumar, R. Ghosh, R. Malik, K. S. Sangwan, and C. Herrmann, "Development of machine learning algorithm for characterization and estimation of energy consumption of various stages during 3D printing," *Procedia CIRP*, vol. 107, pp. 65–70, 2022. doi: 10.1016/j.procir.2022.04.011
[18] A. M. N. C. Ribeiro, P. R. X. do Carmo, I. R. Rodrigues, D. Sadok, T. Lynn, and P. T. Endo, "Short-term firm-level energy-consumption forecasting for energy-intensive manufacturing: A comparison of machine learning and deep learning models," *Algorithms*, vol. 13, no. 11, p. 274, 2020. https://doi.org/10.3390/a13110274

[19] C. M. Yu, C. F. Chien, and C. J. Kuo, "Exploit the Value of Production Data to Discover Opportunities for Saving Power Consumption of Production Tools," *IEEE Transactions on Semiconductor Manufacturing*, vol. 30, no. 4, pp. 345–350, Nov. 2017. doi: 10.1109/TSM.2017.2750712
[20] A. Reger, C. Oette, A. P. Aires, and R. Steinhilper, "Pattern recognition in load profiles of electric drives in manufacturing plants," *2015 5th International Electric Drives Production Conference (EDPC)*, pp. 1–10, 2015. doi: 10.1109/EDPC.2015.7323209.
[21] S. D. Supekar *et al.*, "A framework for quantifying energy and productivity benefits of smart manufacturing technologies," *Procedia CIRP*, vol. 80, pp. 699–704, 2019. doi: 10.1016/j.procir.2019.01.095
[22] I. Smajla, D. K. Sedlar, D. Vulin, and L. Jukić, "Influence of smart meters on the accuracy of methods for forecasting natural gas consumption," *Energy Reports*, vol. 7, pp. 8287–8297, Nov. 2021. doi: 10.1016/j.egyr.2021.06.014
[23] J. Huang and D. D. Koroteev, "Artificial intelligence for planning of energy and waste management," *Sustainable Energy Technologies and Assessments*, vol. 47, Oct. 2021. doi: 10.1016/j.seta.2021.101426
[24] S. T. Newman, A. Nassehi, R. Imani-Asrai, and V. Dhokia, "Energy efficient process planning for CNC machining," *CIRP Journal of Manufacturing Science and Technology*, vol. 5, no. 2, pp. 127–136, 2012. doi: 10.1016/j.cirpj.2012.03.007
[25] S. Terry *et al.*, "The Influence of Smart Manufacturing towards energy conservation: A review," *Technologies (Basel)*, vol. 8, no. 2, p. 31, May 2020. doi: 10.3390/technologies8020031
[26] A. Dean and M. O. Agyeman, "A study of the advances in IoT security," *Proceedings of the 2nd International Symposium on Computer Science and Intelligent Control*, pp. 1–5, 2018. doi: 10.1145/3284557.3284560.
[27] H. Mrabet, S. Belguith, A. Alhomoud, and A. Jemai, "A survey of IoT security based on a layered architecture of sensing and data analysis," *Sensors*, vol. 20, no. 13, p. 3625, 2020. doi: 10.3390/s20133625

Chapter 8

Challenges in achieving sustainability during manufacturing

Arun Kumar Bambam and Kishor Kumar Gajrani

Contents

Abbreviations

WHO: World Health Organization
WCED: World Commission on Environment and Development
GDP: Gross Domestic Product
MQL: Minimum Quantity Lubricant
LCA: Life Cycle Assessment
SME: Small and Medium-Sized Enterprise

8.1 Introduction

The upsurge of manufacturing companies around the world has enhanced liveability, but it has also impacted the environment [1]. Therefore, manufacturing firms are under pressure from government and non-government groups to reduce their negative environmental impact and improve worker safety [2]. Thus, addressing these problems and providing effective solutions requires the implementation of sustainable ways in modern manufacturing systems. There is no universal definition of sustainability; however, the most acceptable definition was proposed by the Director-General of the World Health Organization (WHO) and Norway's former Prime Minister, Gro Harlem Bruntland [3], who defined it as "satisfying

DOI: 10.1201/9781003291961-9

the present needs without affecting the needs of future generations". According to Jawahir and Wanigarathne [4], the primary features of sustainability are intended for the environmental, economic, and social orientations in order to attain needs via efficient resource use.

Many firms have begun redesigning their existing manufacturing system to make them more sustainable [5]. However, changing the existing process is projected to have significant obstacles and difficulties on a technological and economic level. However, redesigning an existing manufacturing system does not guarantee sustainable achievement [6]. Therefore, it is essential to consider sustainability in the manufacturing process and drive plans toward long-term development. Sustainability analysis in the production chain helps identify and recognize potential improvements in the manufacturing system. For this purpose, manufacturing sectors are constantly looking to adopt new technologies to improve their manufacturing processes [7]. However, it is not easy to achieve sustainability in the manufacturing system because it includes several practical aspects. Some of the aspects considered are as follows:

- Using non-hazardous and recyclable materials as well as inputs
- Developing and planning manufacturing processes to reduce resource consumption such as energy, material, and water consumption
- Utilizing renewable energy to conserve the natural environment
- Developing new product designs for reusable, remanufacturable, or recyclable goals
- Extending design concepts using lesser resources and implementing easy-to-repair techniques

The implementation procedures to attain the sustainable manufacturing strategy differ depending on the level of difficulties, types of organization, customer's choice, assets involved, etc. These difficulties can be overcome by considering all influencing factors while going for sustainable manufacturing. A structural overview to achieve sustainability in manufacturing is shown in Figure 8.1.

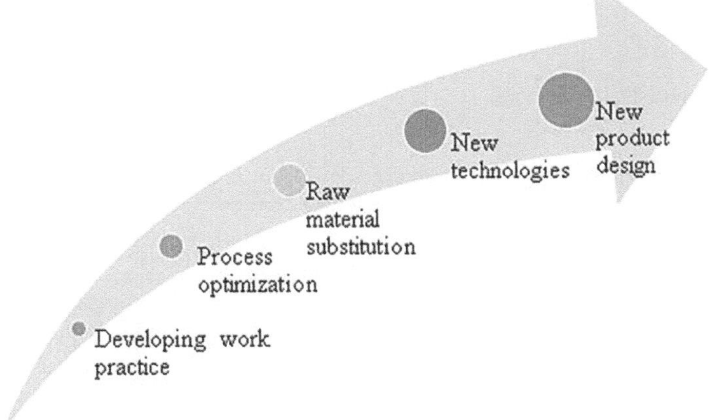

Figure 8.1 Roadmap to achieve sustainable manufacturing [8].

8.2 Sustainability and sustainability dimensions

Sustainability is the concept of improving the effectiveness and efficiency of organizations, products, and systems. Conventionally, sustainability has three dimensions: economics, environment, and social, which have a significant effect on the overall process. Resource conservation benefits both the economy and the environment. Work circumstances, educational conditions, skill levels, and other factors contribute to the social component [9, 10]. The sustainable development program was popularised by the World Commission on Environment and Development (WCED), which encouraged many firms to adapt sustainable ways [11]. The adaptation of sustainability in any firm needs to consider sustainability dimensions, as shown in Figure 8.2.

Sustainability dimensions are an essential part of monitoring and analyzing sustainability and attempting to improve it. Dimensions aid in determining the state of the progress achieved toward an aim, the obstacles and issues encountered when working toward an objective, and the actions that must be implemented to solve the challenges and problems. Sustainability dimensions highlight where the connections between the economy, environmental stewardship, and society are lacking. Also, it proposes as well as priorities solutions [12]. Table 8.1 shows some sustainability factors that can be utilized for the analysis of the environmental, economic, and social dimensions [13].

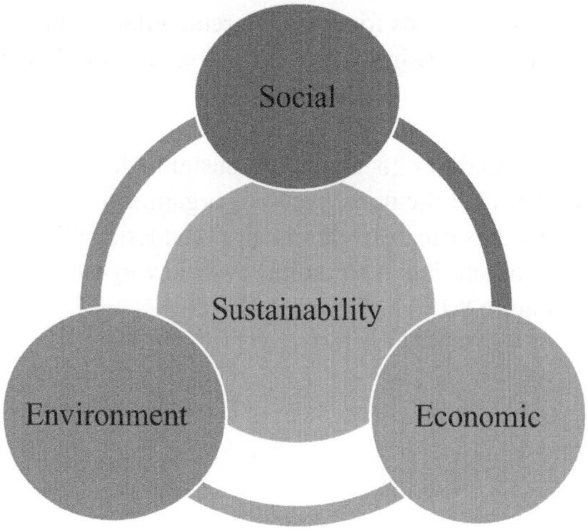

Figure 8.2 Sustainability dimensions.

Table 8.1 Sustainability – Key Performance Indicators

Sustainability dimension	*Key performance indicators*
Environment	Water use rate, renewable energy rate, % of recycled/reusable material, rate of carbon footprint, rate of greenhouse emission, % of waste generated, waste disposal
Social	Equality rate, diversity rate, training hours for each employee, expenses for social initiatives, skilled labor
Economic	Investment in technology, income from recycling programs, inventory and stock cost, return on investment

8.3 Relevance of sustainability in manufacturing

Manufacturing produces essential items and services that enhance human lifestyle and contribute to the global economy. Manufacturing encompasses all industrial development from the customer to the plant and back to the customer, thus including all types of services associated with the production chain. Manufacturing can be classified into three categories: *(i)* discrete manufacturing, *(ii)* process industries, and *(iii)* services [14]. Many studies have been carried out to determine the economic and social significance of manufacturing in various parts of the world. As per the report of the World Bank, in the past decade, the manufacturing industry in India has made a substantial contribution of 13–16% to gross domestic product (GDP) [15] and employed roughly 12% of the total workforce of the countries in 2014. According to researchers, every job created in manufacturing creates an additional two to three jobs in the services sector. Then, returning to the prior model of sustainability, there can be no debate regarding the significance of manufacturing on both the social and economic pillars [8]. However, when it comes to the environment, we cannot avoid the reality that manufacturing has a significant environmental impact. The first major factor is the consumption of raw materials and energy. Furthermore, we must examine the influence of manufacturing on climate change by accounting for the release of greenhouse gases and toxic waste generation (hazardous and solid waste), floating plastic, water emissions, and product end-of-life effects. Last but not least, the working atmosphere must be treated as part of manufacturing sustainability [14].

8.4 Sustainable manufacturing

The new globalization faces difficulty in meeting the growing global demand for resources and consumer products. The industrial sector must be geared toward long-term sustainability in order to maintain the social, cultural, and economic sustainability of human life. The shift to the fourth stage of industrialization is currently shaping the economic value production of the early industrialized countries. This expansion, along with environmental and social aspects, will create an abundance of opportunities for attaining sustainable manufacturing. Sustainable manufacturing can be expressed as a manufacturing approach that reduces waste, environmental impacts, energy consumption, and material costs for greener goods along with meeting the client's needs [16]. These objectives can only be met through employing practices that have an impact on process design, product design, and operating principles that reduces material and energy consumption. Therefore, sustainable manufacturing can also be defined as a system that integrates processes and product design-related issues with production planning and control in order to determine, assess, quantify, and manage the drift of environmental waste with the aim of decreasing environmental impact.

Sustainable manufacturing is based on the 6R method (as shown in Figure 8.3) rather than the 3R approach (reuse, reduce, and recycle) [17]. The term "reduce" in the 6R process refers to minimizing the effort of utilizing resources and fuel/energy consumption during production, resulting in less waste at the use stage. The "reuse" concept is associated with the reuse of previously created items or components after their original lifespan, which leads to reduced resource consumption [17]. "Recycling" refers to the practice of repurposing existing materials that are traditionally regarded as garbage to create new materials or products. When components are collected at the conclusion of their first lifespan, they are dismantled, cleaned, and readied for the next life cycle. The term "redesign" involves redesigning the goods using

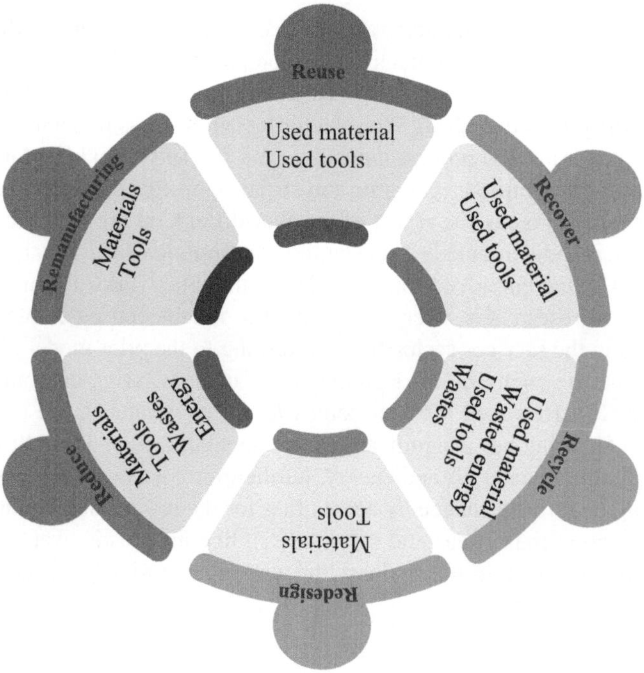

Figure 8.3 6R method in manufacturing to achieve sustainability [20].

methodologies such as "design for the environment" to make them more manageable [18]. In terms of "remanufacture", it covers the reuse of a previously used product, returning it to its original condition via the recycling of as many components as possible while maintaining function. There have been several initiatives to build models for adopting sustainability in the manufacturing industries. Also, various frameworks for sustainable manufacturing, modeling-optimization tools, and supply chains have recently been introduced. Regardless of these frameworks and models, production engineers and designers must acknowledge the need to produce sustainable products and processes. To properly incorporate the sustainability idea into production, a long-term commitment to the whole design approach is essential [7, 19]. The manufacturing industry's sustainability may be addressed in three stages:

- ***Research:*** Needs for early evaluation of explicit sustainability, such as energy, pollution, and material use. Throughout the lifespan of a product, research should anticipate and prevent potential sustainability risks
- ***Development:*** Enhancement of environmental performance with an emphasis on optimal equipment selection and system design utilizing appropriate systems and methodologies
- ***Commercialization:*** The cooperation of consumers, vendors, and suppliers to detect potential issues in the product life cycle and fix it as needed

Modern manufacturing systems are helping to achieve stable, long-term processes, such as additive manufacturing technologies are an example of manufacturing procedures that

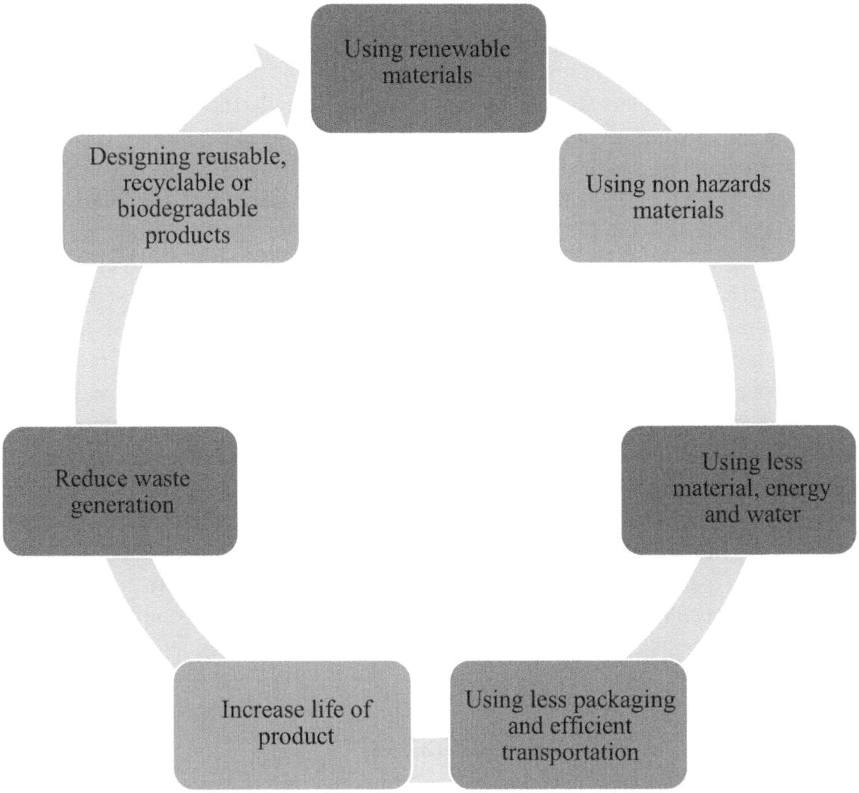

Figure 8.4 Sustainability aspects in manufacturing.

were previously solely employed for quick prototyping. However, additive manufacturing is now opening the door to the additive production of diverse components. Additive manufacturing requires fewer number of production stages, resulting in significant waste and energy savings [21]. Another example of how new technologies contribute to sustainable production is laser-assisted manufacturing/machining, also known for prolonging tool life [22]. It is also known to increase surface wear resistance and prolong product life. The development of these technologies and other techniques, like minimal quantity lubrication (MQL) [23–26], will support efficient large-scale production, which will benefit both environment and the profitability of firms. However, sustainability is not limited to this extent and we have to focus on many other aspects, as shown in Figure 8.4.

8.5 Challenges associated with sustainable manufacturing

In the manufacturing sector, the lack of knowledge, facts, and information on green manufacturing, the threats associated with new kinds of technology, and poor laws are the three most priority problems identified by Mittal and Sangwan [27, 28]. Bhanot et al. [29] compared their views on the issues encountered in sustainable manufacturing using surveys conducted by researchers and industrial employees. From the different issues analyzed for the survey, the perspective of two challenges appeared to match both the industrial employees and the researchers. The first is that the implementation of sustainable manufacturing is hampered by no access or limited to sustainable literature, which limits understanding the ideas and techniques necessary to adopt sustainable manufacturing. The second problem encountered by industrialists in adopting sustainable manufacturing is the delay in gaining economic advantages owing to the high initial cost of implementation. The researchers' and industrialists' perspectives on the two difficulties also differed significantly. The very first challenge is a lack of knowledge about sustainable products among local customers. Researchers believe that there is a large gap in customer knowledge about green products or goods and that it is futile for the industry to follow the goal of manufacturing sustainable or green products/goods until this goal is achieved. Industrial workers believe that there is already a huge demand for green or sustainable processes and products. The second difficulty is the lack of established measurements, on which both industries' and researchers' perspectives differ greatly. Furthermore, the other factors that hamper achieving sustainable manufacturing are a deficit of a standard measure for assessing sustainability, confined progress in identifying the suitable allocation schemes, lack of data for conducting life cycle assessment in a streamlined manner, lack of adequate emphasis on a modular design, lack of specific data for designers to select materials, and a need for dematerialization (particularly for service-related products) [30, 31]. There is a need for more information on the ecological, economic, and social aspects of current manufacturing domains, as well as some concepts for sustainable/greener operations, as shown in Figure 8.5.

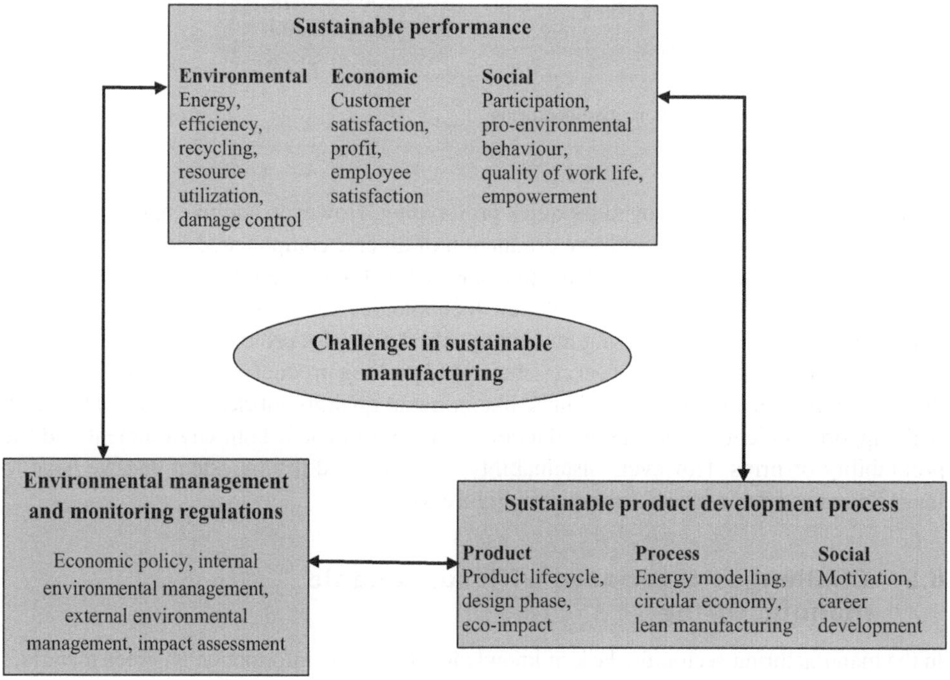

Figure 8.5 Factors influencing sustainability.

8.5.1 Challenges in sustainable product design

The traditional role of a product/part design in manufacturing organizations has been to ensure that the product accomplishes its targeted objectives based on the consumer's choice, considering the performance, economics, aesthetics, and efficiency of the product. However, this does not account for issues such as waste management, end-of-life considerations, or ethical considerations. In other words, the sustainable development parts of a design are usually neglected since they have broader environmental, social, and economic consequences [8, 32]. A product's usage and disposal must be defined by the designer, and they must influence both the production and distribution processes in order to reduce environmental, economic, and social effects. When it comes to disposing of automobiles and technological garbage, modern environmental regulation has placed a heavy focus on recycling and reusing these items [33]. With a greater emphasis on sustainable production, the product designer is burdened with the added duty of evaluating the environmental effect of his/her actions. As a result, it is crucial for the designer to first familiarize himself/herself with the numerous challenges of sustainability. The materials and techniques used have a substantial impact on the environment, economics, and society, so while designing any product, designers have to consider many things. Figure 8.6 depicts the factors that influence the product designer's decisions.

Designers are trying to address most of the issues in the designing stage to achieve sustainability in manufacturing, as shown in Figure 8.7. Design for sustainability is a product design method that considers how a product may impact the environment, economics, society, and its impact on sustainability aspects over the course of its life and aims to mitigate any negative consequences by addressing them throughout the design phase. Addressing all influencing factors for sustainable manufacturing is very challenging for designers.

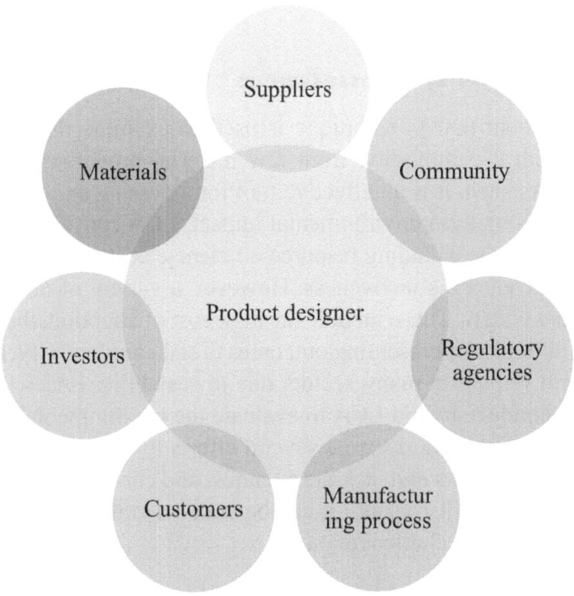

Figure 8.6 Factors affecting designer decisions.

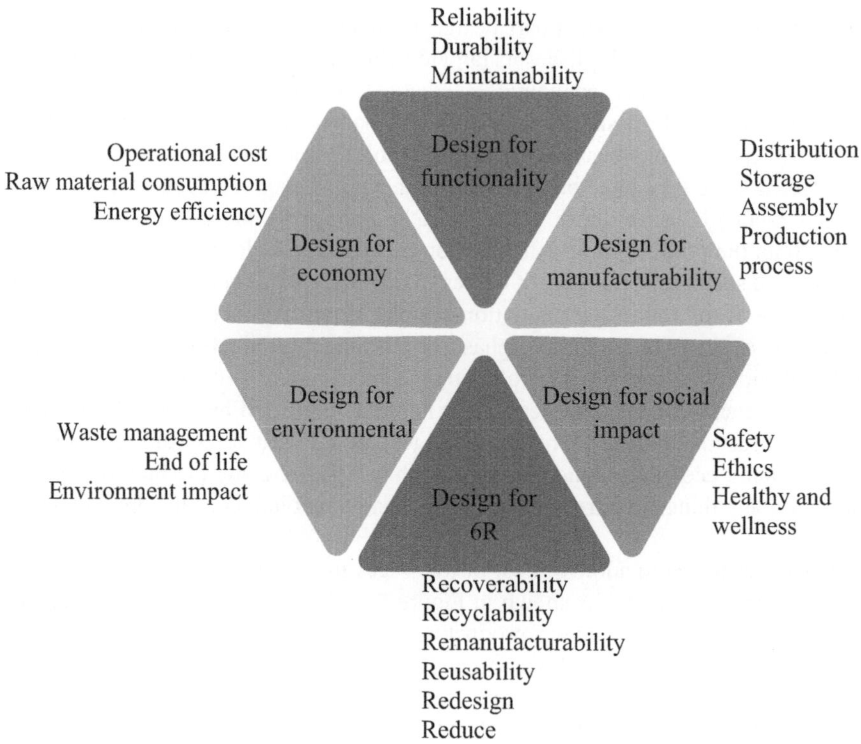

Figure 8.7 Design for sustainability.

8.5.2 *Challenges in life cycle assessment*

The life cycle assessment (LCA) technique is used to examine the many environmental aspects connected with a commodity, from raw material procurement to manufacturing, usage, and disposal. As such, it is an effective tool for assessing and finding possibilities to improve a company's long-term environmental impact. LCA consultants and organizations cite a variety of advantages, including resource efficiency, cost savings, decreased environmental liability, and product distinctiveness. However, a variety of obstacles prevent LCA from being used more widely. These include the high cost of adoption, the difficulties of data collecting, and the difficulty of presenting outcomes to stakeholders. Given these challenges, LCA adoption is still limited in many sectors due to scalability issues [34]. Due to a lack of data, it is hard to conduct a valid LCA to evaluate the environmental impact of developing technology in manufacturing. Despite several efforts to conduct simplified or screened LCAs, all techniques have flaws and lack the openness and consistency essential for accurate assessment. From the above discussion, it can be concluded that LCA of developing technologies revealed some major challenges, such as:

- Difficulty in comparability
- A lack of appropriate and quality data

- Scaling concerns
- Ambiguities and communication uncertainty
- High cost

Hence, it is not possible to include all aspects of sustainability in the sustainable practice, as shown in Table 8.2.

8.5.3 Challenges in material selection

The material selection process is designed to provide one or more materials with attributes that meet the required functionality of products, such as stiffness or strength. Furthermore, it is preferable, but not essential, that the materials effectively manage performance targets, such as expense or ecological effects. Material selection is an interdisciplinary process that needs the collaboration of many stakeholders, such as material scientists, test engineers, product designers, and end-users. As a result, material selection challenges are often open-ended, along with appropriate solutions subject to a continual trade-off between a variety of restrictions and targets [35, 36].

Numerous uncertainties exist in material selection, including those related to design specifications and material attributes. Robust material selection includes a design specification that completely describes a product's functional needs and goals. However, design specifications are often insufficient or poorly specified. Furthermore, the primary failure mechanism is determined by the material properties, and a design specification that applies to one material type may not fit another. Material replacement participants must use caution to ensure that the introduction of new materials does not disclose a latent failure mode that was not specified in the original design specification [37, 38].

Several restrictions and goals often constrain material selection challenges. These may be evaluated in non-comparable units, such as cost and weight, making direct correlation challenging. Another source of uncertainty is material qualities. Material property data may be inadequate or measured using incompatible experimental techniques. Since there are so many uncertainties in material selection (as shown in Table 8.3), they are often based on hunches or precedents. Precedent-based design is difficult since the results of the material

Table 8.2 Structural Overview of Challenges for Life Cycle Assessment

Inventory aspects	*Impact assessment aspects*			*Generic aspect*	*Evolving aspects*
	Ecosystem	*Human health*	*Resources*		
Renewable energy, water use and consumption, delayed emission, allocation, functional unit	Biological invasion, ecotoxicity, land use	Human toxicity, particulate matter, endocrine disruptors, direct health effects, odor, noise	Abiotic resources, biotic resources, changes in soil quality, salinization	Modeling approach: Consequential LCA, rebound effect, weighting, uncertainty analysis, data quality analysis	Positive impact, animal wellbeing, littering

Table 8.3 Key Factors for Material Selection

Environmental	*Economic*	*Society*	*Manufacturability*
Energy consumption Toxic material Resource depletion Greenhouse gas emissions End of life and recyclability	Material cost Manufacturing cost End-of-life cost Operation cost Maintenance cost	Noise, harshness, and vibration Crashworthiness Health and wellness	Formability Jointability Paintability

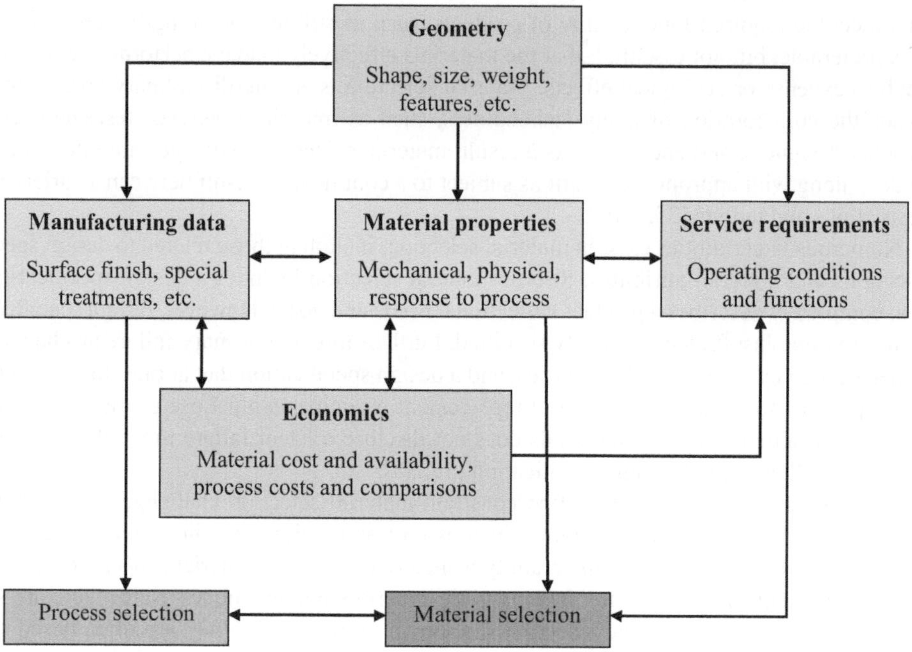

Figure 8.8 Overview of material selection.

selection are sometimes counterintuitive and cannot be immediately addressed. Also, this technique is not acceptable in the context of changing design parameters, like ecological impact and fuel or energy usage [39]. Along with all the above challenges in material selection, some other considerations are shown in Figure 8.8.

8.5.4 Challenges in the selection of manufacturing processes

It is a very challenging decision for any manufacturing industry to select a process that consumes fewer resources and less energy, generates low waste, and helps minimize the total environmental, economic and social impact of the product. Pecas et al. [40] established a model for comparing the life cycle performance of mold manufacturing procedures appropriate for small production volumes. The approach incorporates three distinct but necessary and interdependent performance criteria: economical, technological, and ecological. Two possible technologies were analyzed: one comprising a metal spray shell backfilled with aluminum powder and resin and the other based on aluminum machining. While the first mold provides the greatest environmental and economical solutions, the second has greater technical quality. Bambam et al. [41, 42] performed tribological and machining experiments with four different environmentally friendly lubricants, including two halogen-based ionic liquids mixed with canola oil, one halogen-free ionic liquid mixed with canola oil, and virgin canola oil. Analysis of the result suggests that the reduction in cutting and thrust forces were achieved by 21.70% and 26.80%, respectively, compared to virgin canola oil lubricant. Also, a 32% reduction in surface roughness was achieved in the case of machining with halogen-free ionic liquids mixed with canola oil compared with virgin canola oil. Although the base lubricants were eco-friendly, the influence of different additives on the machining performance was different.

Sustainable manufacturing is a fundamentally difficult subject that is critically dependent on a variety of physical phenomena, like waste generation, heat generation, and energy consumption, during manufacturing processes. Also, new materials and advanced manufacturing technologies require new manufacturing processes to make the process more sustainable in terms of cost, reliability and quality, remanufacturability, energy consumption, product flexibility, and reductions in fossil carbon emissions. Therefore, it is very difficult to select a manufacturing process as various factors directly or indirectly influence the selection process, as shown in Figure 8.9.

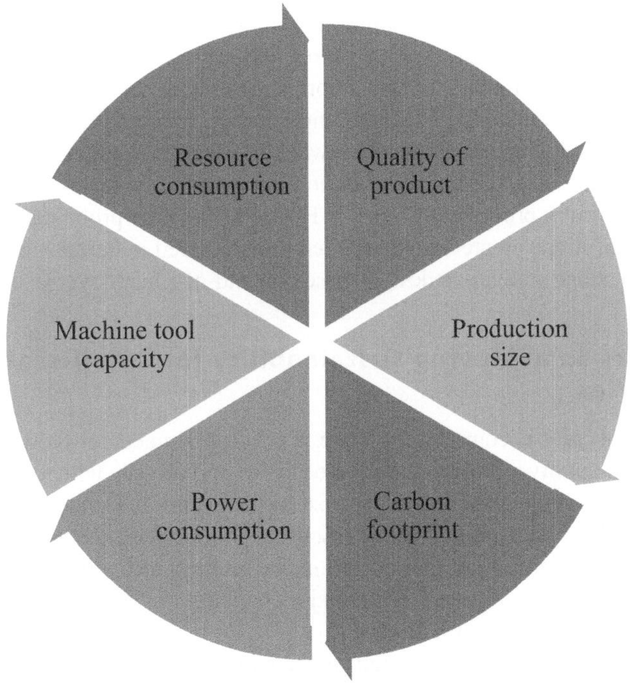

Figure 8.9 Factors affecting the manufacturing process selection.

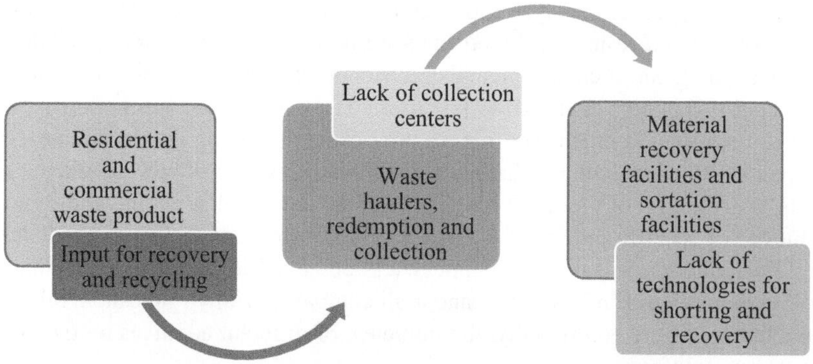

Figure 8.10 Major challenges for recovery and recycling.

8.5.5 *Challenges in recovery and recycling process*

Several obstacles lie in the area of product recovery and recycling, which is a critical step in the application of green concepts in sustainable manufacturing. The process of end-of-life or discarded product collecting and returning to a recycling facility is one of the most complex and expensive recovery fields. Another major challenge with recycling is the shortage of markets for recycled material. Generally, both ferrous and nonferrous metals are ideal targets for sustainable product recycling or recovery because they can be utilized in closed-loop cycles. For example, steel and aluminum from a scrap automotive parts can be used to manufacture a new product. However, recovered and recycled plastic parts are not preferred for commercial use due to the degradation of properties and are also not advisable for closed-loop recovery and recycling [43, 44]. Also, there is a scarcity of systems that can facilitate mixed and refurbished parts, insufficient knowledge, literature, and technology on recycling and recovery, and the lack of smart manufacturing systems for remanufacturing and recycling as well as shorting process, as shown in Figure 8.10. Other problems include the negative perceptions of items created from recovered or recycled materials and manufacturers' unwillingness to adopt innovative designs based on end-of-life recovery.

8.6 Barriers to achieving sustainability for small-scale industries

In today's world of global competition, there is a rising demand for products that are economical, of higher quality and variety, delivered faster, and comply with local and worldwide environmental rules. As a consequence, industrial companies are under enormous pressure to run operations that make optimal use of resources and services. Small- and medium-sized enterprises (SMEs) are failing in two crucial manufacturing areas, process technology, and quality. According to the study, one of the major challenges in small businesses is a lack of awareness of green production initiatives [45].

Furthermore, stakeholders mostly focus on economic challenges rather than addressing environmental and social sustainability. The most significant impediment to achieving sustainable manufacturing in SMEs is a lack of knowledge of the importance of sustainable

Figure 8.11 Barriers for small- and medium-sized enterprises to achieving sustainability.

design, owing to a lack of awareness among stakeholders. People issues (lack of organizational commitment), strategic problems (financial, resources, information, technical, performance measurement, managerial, organizational, etc.), and functional issues (procurement and supply chain function, lack of management structures and processes, etc.) are regarded as barriers to sustainable practices. Developing a culture of sustainability among collaborators or stakeholders is progressing slowly in SMEs, despite the fact that sustainability issues are now incorporated into many government policies and initiatives. Figure 8.11 shows the key barriers to the development of sustainable manufacturing in general. The challenges to addressing the carbon emission aspects of sustainability practices and limiting emissions were absent. Also, SMEs cannot disregard what their shareholders and stakeholders expect since, in today's competitive climate of technological progress, globalization, and inventive rivals, enterprises must continually improve their performance.

8.7 Conclusion

In this work, the study identifies the challenges that must be overcome in order to adopt a sustainable manufacturing system. Sustainable manufacturing techniques are the industry's biggest strengths, but all organizations are not able to adopt sustainable manufacturing due to many reasons, such as aspects of the environment, society, and economics, which may be difficult to achieve because of high investment, stockholder's decision, lack of awareness, etc. Furthermore, the problems highlighted in this work will assist decision-makers in various sectors in implementing sustainable manufacturing practices for high-quality products at a reasonable cost.

Acknowledgments

The authors express gratitude to the Indian Institute of Information Technology, Design and Manufacturing, Kancheepuram, India for financial support through the Institute Seed Grant (IIITDM/ISG/2022/ME/02).

References

1. Abdul-Rashid, S. H., Sakundarini, N., Ghazilla, R. A. R., Thurasamy, R. (2017). The impact of sustainable manufacturing practices on sustainability performance: Empirical evidence from Malaysia. *International Journal of Operations & Production Management*, 37(2), 182–204.
2. Stock, T., Seliger, G. (2016). Opportunities of sustainable manufacturing in industry 4.0. *Procedia CIRP*, *40*, 536–541.
3. Brundtland, G. H., Khalid, M. (1987). *Our common future*. Oxford University Press, Oxford.
4. Jawahir, I. S. (2008). Beyond the 3R's: 6R concepts for next generation manufacturing: Recent trends and case studies. In *Symposium on sustainability and product development, IIT*, Chicago. https://people.utm.my/zulk/wp-content/blogs.dir/916/files/2017/09/Beyond-the-3R-to-6R-concept-Jawahir.pdf
5. Mapar, M., Jafari, M. J., Mansouri, N., Arjmandi, R., Azizinejad, R., Ramos, T. B. (2017). Sustainability indicators for municipalities of megacities: Integrating health, safety and environmental performance. *Ecological Indicators*, *83*, 271–291.
6. Nishitani, K., Kaneko, S., Fujii, H., Komatsu, S. (2011). Effects of the reduction of pollution emissions on the economic performance of firms: an empirical analysis focusing on demand and productivity. *Journal of Cleaner Production*, *19*(17–18), 1956–1964.
7. Rosen, M. A., Kishawy, H. A. (2012). Sustainable manufacturing and design: Concepts, practices and needs. *Sustainability*, *4*(2), 154–174.
8. Kishawy, H. A., Hegab, H., Saad, E. (2018). Design for sustainable manufacturing: Approach, implementation, and assessment. *Sustainability*, *10*(10), 3604.
9. Mesa, J., Esparragoza, I., Maury, H. (2018). Developing a set of sustainability indicators for product families based on the circular economy model. *Journal of Cleaner Production*, *196*, 1429–1442.
10. Ahmad, S., Wong, K. Y., Rajoo, S. (2018). Sustainability indicators for manufacturing sectors: A literature survey and maturity analysis from the triple-bottom line perspective. *Journal of Manufacturing Technology Management*, *30*(2), 312–334.
11. McManus, P. (2014). Defining sustainable development for our common future: A history of the world commission on environment and development (Brundtland Commission). *Australian Geographer*, *45*(4), 559–561.
12. Joung, C. B., Carrell, J., Sarkar, P., Feng, S. C. (2013). Categorization of indicators for sustainable manufacturing. *Ecological Indicators*, *24*, 148–157.
13. Hristov, I., Chirico, A. (2019). The role of sustainability key performance indicators (KPIs) in implementing sustainable strategies. *Sustainability*, *11*(20), 5742.
14. Garetti, M., Taisch, M. (2012). Sustainable manufacturing: Trends and research challenges. *Production Planning & Control*, *23*(2–3), 83–104.
15. World Bank: https://data.worldbank.org/indicator/NV.IND.MANF.ZS?locations=IN (Link Date: 14-06-2022).
16. Gunasekaran, A., Spalanzani, A. (2012). Sustainability of manufacturing and services: Investigations for research and applications. *International Journal of Production Economics*, *140*(1), 35–47.
17. Pini, M., Lolli, F., Balugani, E., Gamberini, R., Neri, P., Rimini, B., Ferrari, A. M. (2019). Preparation for reuse activity of waste electrical and electronic equipment: Environmental performance, cost externality and job creation. *Journal of Cleaner Production*, *222*, 77–89.
18. Milios, L., Esmailzadeh Davani, A., Yu, Y. (2018). Sustainability impact assessment of increased plastic recycling and future pathways of plastic waste management in Sweden. *Recycling*, *3*(3), 33.

19. Zarte, M., Pechmann, A., Nunes, I. L. (2019). Decision support systems for sustainable manufacturing surrounding the product and production life cycle: A literature review. *Journal of Cleaner Production, 219*, 336–349.
20. Bi, Z. M. (2011). Revisit system architecture for sustainable manufacturing. *Journal of Sustainability, 3*(9), 1323–1340.
21. Peng, T., Kellens, K., Tang, R., Chen, C., Chen, G. (2018). Sustainability of additive manufacturing: An overview on its energy demand and environmental impact. *Additive Manufacturing, 21*, 694–704.
22. Banik, S. R., Kalita, N., Gajrani, K. K., Kumar, R., Sankar, M. R. (2018). Recent trends in laser assisted machining of ceramic materials. *Materials Today: Proceedings, 5*(9), 18459–18467.
23. Gajrani, K. K., Suvin, P. S., Kailas, S. V., Rajurkar, K. P., Sankar, M. R. (2021). Machining of hard materials using textured tool with minimum quantity nano-green cutting fluid. *CIRP Journal of Manufacturing Science and Technology, 35*, 410–421.
24. Gajrani, K. K., Prasad, A., Kumar, A. (Eds.). (2022). *Advances in sustainable machining and manufacturing processes*. CRC Press, Boca Raton and London. ISBN: 9781003284574. https://doi.org/10.1201/9781003284574
25. Gajrani, K. K., Suvin, P. S., Kailas, S. V., Sankar, M. R. (2019). Hard machining performance of indigenously developed green cutting fluid using flood cooling and minimum quantity cutting fluid. *Journal of Cleaner Production, 206*, 108–123.
26. Gajrani, K. K., Ram, D., Sankar, M. R. (2017). Biodegradation and hard machining performance comparison of eco-friendly cutting fluid and mineral oil using flood cooling and minimum quantity cutting fluid techniques. *Journal of Cleaner Production, 165*, 1420–1435.
27. Mittal, V. K., Sangwan, K. S. (2014). Prioritizing barriers to green manufacturing: environmental, social and economic perspectives. *Procedia CIRP, 17*, 559–564.
28. Mittal, V. K., Sangwan, K. S. (2014). Development of a model of barriers to environmentally conscious manufacturing implementation. *International Journal of Production Research, 52*(2), 584–594.
29. Bhanot, N., Rao, P. V., Deshmukh, S. G. (2017). An integrated approach for analysing the enablers and barriers of sustainable manufacturing. *Journal of Cleaner Production, 142*, 4412–4439.
30. Bhanot, N., Qaiser, F. H., Alkahtani, M., Rehman, A. U. (2020). An integrated decision-making approach for cause-and-effect analysis of sustainable manufacturing indicators. *Sustainability, 12*(4), 1517.
31. Haapala, K. R., Zhao, F., Camelio, J., Sutherland, J. W., Skerlos, S. J., Dornfeld, D. A., Rickli, J. L. (2013). A review of engineering research in sustainable manufacturing. *Journal of Manufacturing Science and Engineering, 135*(4).
32. Jovane, F., Yoshikawa, H., Alting, L., Boer, C. R., Westkamper, E., Williams, D., Paci, A. M. (2008). The incoming global technological and industrial revolution towards competitive sustainable manufacturing. *CIRP Annals-Manufacturing Technology, 57*(2), 641–659.
33. Das, S., Lee, S. H., Kumar, P., Kim, K. H., Lee, S. S., Bhattacharya, S. S. (2019). Solid waste management: Scope and the challenge of sustainability. *Journal of Cleaner Production, 228*, 658–678.
34. Finkbeiner, M., Ackermann, R., Bach, V., Berger, M., Brankatschk, G., Chang, Y. J., Wolf, K. (2014). Challenges in life cycle assessment: an overview of current gaps and research needs. *Background and Future Prospects in Life Cycle Assessment*, 207–258.
35. Mathiyazhagan, K., Sengupta, S., Mathivathanan, D. (2019). Challenges for implementing green concept in sustainable manufacturing: A systematic review. *Opsearch, 56*(1), 32–72.
36. Wiktorsson, M., Bellgran, M., Jackson, M. (2008). Sustainable manufacturing-challenges and possibilities for research and industry from a Swedish perspective. In *Manufacturing systems and technologies for the new frontier*, 119–122. Springer, London.
37. Deng, Y. M., Edwards, K. L. (2007). The role of materials identification and selection in engineering design. *Materials & Design, 28*(1), 131–139.

38. Mayyas, A. T., Omar, M. (2020). Eco-material selection for lightweight vehicle design. In *Energy efficiency and sustainable lighting-a bet for the future*. IntechOpen. London.
39. Zarandi, M. H. F., Mansour, S., Hosseinijou, S. A., Avazbeigi, M. (2011). A material selection methodology and expert system for sustainable product design. *International Journal of Advanced Manufacturing Technology*, *57*(9), 885–903.
40. Peças, P., Ribeiro, I., Folgado, R., Henriques, E. (2009). A life cycle engineering model for technology selection: A case study on plastic injection moulds for low production volumes. *Journal of Cleaner Production*, *17*(9), 846–856.
41. Bambam, A. K., Dhanola, A., Gajrani, K. K. (2022), Machining of titanium alloys using phosphonium-based halogen-free ionic liquid as lubricant additives. *Industrial Lubrication and Tribology*, *74*(6), 722–728. https://doi.org/10.1108/ILT-03-2022-0083
42. Bambam, A. K., Alok, A., Rajak, A., Gajrani, K. K. (2022). Tribological performance of phosphonium-based halogen-free ionic liquids as lubricant additives. *Proceedings of the Institution of Mechanical Engineers, Part J: Journal of Engineering Tribology*. https://journals.sagepub.com/doi/10.1177/13506501221121898
43. Rahimifard, S., Coates, G., Staikos, T., Edwards, C., Abu-Bakar, M. (2009). Barriers, drivers and challenges for sustainable product recovery and recycling. *International Journal of Sustainable Engineering*, *2*(2), 80–90.
44. Abdelbasir, S. M., El-Sheltawy, C. T., Abdo, D. M. (2018). Green processes for electronic waste recycling: A review. *Journal of Sustainable Metallurgy*, *4*(2), 295–311.
45. Álvarez Jaramillo, J., Zartha Sossa, J. W., Orozco Mendoza, G. L. (2019). Barriers to sustainability for small and medium enterprises in the framework of sustainable development literature review. *Business Strategy and the Environment*, *28*(4), 512–524.

Chapter 9

Sustainability in manufacturing

Future trends

Şenol Şirin

Contents

9.1 Introduction

In the late 20th century, the emergence of harmful consequences exposed the harms of economic growth and led to attempts to raise awareness and possibly correct the consequences of modernization. Important international organizations and academic institutions expressed new approaches to development, and one of them emerged as "sustainability" (Javanmardi et al. 2020). Sustainability resulted from increased awareness of social, environmental, and economic issues on a global scale, along with concerns about poverty, inequality, and a healthy future for humanity (Lindsey 2011). The role of the manufacturing industry in economic growth is substantial. Due to rapid production techniques, the factors causing environmental pollution have increased. Studies continue to reduce or completely eliminate these harmful factors that cause climate change. The concept of sustainable manufacturing, which emerged in the production industry, is aimed to eliminate

DOI: 10.1201/9781003291961-10

harmful factors and environmental concerns or reduce unwanted effects. Sustainable manufacturing aims to protect energy and natural resources, reduce adverse environmental conditions, and produce products under economic conditions (Sarkis 2001). Sustainable manufacturing processes have been popular in recent years (Machado et al. 2019). Since the early 1990s, conventional fluids have been used to improve the processes of machining (Haider and Hashmi 2014). In manufacturing processes (turning, milling, grinding, drilling, etc.), cutting tools account for approximately 4% of production costs, while metal cutting fluids account for approximately 17% (Gajrani et al. 2022). However, when machining difficult-to-machine materials, cutting fluid costs can rise to 20–30% of production costs (Pusavec et al. 2010). Cutting fluids basically have three important functions, lubrication, cooling, and removal chips, from the cutting zone (Figure 9.1). Conventional cutting fluids can reduce the cutting temperature, the friction between the cutting tool, workpiece, and chip interfaces, prolong the cutting tool and increase processing efficiency (Yin et al. 2018; Yang et al. 2021).

Cutting fluids used in metal machining are in the petroleum-based mineral oil group (Benedicto et al. 2017). Since cutting fluids are petroleum-based, they can cause serious diseases in terms of human health, such as cancer, skin disease, respiratory tract disorders, etc. (Şirin and Şirin 2014). Studies on methods that will minimize or completely eliminate the adverse effects of cutting fluids on ecological and occupational health in sustainable manufacturing processes have increased in recent years. The best-known of these methods is minimum quantity lubrication (MQL-also known as near-dry machining). MQL represents a system that converts compressed air into an aerosol mist with a small amount of oil. Different cutting fluids are used in the MQL system. In studies on the MQL system, researchers examined the performance of different oils

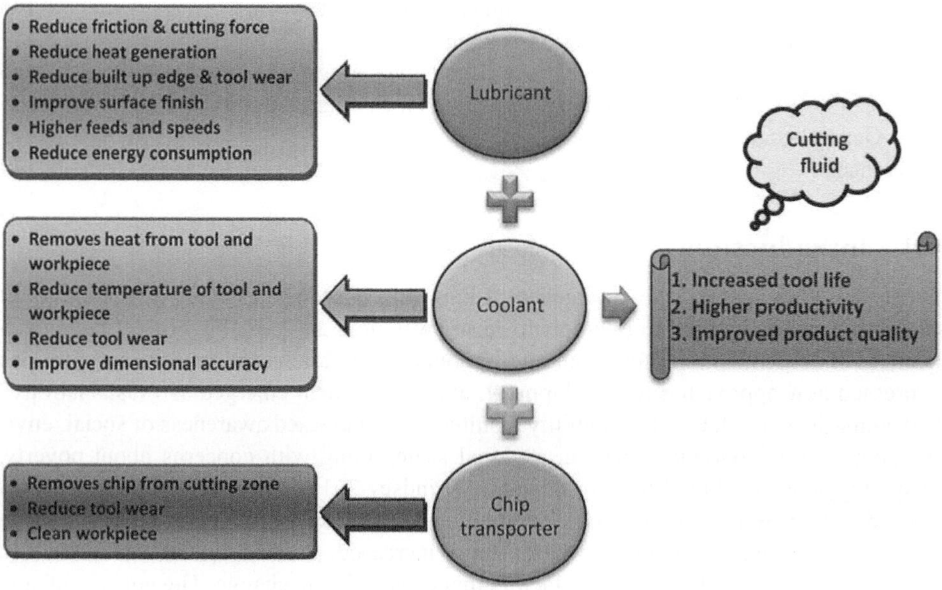

Figure 9.1 Role of the cutting fluids.

(i.e., vegetable-based, synthetic, fatty, semi-synthetic, and mineral) (Gajrani and Sankar 2020). In the results obtained by the researchers, they stated that vegetable-based oils showed the best performance in the MQL system. Researchers have tried different vegetable-based oils in the MQL system, for example, coconut oil (Fernando et al. 2020), rice bran oil (Bedi et al. 2020), jojoba oil (Gaurav et al. 2020), sunflower oil (Anand and Mathew 2020), canola oil (Bambam et al. 2022), soybean oil (Okafor and Nwoguh 2020), rapeseed oil (Subramani et al. 2022), and corn oil (Arsene et al. 2021). Due to the non-toxicity and biodegradability properties of these vegetable-based oils, their negative effects on human health are extremely low compared to conventional cutting fluids. Due to all its advantages, although vegetable-based oils are used in the MQL system, it lags behind conventional cutting fluids with regard to machining efficient (Bhuyan et al. 2018). The researchers used nano-sized particles with high lubricant performance and thermal conductivity coefficient (Bandarra et al. 2014) to increase the performance of the MQL system. The mixture formed with the addition of small sized particles is called nanofluid. Nanofluid mixtures are prepared by adding nano sized (usually 1–100 nm) particles to the cutting oil with different mixing processes and mixing ratios (wt.% or vol.%). In the nanofluid preparation process, two methods, one-step and two-step, are preferred. However, previous studies show that the two-step method is preferred due to its ease of application and cost-effectiveness. Commercially nanoparticles are doped directly into the cutting oil in two step method. More detailed information about the nanofluid mixing processes are detailed in other sections. The prepared mixtures are included in the MQL system and aerosol mist is created with compressed air. The pulverized nanofluids (Figure 9.2) are sent to the cutting zone under pressure. Nanofluids provide satisfactory results not only because of their lubricating and thermal conductivity properties but also the mechanisms in the cutting zone. Nanofluids outperform due to the rolling, mending, filling and polishing etc. in the cutting zone compared to MQL (Gajrani et al. 2019).

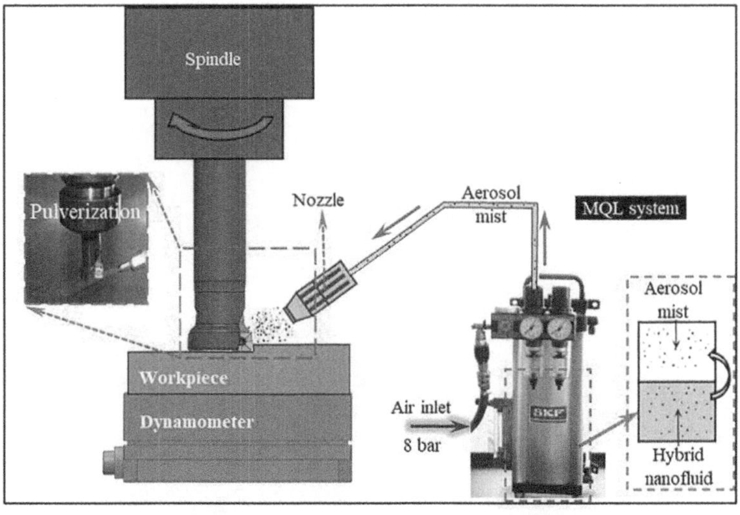

Figure 9.2 Nanofluid aerosol mist and pulverization (Şirin and Kıvak 2021).

Nanoparticles are more preferred in nanofluid mixtures, such as ceramic-based (e.g., hBN, graphite, and graphene nanoplatelets (GnP)) and metal-based (Au, Cu, Fe, Zn, and Ti). Nanoparticles have different shapes (layer, spherical, fiber, etc.), thermal conductivity coefficients, density, and dimensions (diameter, thickness, length, etc.). The different superior properties of nanoparticles have provided researchers with different investigation opportunities (Ağbulut et al. 2021). When two or more nanoparticles are doped to the same mixture, it is called a hybrid nanofluid (Junankar et al. 2021). It was found few studies in the literature on the performance evaluation and/or characterization of hybrid nanofluids in the machining process. Agglomeration and sedimentation should be avoided in mono or hybrid nanofluid mixtures. In the literature, researchers used different mixing processes, such as ultrasonication, vibration, mechanical, and magnetic, for homogeneous mono or hybrid nanofluids. It has been stated that the use of hybrid nanoparticles is more effective in providing ideal cooling/lubrication owing to the synergistic effect of nanoparticles in the MQL method. Surfactants are added to the nanofluid mixture to increase the shelf life of mono or hybrid nanofluids and to relatively prevent aggregation and sedimentation problems. However, in some studies, it has been mentioned that the added surfactants have some disadvantages, such as foaming, reducing the thermal transfer coefficient, and increasing the viscosity (Babita et al. 2016). Studies in which mono or hybrid nanofluid mixtures prepared with adding surfactants have tribological effects are limited. Another cooling method used in sustainable manufacturing is cryogenic cooling, using different cryogen gases. Gases such as nitrogen (N_2), carbon dioxide (CO_2), argon (Ar), and helium (He) are used in cryogenic cooling, also known as subzero cooling (Yildiz and Nalbant 2008). Cryogens used in liquid or gaseous form are non-toxic and extremely effective in lowering the cutting temperature. However, it has been reported in some literature studies that the chips have difficulty leaving the workpiece, as it reduces the cutting temperature to low levels (Figure 9.3) (Yıldırım et al. 2020).

Hybrid cooling/lubrication, which combines the lubrication of nanofluid method and the cooling properties of cryogenic method, has been a subject of interest for researchers in recent years. Studies continue on the sustainable hybrid cooling/lubrication method, which has just begun to be studied in the literature.

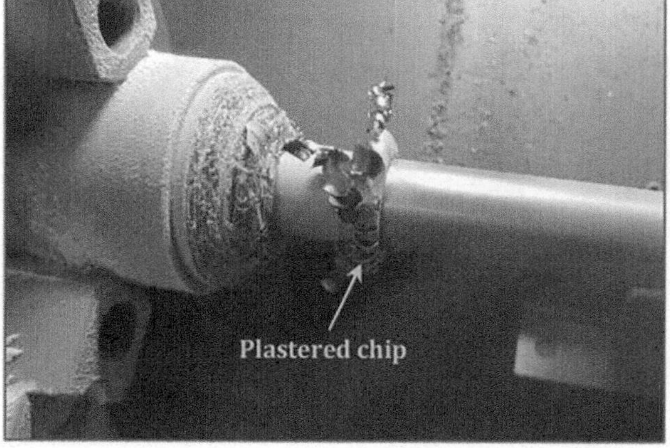

Figure 9.3 Application of cryogenic cooling (Yıldırım et al. 2020).

This book chapter aims to examine the cooling/lubrication methods to be used in the future for sustainable manufacturing. For this purpose, MQL machining, nanofluid machining, hybrid nanofluid machining, cryogenic cooling machining, and hybrid cooling/lubrication machining methods were investigated. The chapter also provides the reader with suggestions for improving the cooling/lubrication methodology.

9.2 Results and discussions

9.2.1 *MQL machining*

The first use of "minimum quantity lubrication" in the literature seems to have been made by Weck and Koch in 1993 regarding the bearings (Boswell et al. 2017). MQL is a method in which a small scale of liquids (20–100 mL/h) is sent directly to the cutting zone as an aerosol mist by combining with compressed air (Şirin and Kıvak 2021). The use of the MQL method, also called semi or near-dry processing, is increasing day by day in industrial applications (Sartori et al. 2018). The MQL method is described as the pressurized spraying of a little amount of biodegradable liquid particles mixed with compressed air onto the cutting zone (Şirin and Kıvak 2019). The effective lubrication film layer formed in cutting tool-workpiece surfaces can reduce cutting temperature and tool wear mechanisms. MQL is an alternative method that minimizes the negative effect on employee health, uses biodegradable oil, does not contain toxic substances, reduces cutting forces, prolongs tool life, and increases product quality compared to wet cutting consist of large amount of cutting fluids (Mia et al. 2019; Makhesana and Patel 2021). Although the MQL method is one of the solutions that eliminates the adverse effects of both petroleum-based traditional fluid and dry machining, it can be used efficiently in traditional processes, such as turning, milling, grinding, and drilling (Şirin and Şirin 2021; Fratila and Caizar 2011).

Due to sustainability and low oil consumption requirements, it is important that MQL fluids have the following characteristics (Boswell et al. 2017):

- biodegradable,
- high lubrication, and
- high stability.

In the MQL method, the pressurized oil mixture is applied to the cutting tool and workpiece zones in two different forms, from inside the cutting tool or outside through the nozzle (Figure 9.4). In the internal application system, air and oil mix together and reach a nozzle through a single/double channel to the cutting tool (Figure 9.5).

In addition, two different nozzles, ejector-type and conventional-type, can be used in an external MQL application. The air/lubricant aerosol mist sent to the cutting zone can be mixed before in an MQL system, and it can also be mixed in the nozzle. External MQL applications are given in Figure 9.6.

9.2.1.1 *MQL machining case study 1*

Dhar et al. turned AISI 4340 material under different conditions (dry, wet, and MQL). According to experimental results, researchers claimed that the MQL condition reduced tool flank wear value (Figure 9.7 (a)). In addition, according to the surface roughness results,

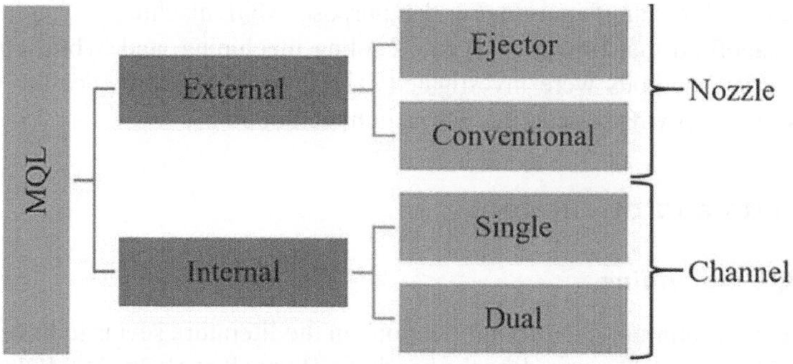

Figure 9.4 Application of MQL system.

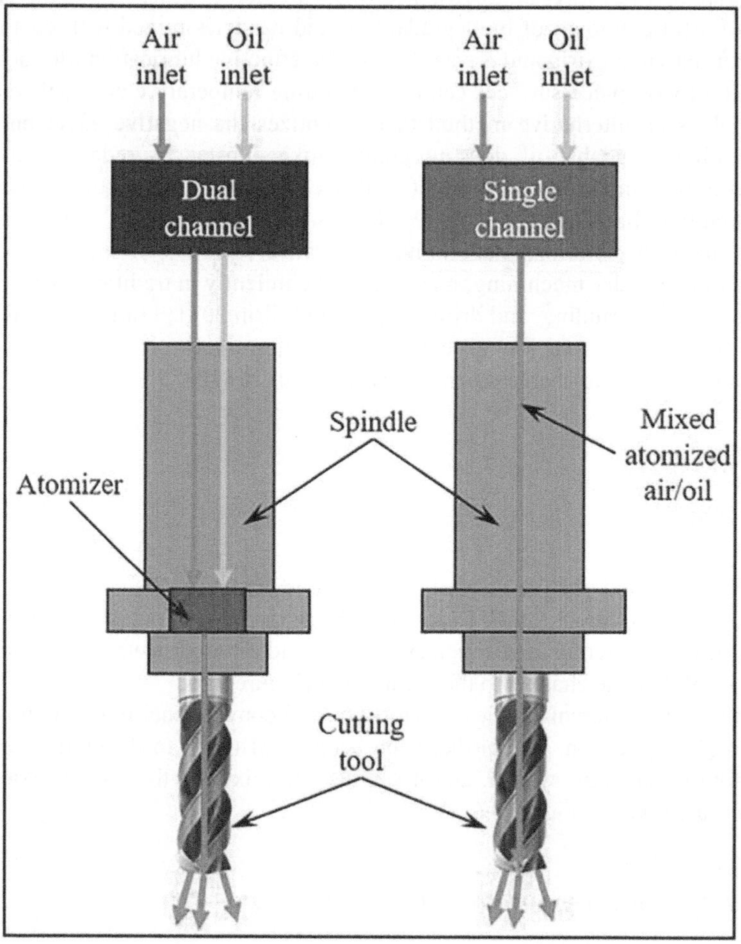

Figure 9.5 Internal channel application of MQL system.

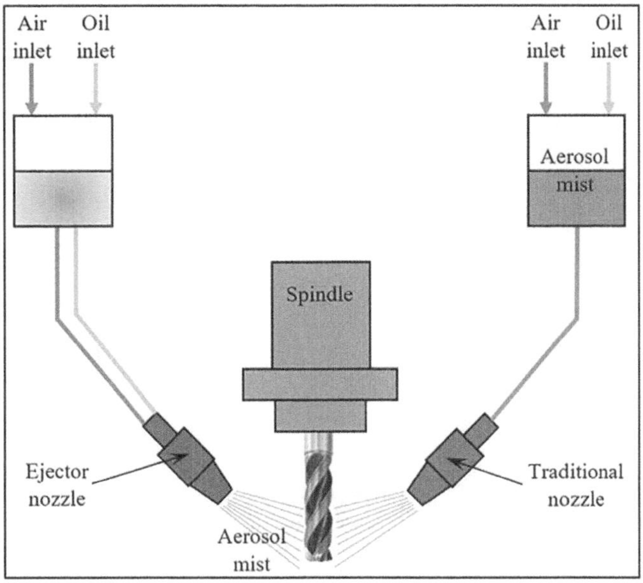

Figure 9.6 External application of MQL system.

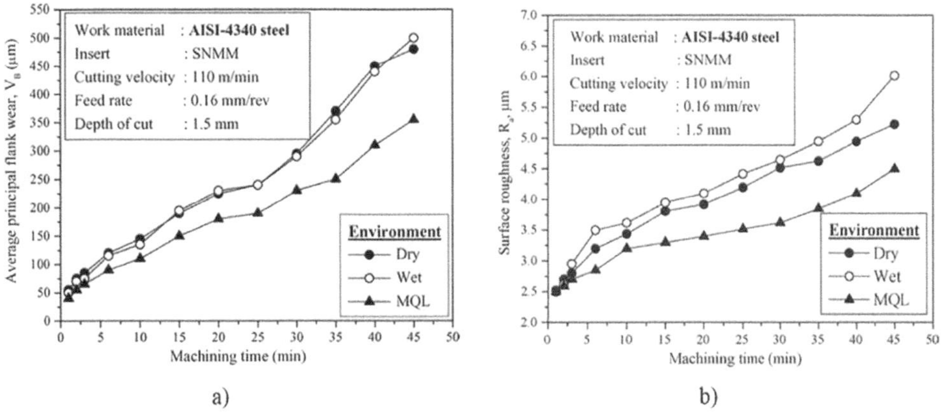

Figure 9.7 Experimental result: (a) flank wear and (b) surface roughness (Dhar et al. 2006).

it was stated that MQL machining showed the best performance (Figure 9.7 (b)). MQL machining was followed by wet and dry machining in tool wear and surface quality results, respectively (Dhar et al. 2006).

9.2.1.2 MQL machining case study 2

Hadad and Sadeghi turned AISI 4140 material under dry, wet, and MQL conditions. Machining forces and surface quality are investigated as machining performance in the

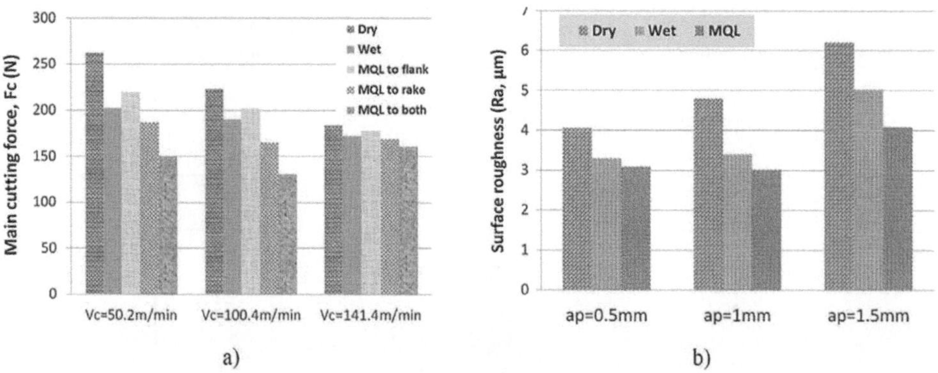

Figure 9.8 Experimental results: (a) cutting force and (b) surface quality (Hadad and Sadeghi 2013).

study. In addition, the MQL nozzle is positioned as flank, rake, and both faces. As shown in Figure 9.8 (a), MQL condition and both nozzle (flank + rake face) positions provided the best performance. As shown in Figure 9.8 (b), MQL condition performed better compared to dry and wet conditions (Hadad and Sadeghi 2013).

9.2.2 Nanofluid machining

There are different types of nanoparticles in terms of physical, chemical, size, and morphology properties (Khan et al. 2019). Nanoparticles can be divided into six groups: carbon, ceramic, metal, semiconductor, polymeric, and lipid-based nanoparticles. Nanoparticles have an average size of 1–100 nm, which is equal to one billionth of a meter (0.000000001 m), and have shapes such as star, spherical, triangular, cubic, hexagonal, oval, helical, prism, rod, cube, sphere, plate, and fiber. Nanoparticles can be produced from chemically stable metals, ceramics, metal carbides, metal carbons, and nitride. Nanoparticles in a mono state have some advantages: lubricity, high thermal conductivity, mechanisms exhibited, etc. When nanoparticles are added to a liquid to form a mixture, even in very small quantities, nanoparticles significantly increase the heat transfer capacity of the resulting mixture (Rajak et al. 2022). Thanks to the mechanisms of nanoparticles, it provides the mixture with properties such as lubricating, reducing friction, and increasing its tribological performance. Nanoparticles performed the mechanisms in the nanofluid mixture such as mending, protective thin film, rolling, and polishing (He et al. 2019). Some of the mechanisms that nanoparticles showed in the nanofluid mixture are given in Figure 9.9.

When nanoparticles are added to a fluid, they are also called named nanofluids. Nanofluid was first invented by Choi in 1995 to describe a stable colloid mixture or suspension between nanometer (10^{-9} m)-sized solid particles into a liquid-based fluid (Suresh et al. 2011; Minea 2017). Since nanoparticles are hydrophobic, extreme care should be taken while preparing the mixture. There are two main methods of nanofluid preparation: one-step and two-step methods. While the two-step method covers the production and distribution operations, the one-step method depends on both being done at the same time. The two-step method is simpler to implement compared to other methods. Therefore, researchers were recommend the two step method for preparing a nanofluid suspension (Şirin and Şirin 2021; Sarıkaya et al. 2021). Some studies conducted in the literature, researchers use

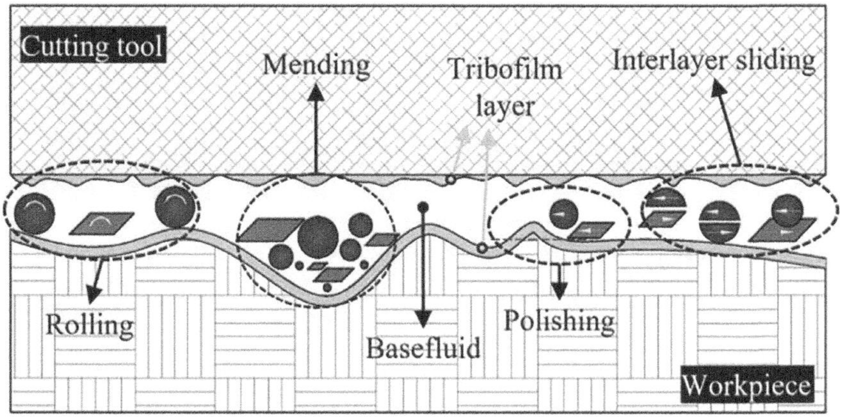

Figure 9.9 Nanoparticle mechanisms.

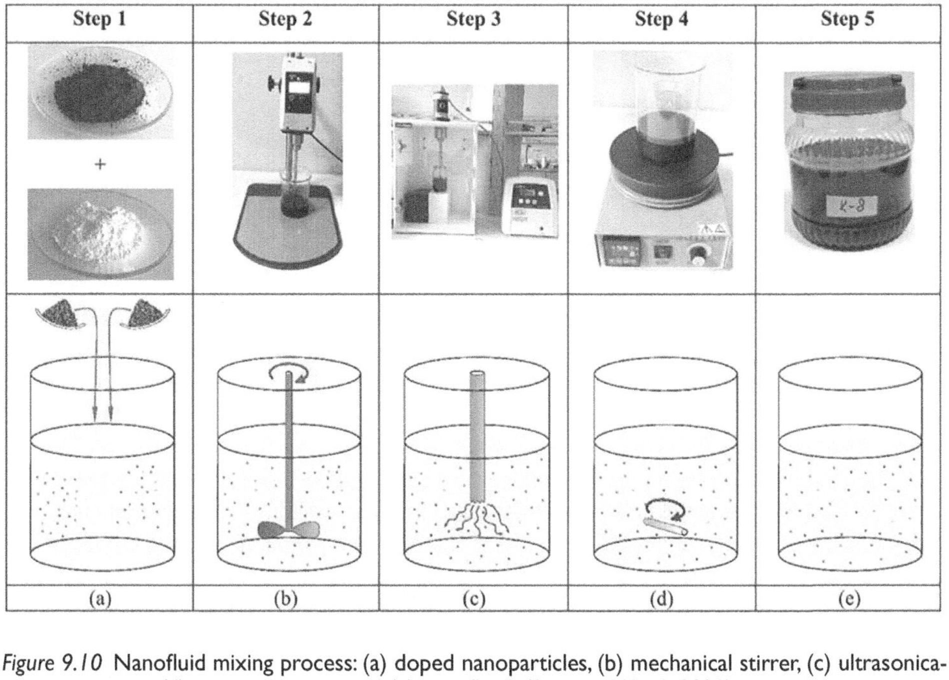

Figure 9.10 Nanofluid mixing process: (a) doped nanoparticles, (b) mechanical stirrer, (c) ultrasonication, (d) magnetic stirrer, and (e) nanofluids (Şirin and Kıvak 2021).

mixing techniques such as mechanical stirrer, ultrasonic homogenizer, magnetic stirrer, and ultrasonic bath (Şirin 2022). The mixing processes applied while preparing the nanofluid are given in Figure 9.10.

Increasing the shelf life of nanofluid mixtures after the mixing process is still a matter of interest. Therefore, it is aimed to increase the shelf life of nanofluid mixtures and to provide a homogeneous suspension mixture in the studies. Agglomeration and sedimentation are the biggest problems in nanofluid suspension mixtures. The sedimentation rate can change

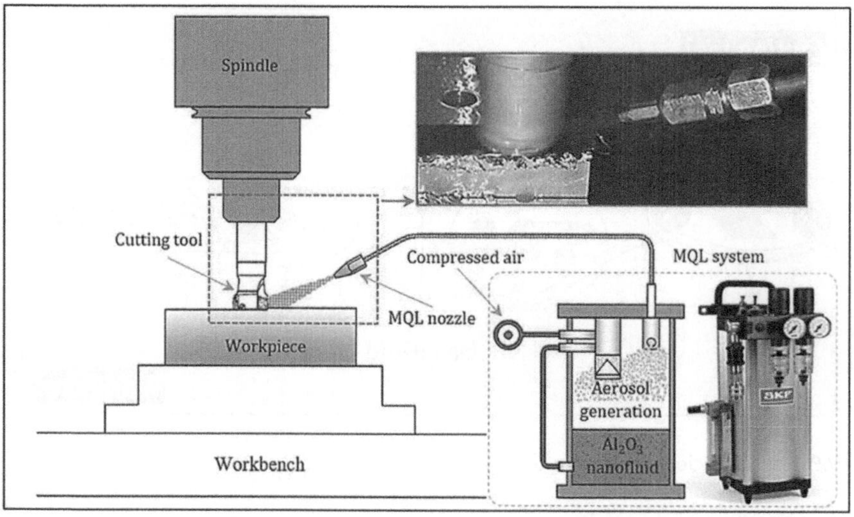

Figure 9.11 Application of nanofluid machining (Günan et al. 2020).

proportional to the square of nanoparticle radius according to Stokes law given in Eq. (9.1) (Singh et al. 2020):

$$V_s = \frac{2R^2}{9\mu_m}\left(p_p - p_m\right)g \tag{9.1}$$

In the equation, V_s: sedimentation rate, R: average radius of the nanoparticle, μ: viscosity of the liquid, p_p: density of the nanoparticle, and p_m: density of the liquid. In addition, using smaller radius nanoparticles can reduce sedimentation rate. Increasing the nanoparticles surface energy can cause nanoparticle aggregation. Therefore, choosing an optimal rate of the nanoparticle size is necessary for homogeneous mixture.

The long shelf life, low stability and tendency to agglomerate of nanofluid mixtures make their future use difficult on a large scale. Nanofluids may lose their heat transfer capacity and homogeneous because of sedimentation and change their thermophysical properties (Fuskele and Sarviya 2017). The most commonly used techniques for nanofluid stability are: surfactant addition (E. Şirin et al. 2021), ultrasonic mixing, and pH control of suspensions (Nobrega et al. 2022).

Prepared homogeneous nanofluids were transferred to the cutting zone by MQL method. It has been stated in the literature study that nanofluids gave prolonged cutting tool life (Şirin and Şirin 2021; Günan et al. 2020), and improve the surface quality (E. Şirin et al. 2021). The image in which the nanofluids are sent to the cutting zone is given in Figure 9.11.

9.2.2.1 *Nanofluid machining case study 1*

Yücel et al. turned AA 2024 T3 aluminum alloy under dry, MQL, and MoS_2 nanofluid conditions. Researchers examined surface quality, cutting temperature, and cutting tool wear mechanisms as performance criteria in their studies. According to experimental results, nanofluid machining improved the surface quality (Figure 9.12 (a)), reduced the cutting temperature (Figure 9.12 (b)), and delayed tool wear compared to dry and MQL conditions (Yücel et al. 2021).

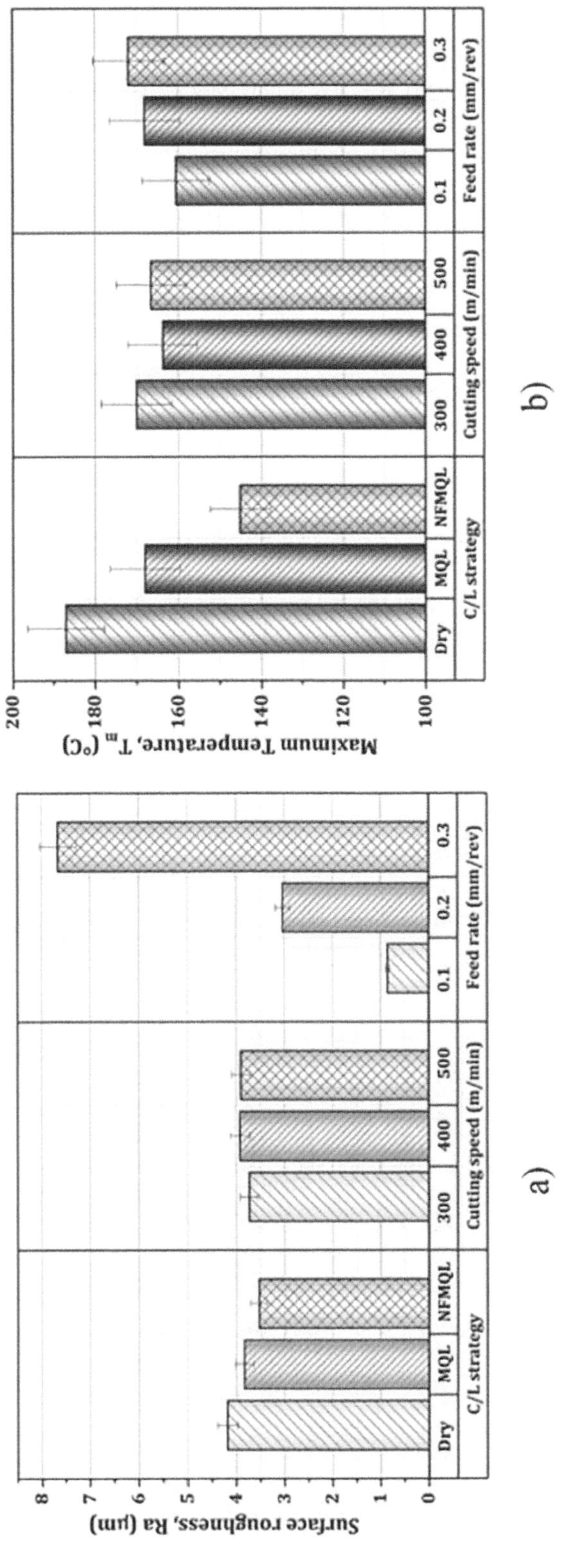

Figure 9.12 Experimental result: (a) surface roughness and (b) cutting temperature (Yücel et al. 2021) (Open-access).

9.2.2.2 *Nanofluid machining case study 2*

Gaurav et al. (2020) turned hardened Ti6-Al-4V material under dry, MQL (vegetable-based jojoba oil and mineral-based LRT30), 0.1, 0.5, and 0.9 wt.% doped MoS_2 to jojoba oil, 0.1, 0.5 and 0.9 wt.% doped MoS_2 to LRT 30 nanofluid conditions. Researchers preferred machining performance criteria: cutting forces, surface roughness, and tool wear. At the end of tests, jojoba 0.1 wt.% doped MoS_2 showed better performance in cutting forces (Figure 9.13 (c)), surface roughness (Figure 9.13 (a)), and tool flank wear (Figure 9.13 (b)) (Gaurav et al. 2020)

9.2.3 *Hybrid nanofluid machining*

In the previous sections, it was stated that a hybrid nanofluid was obtained by mixing two or more nano-sized particles. In other words, hybrid nanofluids are liquids containing different types of nanoparticles. The nanofluid preparation process is extremely important because

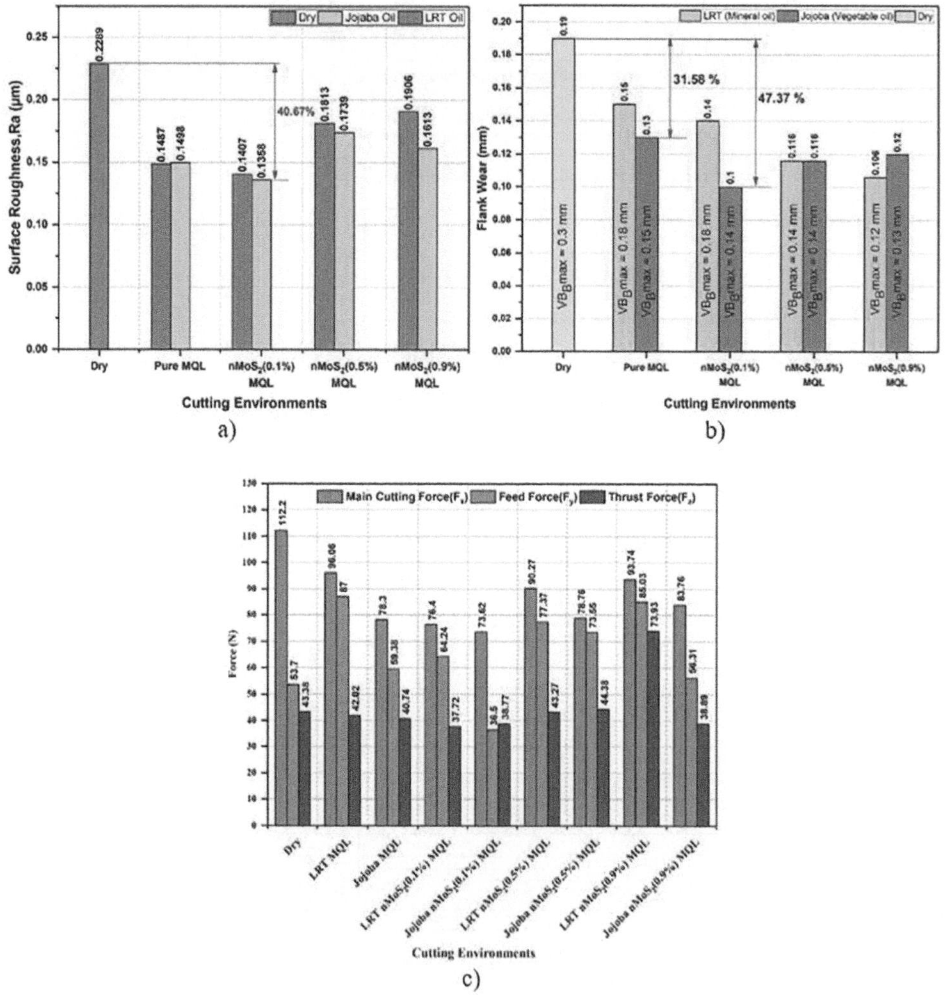

Figure 9.13 Experimental results: (a) surface roughness, (b) flank wear, and (c) cutting forces (Gaurav et al. 2020).

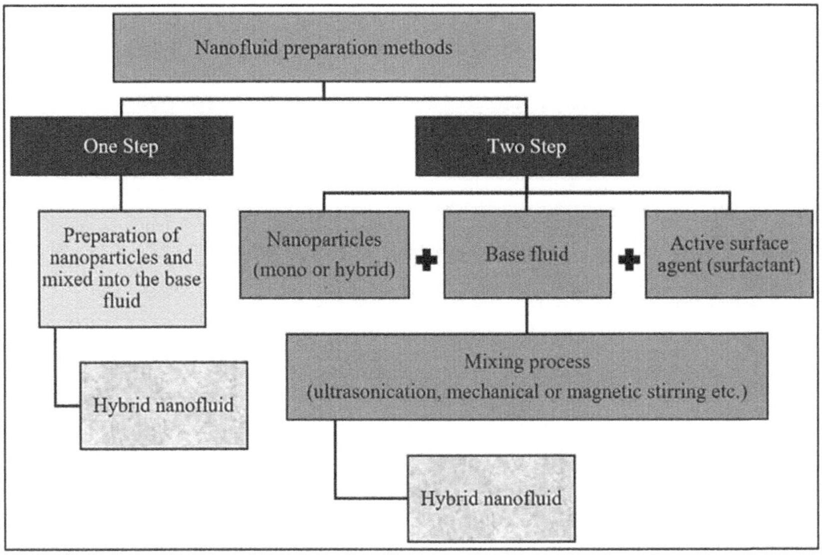

Figure 9.14 Preparation of hybrid nanofluids.

the Van der Waals and cohesive forces of different nanoparticles are different. One-step or two-step methods are utilized in the preparation of the hybrid nanofluids, such as mono-type nanofluids. Mono or hybrid nanofluid preparation methods are given in Figure 9.14.

Nanoparticles with different densities, shapes, and thermophysical properties can show different mechanism effects in the same fluid. In the section on nanofluid machining, it was stated that mono nanoparticles showed the mechanisms of mending, protective thin film, rolling, and polishing effects. Nanoparticles in the hybrid nanofluids show some additional mechanisms (synergistic effect (Yıldırım et al. 2022), interlayer sliding effect (Şirin and Kıvak 2021), and hybrid particle effect (Şirin 2022). Some mechanisms of hybrid nanoparticles are given in Figure 9.15.

The long-term stability of nanoparticles can be a challenge, especially in hybrid nanofluids. This situation can be expressed by the colloidal suspension behavior of two different nanoparticles against mono nanoparticles. The selection of hybrid particles is important to obtain a synergetic effect, superior tribological performance, and better heat transfer performance. Different base fluids and nanoparticles are used to prepare hybrid nanofluid mixtures. Some base fluids and nanoparticles used in nanofluid mixtures are shown in Figure 9.16.

In the literature research, it has been seen that there are not enough studies on hybrid nanofluids. In a few studies, it has been stated that hybrid nanofluids are more advantageous than mono nanofluids. The advantages of hybrid nanofluids include a high heat transfer coefficient, better thermophysical properties, high tribological performances, effective lubrication during machining, and better wettability characteristics.

9.2.3.1 Hybrid nanofluid machining case study 1

Yıldırım et al. turned AISI 420 hardened steel under dry, MQL, mono/hybrid MWCNT, Al_2O_3 and MoS_2 nanofluid conditions. The researchers prepared the hybrid nanofluids by mixing at 0.6 vol.% in ratios of 1:1, 1:2, and 2:1. As performance criteria, surface quality, cutting temperature, and cutting tool wear mechanisms were examined in the study.

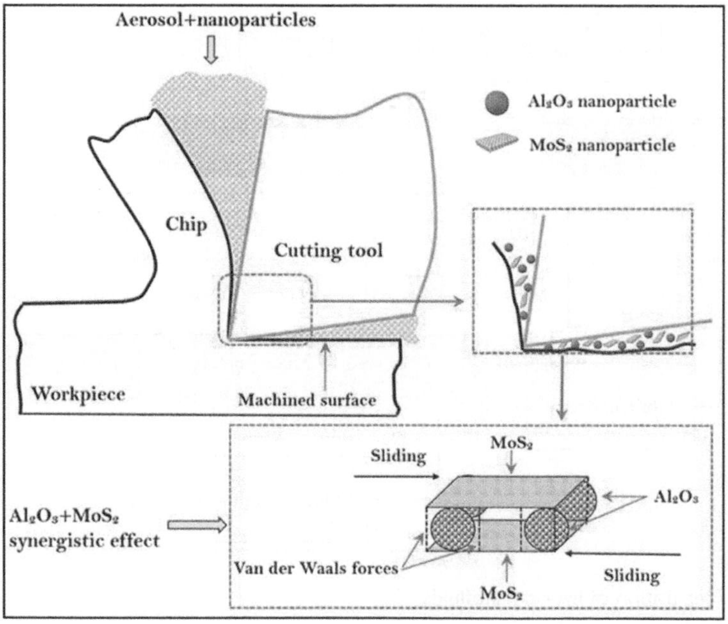

Figure 9.15 Hybrid nanoparticles mechanisms (Yıldırım et al. 2022).

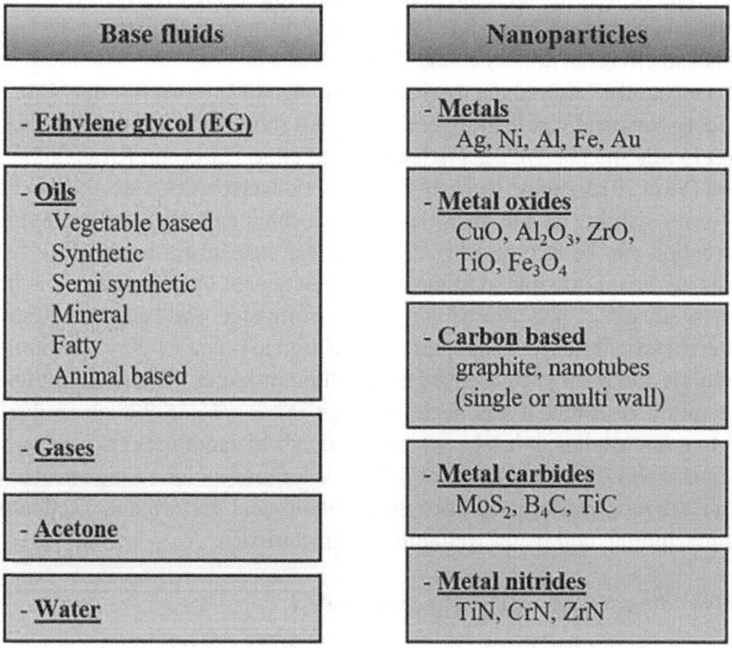

Figure 9.16 Base fluid and nanoparticles for preparing mono or hybrid nanofluids.

Researchers reached the following conclusions: Al_2O_3:MoS_2 2:1 hybrid nanofluid showed the best performance in terms of surface roughness (Figure 9.17 (a)), Al_2O_3:MoS_2 1:1 hybrid nanofluid showed the best performance in terms of cutting temperature (Figure 9.17 (b)), and Al_2O_3:MoS_2 2:1 hybrid nanofluid exhibited the best performance in terms of flank wear (Figure 9.17 (c)) (Yıldırım et al. 2022).

9.2.3.2 Hybrid nanofluid machining case study 2

Khan et al. turned AISI 52100 steel under MQL, Al_2O_3 nanofluid, and hybrid nanofluid conditions. Mono and hybrid nanofluids were prepared at 0.2 vol.%, 0.75 vol.%, and 1.20 vol.%. The researchers chose surface quality, power-energy consumption, and cutting tool life as performance characteristics. The outstanding results in the experimental study: Al-GnP hybrid nanofluid with 1.20 vol.% showed the best performance in terms of surface roughness (Figure 9.18 (a)), cutting power, and machine tool power (Figure 9.18 (b)) (Khan et al. 2020).

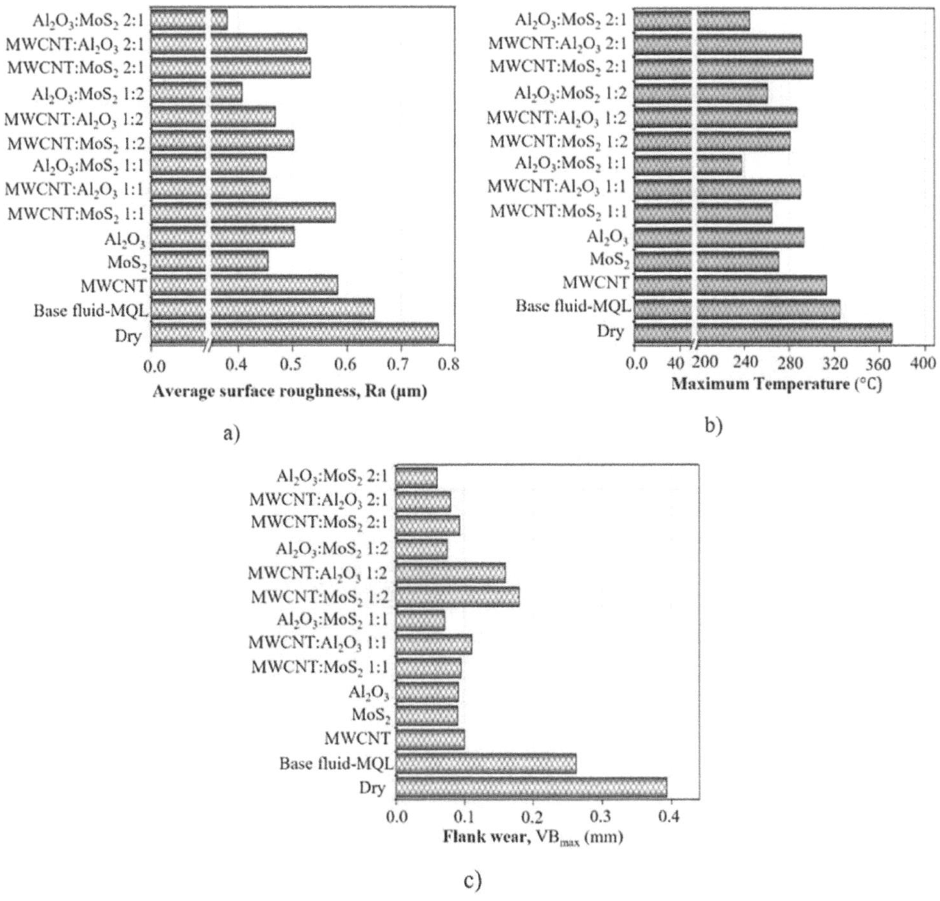

Figure 9.17 Experimental results: (a) surface roughness, (b) maximum cutting temperature, and (c) flank wear (Yıldırım et al. 2022).

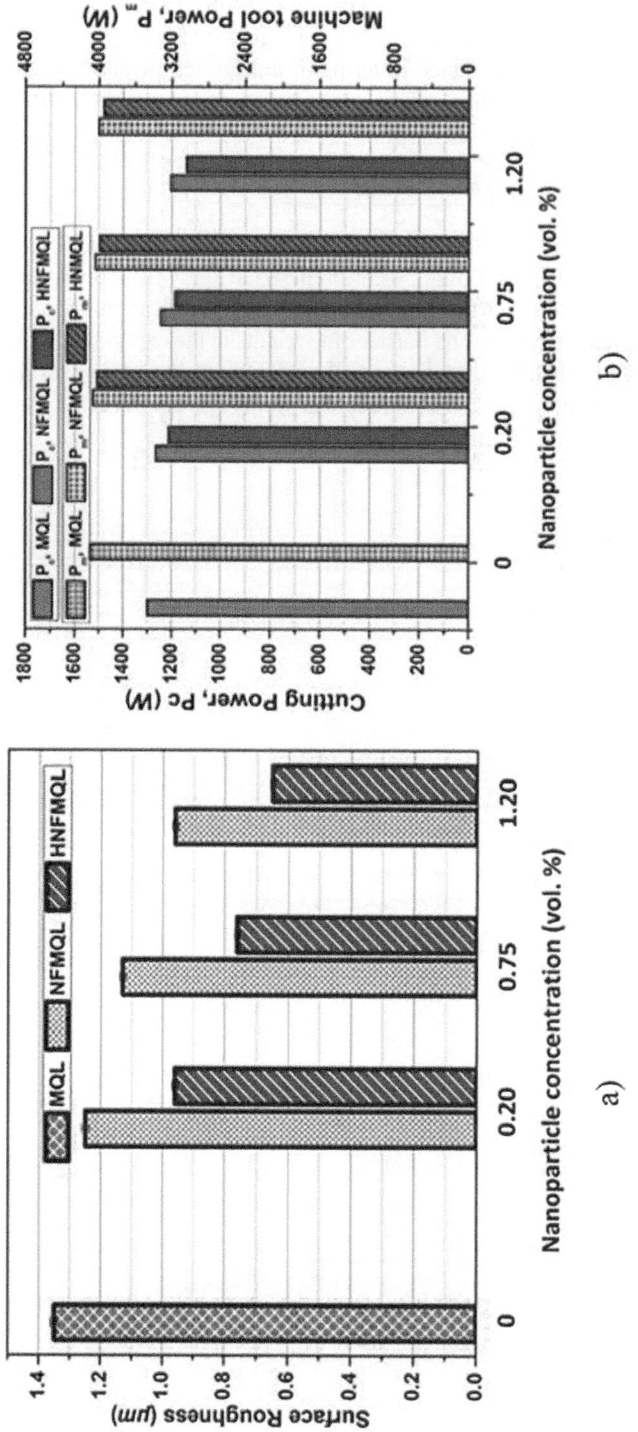

Figure 9.18 Experimental results: (a) surface roughness and (b) cutting and machine tool power (Khan et al. 2020).

9.2.4 Cryogenic cooling machining

Another process used in sustainable manufacturing is cryogenic cooling (also called sub-zero cooling) machining. Cryogenic cooling is an environmentally friendly machining process and was first used by Uehara and Kumagai in 1969 (Sivaiah and Chakradhar 2020). Cryogenic cooling machining, one of the most effective methods of controlling temperature during manufacturing processes (i.e., turning, milling, grinding, drilling, etc.), is performed using a gaseous-based cooler. The most preferred cryogens for cooling are air, nitrogen (N_2), helium (He), carbon dioxide (CO_2), and argon (Ar) (Yildiz and Nalbant 2008; Şirin 2022). Recently, these cryogens have been used in the gas or liquid phase to control cutting temperature by researchers (Jebaraj and Kumar 2019). However, in cryogenic cooling, CO_2 and LN_2 are often preferred because they are anti-corrosive, anti-flammable, nontoxic, and odorless (Gupta et al. 2021). When both gases are used as refrigerants, low-temperature generation needs to be distinguished. CO_2 and LN_2 phase diagrams are shown in Figure 9.19.

The gases used in cryogenic cooling must be stored in a storage tank. The gases pressurized in the storage tanks produced in different capacities are sent to the cutting zone with a cryogenic nozzle (Figure 9.20). To minimize thermal losses, researchers prefer a vacuum-jacketed hose for cryogen transfer (Yıldırım 2019).

9.2.4.1 Cryogenic cooling machining case study 1

Gupta et al. (2021) turned α-β titanium alloy material under dry, CO_2, and LN_2 conditions. In the study, the researchers preferred flank wear, surface roughness, cutting temperature, and cutting force as performance criteria. Researchers claimed that LN_2 showed better performance in terms of flank wear (Figure 9.21 (a)) and cutting force, while CO_2 showed better performance in terms of surface roughness (Figure 9.21 (b)) and cutting temperature (Figure 9.21 (c)) (Gupta et al. 2021).

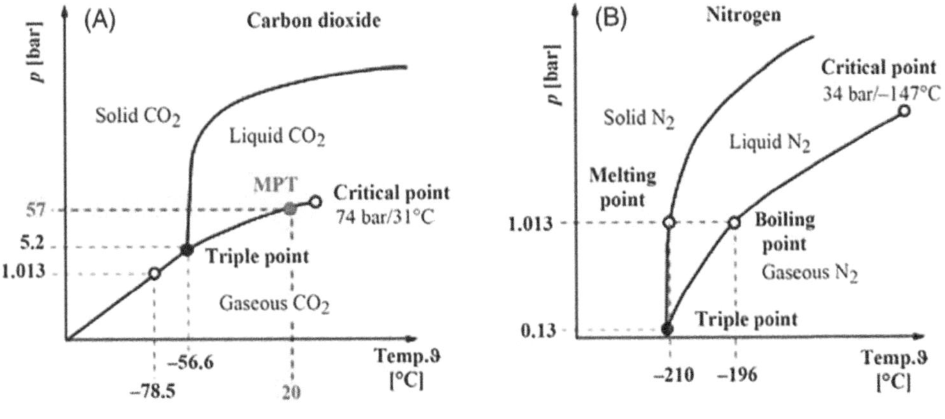

Figure 9.19 Phase diagram for (a) CO_2 and (b) LN_2 (Blau et al. 2015).

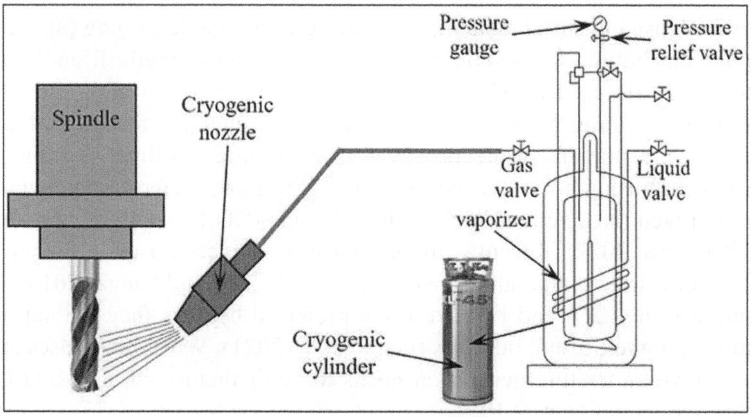

Figure 9.20 Schematic representation of cryogenic cooling machining.

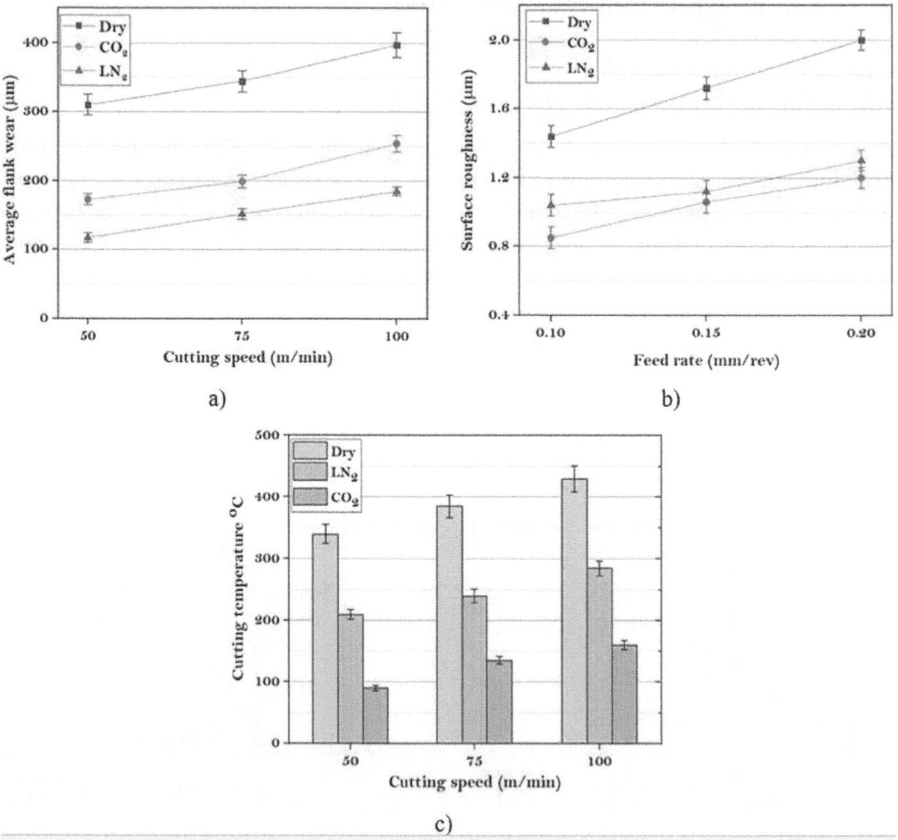

Figure 9.21 Experimental results: (a) flank wear, (b) surface roughness, and (c) cutting temperature (Gupta et al. 2021).

9.2.4.2 *Cryogenic cooling machining case study 2*

Uçak and Çiçek drilled Ni alloy 718 with TiAlN coated and uncoated drills under dry, wet, and cryogenic cooling/lubrication. The authors claimed that wet conditions showed the best performance in surface quality (Figure 9.22 (a)). LN_2 condition showed the best performance at the cutting temperature (Figure 9.22 (b)) (Uçak and Çiçek 2018).

9.2.5 *Hybrid cooling/lubrication machining*

Hybrid cooling/lubrication is one of the most curious topics, especially in sustainable manufacturing. In this method, a combination of cryogen cooling properties and MQL applications such as mono/hybrid nanofluid properties are employed. However, it has been understood that the studies in the literature on hybrid cooling/lubrication are insufficient. Hybrid cooling/lubrication methods include MQL + cryogenic, nanofluid + cryogenic, and hybrid nanofluid + cryogenic. MQL + cryogenic hybrid cooling/lubrication systems are shown in Figure 9.23.

Hybrid cooling/lubrication machining has many advantages over other cooling/lubrication methods. Controlling the temperature occurs in deformation regions (primary, secondary, and tertiary) of the cutting zone with cryogenic cooling, efficient lubrication with MQL, and synergistic effect with mono and hybrid nanofluids can be given as examples for hybrid cooling/lubrication. The effects of different cooling/lubrication efficient are given in Table 9.1. In particular, hybrid cooling/lubrication machining results show promise in cooling, lubrication, and product quality.

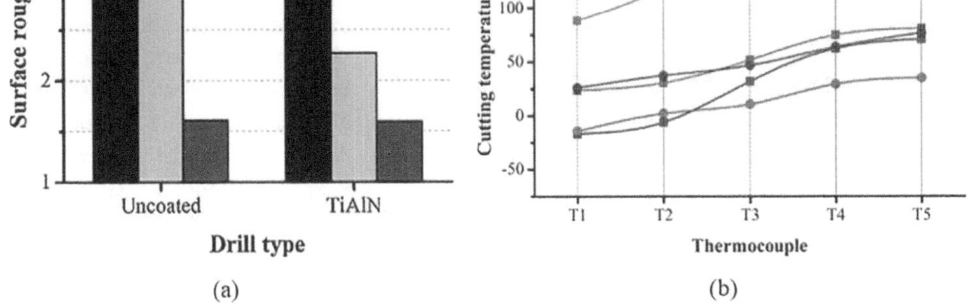

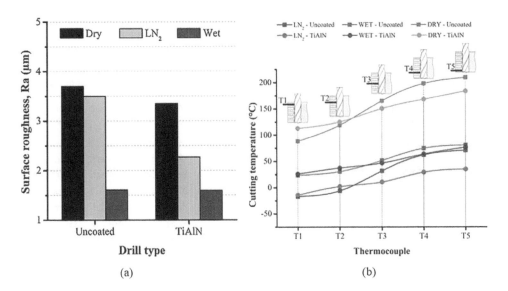

Figure 9.22 Experiment results: (a) surface roughness and (b) cutting temperature (Uçak and Çiçek 2018).

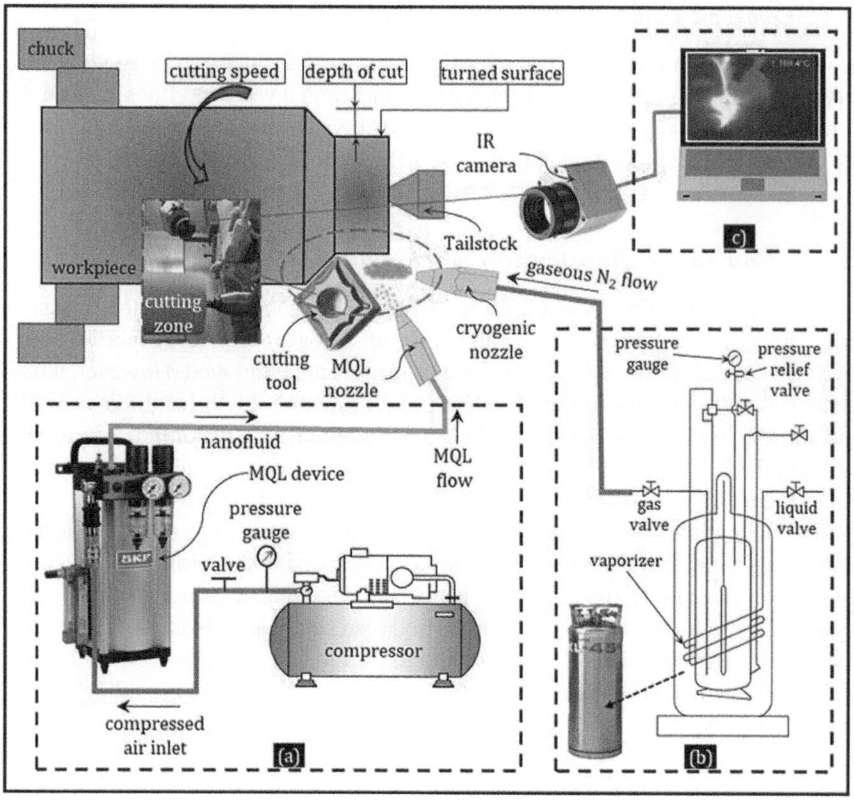

Figure 9.23 Application of hybrid cooling/lubrication machining: (a) MQL machining, (b) cryogenic machining, and (c) cutting temperature (Şirin 2022).

Table 9.1 Application and Efficiency of Various Cooling/Lubrication Machining (Jawahir et al. 2016)

	Effects of the cooling and lubricating strategy	*Food (emulsion/ oil)*	*Dry (compressed air)*	*MQL (oil)*	*Cryogenic (LN2)*	*Hybrid (LN2 + MQL)*
Primary	*Cooling*	Good	Poor	Marginal	Excellent	Excellent
	Lubrication	Excellent	Poor	Excellent	Marginal	Excellent
	Chip Removal	Good	Good	Marginal	Good	Good
Secondary	*Machine Cooling*	Good	Poor	Poor	Marginal	Marginal
	Workpiece Cooling	Good	Poor	Poor	Good	Good
	Dust/Particle Control	Good	Poor	Marginal	Marginal	Good
	Product Quality (Surface Integrity)	Good	Poor	Marginal	Excellent	Excellent
	Sustainability Concerns	*Water pollution, microbial infestation, and high cost*	*Poor surface integrity due to thermal damage*	*Harmful oil vapor*	*Initial cost*	*Initial cost and oil vapor*

(Source: Data from Brinksmeier, Jawahir, I. S., H. Attia, D. Biermann, J. Duflou, F. Klocke, D. Meyer, S. T. Newman, et al. "Cryogenic Manufacturing Processes" *CIRP Annals* 65, no 2 (2016): 713–736.)

9.2.5.1 Hybrid cooling/lubrication machining case study 1

Yıldırım et al. turned Inconel 625 material under different cooling/lubrication methods (i.e., MQL, cryogenic and hybrid). The researchers used liquid LN_2 for cryogenic cooling. The cutting temperature (Figure 9.24 (a)), surface roughness (Figure 9.24 (b)), and tool wear (Figure 9.24 (c)) results showed that cryogenic + MQL hybrid cooling/lubrication machining performed better than dry and MQL machining. It has even been claimed that MQL machining outperforms tool wear in cryogenic conditions. It was also emphasized that in cryogenic machining, the chip does not break regularly and plasters to the workpiece (Yıldırım et al. 2020).

9.2.5.2 Hybrid cooling/lubrication machining case study 2

Şirin turned 25 superalloy materials under dry, MQL, nanofluids (GnP and MWCNT), hybrid nanofluid, cryogenic (gaseous N_2), and hybrid cooling/lubrication conditions. The researcher stated that at the cutting temperature of the cryogenic condition (Figure 9.25 (b)), the MWCNT/GnP + cryogenic hybrid cooling lubrication condition performed better in terms of surface roughness (Figure 9.25 (a)) and tool wear (Figure 9.25 (c)) compared to all other conditions (Şirin 2022).

9.3 General conclusion and future trends

The aerospace, weapons, petrochemical, nuclear energy, and medical industries are among the rapidly developing global manufacturing industries. The specifications such as high strength, toughness, and high-temperature resistance, etc. required by these industries have led to the emergence of superior materials (Kim et al. 2020). Examples of superior materials are high-strength steel, titanium alloys, and superalloy materials. Although the superior properties of these materials are advantageous, difficulties are encountered during processing. Therefore, such materials are known to be difficult to machine materials. When machining these materials, high friction occurs in elastic-plastic deformation. Friction can cause undesirable high-temperature formations in tool-chip-workpiece interfaces. Cutting fluids are an alternative method to reduce the temperature in tool-chip-workpiece interfaces. However, petroleum-based fluids cause important pollution in the environment, and micro-scale mist particles evaporation of the heat damage the health of the employees (Chetan et al. 2015). It is necessary to reduce the environmental pollution caused by cutting fluids and employ clean, sustainable manufacturing according to environmental protection and sustainable improving requirements (Araújo et al. 2017). In recent years, the development of environmental-friendly sustainable manufacturing technology relying on sustainable science and engineering theory has become an important research point in industrial applications (Mia et al. 2018). The most preferred and future-trending methods in sustainable manufacturing are MQL machining, nanofluid machining, hybrid nanofluid machining, cryogenic cooling machining, and hybrid cooling/lubrication machining. In this chapter, sustainability in manufacturing and future trends in cooling/lubrication techniques were examined. As a result of the investigations, the following conclusions were reached:

i) MQL machining: Although MQL machining performs better for surface roughness and reducing the cutting temperature, it lags behind conventional metal cutting fluids.
ii) Nanofluid machining: It has increased the thermal transmission coefficient and machining efficiency of the vegetable-based oil in the MQL system, thanks to the nanoparticle mechanisms.

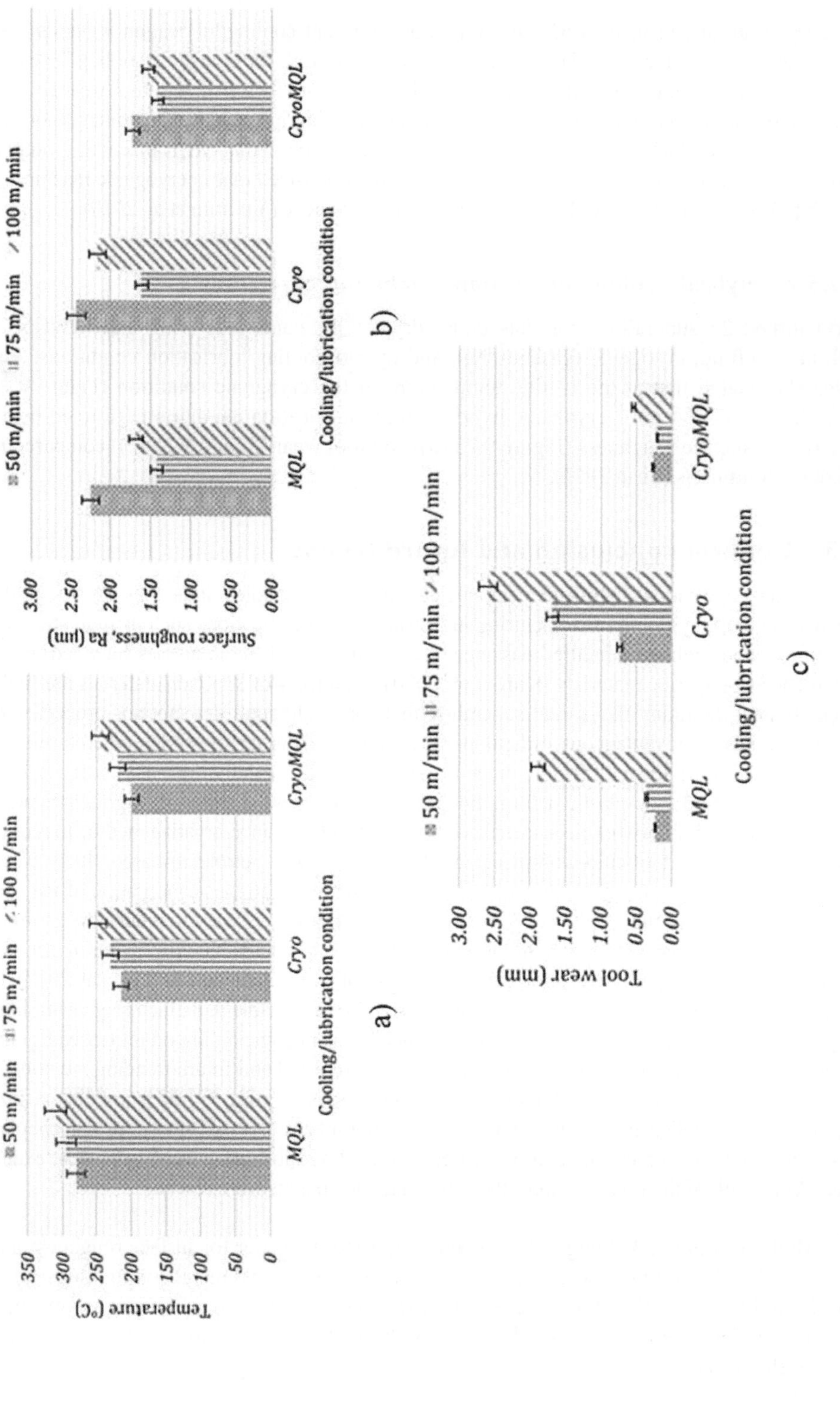

Figure 9.24 The effect of cooling/lubrication machining results: (a) temperature, (b) surface roughness, and (c) tool wear (Yıldırım et al. 2020).

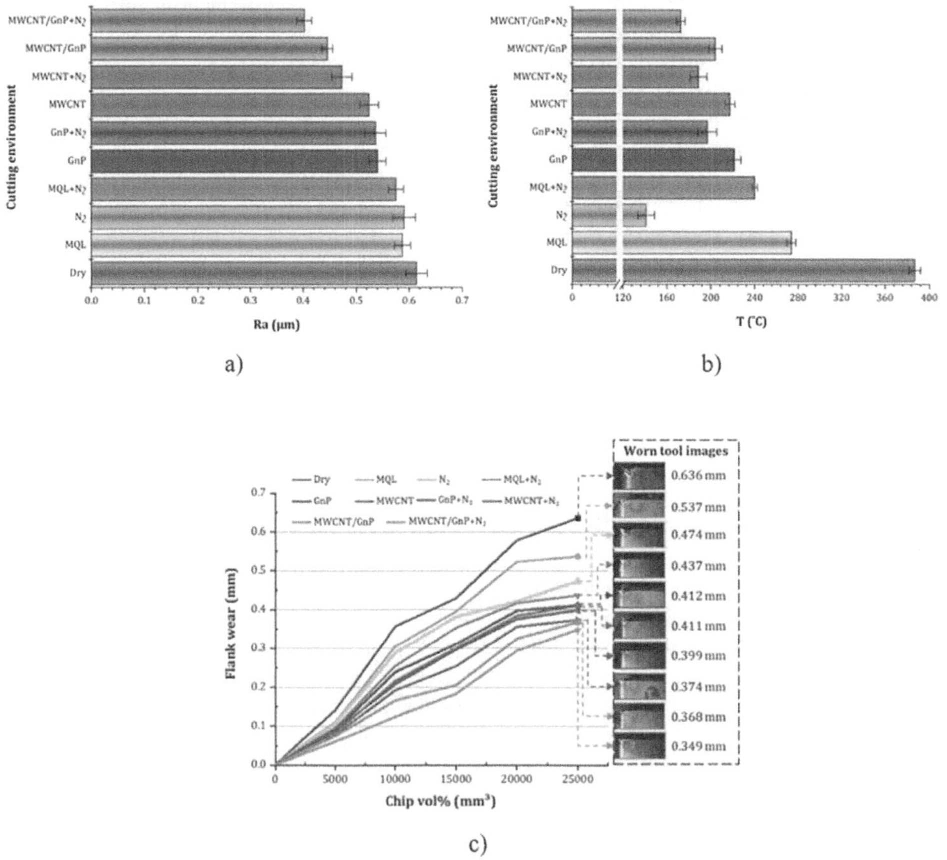

Figure 9.25 Experimental results: (a) surface roughness, (b) temperature, and (c) flank wear (Şirin 2022).

iii) Hybrid nanofluid machining: In the technique, which shows a synergistic effect with the combination of different nanoparticles, substantial improvements in tool life have been observed.

iv) Cryogenic machining: In the technique, which used CO_2 and N_2 gases, the cutting temperature values, especially in the cutting zone, are reduced.

v) Hybrid cooling/lubrication machining: With the combination of MQL, nanofluid and hybrid nanofluid and cryogenic gases, both the processing efficiency has increased, and the temperature values have been under controlled in the cutting zone.

References

Ağbulut, Ü., F. Polat, and S. Sarıdemir. 2021. A comprehensive study on the influences of different types of nano-sized particles usage in diesel-bioethanol blends on Combustion, performance, and environmental aspects. *Energy* 229: 120548. doi:10.1016/J.ENERGY.2021.120548

Anand, K. N., and J. Mathew. 2020. Evaluation of size effect and improvement in surface characteristics using sunflower oil-based MQL for sustainable micro-end milling of Inconel 718. *Journal of the Brazilian Society of Mechanical Sciences and Engineering* 42(4): 1–13. doi:10.1007/S40430-020-2239-0/FIGURES/14

Araújo J., A. Santos, W. F. Sales, R. B. d. Silva, E. S. Costa, and Á. R. Machado. 2017. Lubri-cooling and tribological behavior of vegetable oils during milling of AISI 1045 steel focusing on sustainable manufacturing. *Journal of Cleaner Production* 156: 635–647. doi:10.1016/J.JCLEPRO.2017.04.061

Arsene, B., C. Gheorghe, F. A. Sarbu, M. Barbu, L. I. Cioca, and G. Calefariu. 2021. MQL-assisted hard turning of AISI D2 steel with corn oil: Analysis of surface roughness, tool wear, and manufacturing costs. *Metals, 11:2058* 11. doi:10.3390/MET11122058

Babita, S., K. Sharma, and S. M. Gupta. 2016. Preparation and evaluation of stable nanofluids for heat transfer application: a review. *Experimental Thermal and Fluid Science* 79: 202–212. doi:10.1016/j.expthermflusci.2016.06.029

Bambam, A. K., Dhanola, A., and Gajrani, K. K. (2022), Machining of titanium alloys using phosphonium-based halogen-free ionic liquid as lubricant Additives. *Industrial Lubrication and Tribology* 74(6): 722–728. doi:10.1108/ILT-03-2022-0083

Bandarra F., E. Pedone, O. S. H. Mendoza, C. L. L. Beicker, A. Menezes, and D. Wen. 2014. Experimental investigation of a silver nanoparticle-based direct absorption solar thermal system. *Energy Conversion and Management* 84: 261–267. doi:10.1016/j.enconman.2014.04.009

Bedi, S. S., G. C. Behera, and S. Datta. 2020. Effects of cutting speed on MQL machining performance of AISI 304 stainless steel using uncoated carbide insert: application potential of coconut oil and rice bran oil as cutting Fluids. *Arabian Journal for Science and Engineering* 45(11): 8877–8893. doi:10.1007/S13369-020-04554-Y/FIGURES/21

Benedicto, E., D. Carou, and E. M. Rubio. 2017. Technical, economic and environmental review of the lubrication/cooling systems used in machining processes. *Procedia Engineering* 184: 99–116. doi:10.1016/J.PROENG.2017.04.075

Bhuyan, M., Sarmah, A., Gajrani, K. K., Pandey, A., Thulkar, T. G., and Sankar, M. R. 2018. State of art on minimum quantity lubrication in grinding process. *Materials Today: Proceedings* 5(9): 19638–19647.

Blau, P., K. Busch, M. Dix, C. Hochmuth, A. Stoll, and R. Wertheim. 2015. Flushing strategies for high performance, efficient and environmentally friendly cutting. *Procedia CIRP* 26: 361–366. doi:10.1016/J.PROCIR.2014.07.058

Boswell, B., M. N. Islam, I. J. Davies, Y. R. Ginting, and A. K. Ong. 2017. A review identifying the effectiveness of minimum quantity lubrication (MQL) during conventional machining. *The International Journal of Advanced Manufacturing Technology* 92(1): 321–340. doi:10.1007/S00170-017-0142-3

Brinksmeier, Jawahir, I. S., H. Attia, D. Biermann, J. Duflou, F. Klocke, D. Meyer, S. T. Newman, et al. 2016. Cryogenic manufacturing processes. *CIRP Annals* 65(2): 713–736.

Chetan, S. G., and P. V. Rao. 2015. Application of sustainable techniques in metal cutting for enhanced machinability: A review. *Journal of Cleaner Production* 100: 17–34. doi:10.1016/J.JCLEPRO.2015.03.039

Dhar, N. R., M. Kamruzzaman, and M. Ahmed. 2006. Effect of minimum quantity lubrication (MQL) on tool wear and surface roughness in turning AISI-4340 steel. *Journal of Materials Processing Technology* 172(2): 299–304. doi:10.1016/j.jmatprotec.2005.09.022

Fernando, W. L. R., N. Sarmilan, K. C. Wickramasinghe, H. M. C. M. Herath, and G. I. P. Perera. 2020. Experimental investigation of minimum quantity lubrication (MQL) of coconut oil based metal working fluid. *Materials Today: Proceedings* 23: 23–26. doi:10.1016/J.MATPR.2019.06.079

Fratila, D., and C. Caizar. 2011. Application of Taguchi method to selection of optimal lubrication and cutting conditions in Face milling of $AlMg_3$. *Journal of Cleaner Production* 19(6–7): 640–645. doi:10.1016/j.jclepro.2010.12.007

Fuskele, V., and R. M. Sarviya. 2017. Recent developments in nanoparticles synthesis, preparation and Stability of nanofluids. *Materials Today: Proceedings* 4(2): 4049–4060. doi:10.1016/J.MATPR.2017.02.307

Gajrani, K. K., A. Prasad, and A. Kumar. (Eds.). 2022. *Advances in sustainable machining and manufacturing processes*. CRC Press, Boca Raton and London. ISBN: 9781003284574. doi:10.1201/9781003284574

Gajrani, K. K., and M. R. Sankar. 2020. Role of eco-friendly cutting fluids and cooling techniques in machining. In *Materials forming, machining and post processing*, 159–181. Springer, Cham. ISBN: 978-3-030-18853-5. doi:10.1007/978-3-030-18854-2_7

Gajrani, K. K., P. S. Suvin, S. V. Kailas, and R. S. Mamilla. 2019. Thermal, rheological, wettability and hard machining performance of MoS_2 and CaF_2 based minimum quantity hybrid nano-green cutting fluids. *Journal of Materials Processing Technology* 266: 125–139.

Gaurav, G., A. Sharma, G. S. Dangayach, and M. L. Meena. 2020. Assessment of jojoba as a pure and nano-fluid base oil in minimum quantity lubrication (MQL) hard-turning of Ti-6Al-4V: A step towards sustainable machining. *Journal of Cleaner Production* 272: 122553. doi:10.1016/J.JCLEPRO.2020.122553

Gupta M. K., Q. Song, Z. Liu, M. Sarıkaya, M. Mia, M. Jamil, A. K. Singla, A. Bansal, Y. Pimenov, and M. Kuntoğlu. 2021. Tribological performance based machinability investigations in cryogenic cooling assisted turning of α-β titanium alloy. *Tribology International* 160: 107032. doi:10.1016/j.triboint.2021.107032

Günan, F., T. Kıvak, Ç. V. Yıldırım, and M. Sarıkaya. 2020. Performance evaluation of MQL with Al_2O_3 mixed nanofluids prepared at different concentrations in milling of Hastelloy C276 Alloy. *Journal of Materials Research and Technology* 9(5): 10386–10400. doi:10.1016/J.JMRT.2020.07.018

Hadad, M., and B. Sadeghi. 2013. Minimum quantity lubrication-MQL turning of AISI 4140 steel alloy. *Journal of Cleaner Production* 54: 332–343. doi:10.1016/J.JCLEPRO.2013.05.011

Haider, J., and M. S. J. Hashmi. 2014. Health and environmental impacts in metal machining processes. *Comprehensive Materials Processing* 8: 7–33. doi:10.1016/B978-0-08-096532-1.00804-9

He, J., J. Sun, Y. Meng, and X. Yan. 2019. Preliminary investigations on the tribological performance of hexagonal boron nitride nanofluids as lubricant for steel/steel friction pairs. *Surface Topography: Metrology and Properties* 7(1): 15022. doi:10.1088/2051-672X/ab0afb

Javanmardi, E., S. Liu, and N. Xie. 2020. Exploring grey systems theory-based methods and applications in sustainability studies: A systematic review approach. *Sustainability* 12(11): 4437. doi:10.3390/SU12114437

Jawahir, I. S., H. Attia, D. Biermann, J. Duflou, F. Klocke, D. Meyer, S. T. Newman, et al. 2016. Cryogenic manufacturing processes. *CIRP Annals* 65(2): 713–736. doi:10.1016/j.cirp.2016.06.007

Jebaraj, M., and M. P. Kumar. 2019. Effect of cryogenic CO2 and LN2 coolants in milling of aluminum alloy. *Materials and Manufacturing Processes* 34(5): 511–520. doi:10.1080/10426914.2018.1532591

Junankar, A. A., S. R. Parate, P. K. Dethe, N. R. Dhote, D. G. Gadkar, D. D. Gadkar, and S. A. Gajbhiye. 2021. A review: Enhancement of turning process performance by effective utilization of hybrid nanofluid and MQL. *Materials Today: Proceedings* 38: 44–47. doi:10.1016/J.MATPR.2020.05.603

Khan, A. M., M. K. Gupta, H. Hegab, M. Jamil, M. Mia, N. He, Q. Song, Z. Liu, and C. I. Pruncu. 2020. Energy-based cost integrated modelling and sustainability assessment of Al-GnP hybrid nanofluid assisted turning of AISI52100 steel. *Journal of Cleaner Production* 257: 120502. doi:10.1016/J.JCLEPRO.2020.120502

Khan, I., K. Saeed, and I. Khan. 2019. Nanoparticles: Properties, applications and toxicities. *Arabian Journal of Chemistry* 12(7): 908–931. doi:10.1016/J.ARABJC.2017.05.011

Kim, J. H., E. J. Kim, and C. M. Lee. 2020. A study on the heat affected zone and machining characteristics of difficult-to-cut materials in laser and induction assisted machining. *Journal of Manufacturing Processes* 57: 499–508. doi:10.1016/J.JMAPRO.2020.07.013

Lindsey, T. C. 2011. Sustainable principles: Common values for achieving sustainability. *Journal of Cleaner Production* 19(5): 561–565. doi:10.1016/J.JCLEPRO.2010.10.014

Machado, C. G., M. P. Winroth, and E. H. D. R. da Silva. 2019. Sustainable manufacturing in industry 4.0: An emerging research Agenda. *International Journal of Production Research* 58(5): 1462–1484. doi:10.1080/00207543.2019.1652777

Makhesana, M. A., and K. M. Patel. 2021. Performance assessment of vegetable oil-based nanofluid in minimum quantity lubrication (MQL) during machining of Inconel 718. *Advances in Materials and Processing Technologies*: 1–17. doi:10.1080/2374068X.2021.1945305

Mia, M., M. A. Khan, and N. Dhar. 2019. Study of surface roughness and cutting forces using ANN, RSM, and ANOVA in turning of Ti-6Al-4V under cryogenic jets applied at flank and rake faces of coated WC tool. *The International Journal of Advanced Manufacturing Technology* 93(1–4): 975–991. doi:10.1007/s00170-017-0566-9

Mia, M., M. K. Gupta, G. Singh, G. Królczyk, and D. Y. Pimenov. 2018. An approach to cleaner production for machining hardened steel using different cooling-lubrication conditions. *Journal of Cleaner Production* 187: 1069–1081. doi:10.1016/J.JCLEPRO.2018.03.279

Minea, A. A. 2017. Hybrid nanofluids based on Al_2O_3, TiO_2 and SiO_2: Numerical evaluation of different approaches. *International Journal of Heat and Mass Transfer* 104: 852–860. doi:10.1016/J.IJHEATMASSTRANSFER.2016.09.012

Nobrega, G., R. R. de Souza, I. M. Gonçalves, A. S. Moita, J. E. Ribeiro, and R. A. Lima. 2022. Recent developments on the thermal properties, stability and applications of nanofluids in Machining, solar energy and biomedicine. *Applied Sciences* 12(3): 1115. doi:10.3390/APP12031115

Okafor, A. C., and T. O. Nwoguh. 2020. Comparative evaluation of soybean oil–based MQL flow rates and emulsion flood cooling strategy in high-speed face milling of Inconel 718. *International Journal of Advanced Manufacturing Technology* 107(9–10): 3779–3793. doi:10.1007/S00170-020-05248-3/FIGURES/12

Pusavec, F., D. Kramar, P. Krajnik, and J. Kopac. 2010. Transitioning to sustainable production—part ii: Evaluation of sustainable machining technologies. *Journal of Cleaner Production* 18(12): 1211–1221. doi:10.1016/J.JCLEPRO.2010.01.015

Rajak, U., Ü. Ağbulut, I. Veza, A. Dasore, S. Sarıdemir, and T. N. Verma. 2022. Numerical and experimental investigation of ci engine behaviours supported by zinc oxide nanomaterial along with diesel fuel. *Energy* 239: 122424. doi:10.1016/J.ENERGY.2021.122424

Sarıkaya, M., Ş. Şirin, Ç. V. Yıldırım, T. Kıvak, and M. K. Gupta. 2021. Performance evaluation of whisker-reinforced ceramic tools under nano-sized solid lubricants assisted MQL turning of co-based Haynes 25 superalloy. *Ceramics International* 47(11): 15542–15560. doi:10.1016/J.CERAMINT.2021.02.122

Sarkis, J. 2001. Manufacturing's role in corporate environmental sustainability concerns for the new millennium. *International Journal of Oprations & Production Management* 21(6): 666–686. www.emerald-library.com/ft

Sartori, S., A. Ghiotti, and S. Bruschi. 2018. Solid lubricant-assisted minimum quantity lubrication and cooling strategies to improve Ti6Al4V machinability in finishing turning. *Tribology International* 118: 287–294. doi:10.1016/j.triboint.2017.10.010

Singh, G., M. K. Gupta, H. Hegab, A. M. Khan, Q. Song, Z. Liu, M. Mia, M. Jamil, V. S. Sharma, M. Sarikaya, and C. I. Pruncu. 2020. Progress for sustainability in the mist assisted cooling techniques: A critical review. *The International Journal of Advanced Manufacturing Technology* 109(1): 345–376. doi:10.1007/S00170-020-05529-X

Şirin, E., T. Kıvak, and Ç. V. Yıldırım. 2021. Effects of mono/hybrid nanofluid strategies and surfactants on machining performance in the drilling of Hastelloy X. *Tribology International* 157: 106894. doi:10.1016/j.triboint.2021.106894

Şirin, E., and Ş. Şirin. 2014. The negative effects on the human health of metal working fluids used in central cooling system. *Duzce University Journal of Science and Technology* 2(2): 444–457. https://dergipark.org.tr/en/pub/dubited/issue/4808/66230

Şirin, E., and Ş. Şirin. 2021. Investigation of the performance of ecological cooling/lubrication methods in the milling of AISI 316L stainless steel. *Manufacturing Technologies and Applications* 2(1): 75–84. https://dergipark.org.tr/en/pub/mateca/901460

Şirin, Ş. 2022. Investigation of the performance of cermet tools in the turning of Haynes 25 superalloy under gaseous N2 and hybrid nanofluid cutting environments. *Journal of Manufacturing Processes* 76: 428–443. doi:10.1016/J.JMAPRO.2022.02.029

Şirin, Ş., and T. Kıvak. 2019. Performances of different eco-friendly nanofluid lubricants in the milling of Inconel X-750 superalloy. *Tribology International* 137: 180–192. doi:10.1016/j.triboint.2019.04.042.

Şirin, Ş., and T. Kıvak. 2021. Effects of hybrid nanofluids on machining performance in MQL-milling of Inconel X-750 superalloy. *Journal of Manufacturing Processes* 70: 163–176. doi:10.1016/J.JMAPRO.2021.08.038

Sivaiah, P., and D. Chakradhar. 2020. Identifying the effectiveness of manner of cryogenic coolant supply in different cryogenic cooling techniques in turning process: A review. *Machining Science and Technology* 24(6): 948–999. doi:10.1080/10910344.2020.1815039

Subramani, S., S. N. Muthu, and N. L. Gajbhiye. 2022. A numerical study on the influence of minimum quantity lubrication parameters on spray characteristics of rapeseed oil As cutting fluid. *Industrial Lubrication and Tribology* 74(2): 197–204. doi:10.1108/ILT-08-2021-0305/FULL/PDF

Suresh, S., K. P. Venkitaraj, P. Selvakumar, and M. Chandrasekar. 2011. Synthesis of Al_2O_3–Cu/water hybrid nanofluids using two step method and its thermo physical properties. *Colloids and Surfaces A: Physicochemical and Engineering Aspects* 388(1–3): 41–48. doi:10.1016/J.COLSURFA.2011.08.005

Uçak, N., and A. Çiçek. 2018. The effects of cutting conditions on cutting temperature and hole quality in drilling of Inconel 718 using solid carbide drills. *Journal of Manufacturing Processes* 31: 662–673. doi:10.1016/j.jmapro.2018.01.003

Yang, M., C. Li, Z. Said, Y. Zhang, R. Li, S. Debnath, H. M. Ali, T. Gao, and Y. Long. 2021. Semiempirical heat flux model of hard-brittle bone material in ductile microgrinding. *Journal of Manufacturing Processes* 71: 501–514. doi:10.1016/J.JMAPRO.2021.09.053

Yildiz, Y., and M. Nalbant. 2008. A review of cryogenic cooling in machining processes. *International Journal of Machine Tools and Manufacture* 48(9): 947–964. doi:10.1016/j.ijmachtools.2008.01.008

Yin, Q., C. Li, L. Dong, X. Bai, Y. Zhang, M. Yang, D. Jia, Y. Hou, Y. Liu, and R. Li. 2018. Effects of the physicochemical properties of different nanoparticles on lubrication performance and experimental evaluation in the NMQL milling of Ti-6Al-4V. *The International Journal of Advanced Manufacturing Technology* 99: 3091–3109. doi:10.1007/s00170-018-2611-8

Yıldırım, Ç. V. 2019. Experimental comparison of the performance of nanofluids, cryogenic and hybrid cooling in turning of Inconel 625. *Tribology International* 137: 366–378. doi:10.1016/j.triboint.2019.05.014

Yıldırım, Ç. V., T. Kıvak, M. Sarıkaya, and Ş. Şirin. 2020. Evaluation of tool wear, surface roughness/topography and chip morphology when machining of Ni-based alloy 625 under MQL, cryogenic cooling and CryoMQL. *Journal of Materials Research and Technology* 9(2): 2079–2092. doi:10.1016/j.jmrt.2019.12.069

Yıldırım, Ç. V., Ş. Şirin, T. Kıvak, and M. Sarıkaya. 2022. A comparative study on the tribological behavior of mono&proportional hybrid nanofluids for sustainable turning of AISI 420 hardened steel with cermet tools. *Journal of Manufacturing Processes* 73: 695–714. doi:10.1016/J.JMAPRO.2021.11.044

Yücel, A., Ç. V. Yıldırım, M. Sarıkaya, Ş. Şirin, T. Kıvak, M. K. Gupta, and İ. V. Tomaz. 2021. Influence of MoS_2 based nanofluid-MQL on tribological and machining characteristics in turning of AA 2024 T3 aluminum alloy. *Journal of Materials Research and Technology* 15: 1688–1704. doi:10.1016/J.JMRT.2021.09.007

Part II

Sustainable materials and processing

Chapter 10

Role of materials and manufacturing processes in sustainability

M. B. Kiran and V. J. Badheka

Contents

10.1 Introduction

Environmental sustainability means conserving natural resources so that the health and well-being of the present and future ecosystems can be protected. The well-being of human beings is closely related to environmental sustainability. Clean air and fresh water are the need of the hour. Rapid industrialization is deteriorating the health of our planet. The increased emission of carbon dioxide, methane, carbon monoxide, etc., from industries has worsened the atmosphere. Global warming is causing ice to melt and has resulted in floods, unseasonal rainfall, and variation in climatic conditions. Population increase is another issue that leads to increased farming, resulting in increased carbon emissions. The variation in climate has resulted in detrimental agricultural production. Thus, many state and central governments of several countries worldwide have developed policies and execution methodologies for checking global warming. The government and its policies alone cannot achieve environmental sustainability. Manufacturing companies are also equally responsible for achieving environmental sustainability. Efficient usage of electricity also contributes to ecological sustenance. Companies will have to adopt practices that make them socially responsible and environmentally sustainable. Making use of clean energy is one such example. Over-dependency on fossil fuels is another issue plaguing several countries around the world. Increased use of clean energy, such as solar energy, geothermal energy, wind energy, etc.,

DOI: 10.1201/9781003291961-12

can bring the green gas emission under control. Thus, global warming can be brought under control. Vehicular pollution is also increasing around the world. This will also result in carbon dioxide, carbon monoxide, methane, etc. This will also contribute to global warming to a large extent. This clearly shows the harmful effects of global warming and signifies the need for companies to follow practices that would lead to environmental sustenance and become socially responsible.

10.1.1 Scenario in India

The National Clean Air Programme (NACP) policy floated by the Government of India targets to achieve a 20% to 30% reduction in PM_{10} and $PM_{2.5}$ concentrations by 2024. Pollutants can be in solid, liquid, or gaseous forms. In 2018, the Government of India came out with a comprehensive action plan for implementing the policy, with clear timeframes. It also includes the names of agencies responsible for implementation. The Government of India also floated the Environment Protection Act in 1986. Table 10.1 shows the details of the air pollution act.

The act aims to protect the environment by implementing specific practices, to tackle specific environmental problems in different parts of the country. Specific policies for air, water, and noise have been floated by the Central Pollution Control Board, Government of India. This clearly shows the effort by the government to provide a clean environment to its citizens. The Water Act came in 1974 to prevent and control pollution. Amendment to this act came in 1988. A cess was collected from citizens for performing activities related to the prevention and control of corruption. The act was again amended in 2003 (G.S.R.860(E), G.S.R.840(E), G.S.R.830(E), G.S.R.378(E), G.S.R.58(E)). Table 10.2 shows the details of water management rules.

Hazardous waste management includes proper disposal of electronic waste, nuclear waste, etc. Hazardous waste management is a very significant step in achieving environmental sustainability. For hazardous waste management, the Government of India has initiated rules to ensure the safe handling, generation, storage, and transportation of hazardous waste. The act

Table 10.1 Air Pollution Act

Act	*Description*
No. 14 of 1981, [29/3/1981]—The Air (Prevention and Control of Pollution) Act 1981, amended 1987	The Act aims to prevent, control, and reduce air pollution.
No. 14 of 1981, [29/3/1981]—The Air (Prevention and Control of Pollution) amended 1987	The Act aims to prevent, control, and reduce air pollution.

Table 10.2 Water Management Rules

Act	*Description*
No. 06 of 1974, [23/03/1974]	The Water (Prevention and Control of Pollution) Act, 1974
The Water (Prevention and Control of Pollution) Cess Act, 1977	The Act provides authority for collecting the levy or cesses on water consumption.

Table 10.3 Hazardous Waste Management Rules

Act	*Date*
First Amendment Rules	06-07-2016
Second Amendment Rules	28-02-2017
Third Amendment Rules	11-06-2018
Fourth Amendment Rules	01-03-2019

came in 1989. Finally, it got amended in 2008. Table 10.3 shows the details of the hazardous waste management rules.

The book chapter addresses the different techniques proposed by several researchers for environmental sustenance. The chapter has been organized into the following sections. Section 10.2.1 will address schemes that ensure environmental sustenance through materials. Section 10.2.2 will address strategies that provide environmental sustenance through manufacturing processes. Section 10.2.3 will address the techniques that provide environmental sustenance post-product manufacturing stage. Section 10.3 will elaborate on the conclusions of the present research work.

10.2 Literature review

10.2.1 Sustainable materials

10.2.1.1 Manufacturing

Aircraft components demand high heat resistance, corrosion resistance, and toughness. Ti6Al4V has a very high strength-to-weight ratio (Allwood et al. 2012). This is one of the most commonly used alloys in the aerospace industry. These alloys are costly. Ni alloys are also being used in aerospace industries. Aluminum alloys are also being used in aerospace companies because of their lightweight. Aluminum alloys will have low toughness. Many researchers have studied aluminum alloys, especially defects and quality (Rodrigues et al. 2019). Automobile companies are trying hard for environmental sustainability (Orucu et al. 2022). They observed that very little information was available about the effect of the manufacturing of car parts on the environment. This has motivated them to study the impact of car mats made of rubber on the environment. They performed a sensitivity analysis to figure out the use of alternate raw materials and energy sources. They have concluded that car mat manufacturing is the main contributor to the high-impact group (between 47% and 83%). This is mainly due to the high energy consumption during car mat manufacturing. Energy is consumed during the long mixing and hot-pressing process. They have compared both types of mats-rubber and carpet. They concluded that carpet mats would cause higher environmental impacts (26% to 92%). This study has helped automobile companies and the customers identify car parts that will have the most negligible effect on the environment.

Cutting fluids are being used in manufacturing products to improve efficiency in machining and the surface finish of the products. It does this by providing lubrication, cooling, and also preventing rusting. During machining, large amounts of heat will be generated because of the friction between the chip and the tool, plastic deformation of material at the shear zone, and the rubbing between the tool and the work. If not dissipated from the cutting zone properly, the heat would result in a poor surface finish on the component. Cutting fluids would also help enhance the tool life by reducing the tool wear. The cutting fluids would also improve the stability of the cutting tool during machining. This would improve the accuracy of machining. The cutting fluids would also help clean the tool and the component during machining. Cutting fluids can be either water-based or oil-based. Tang et al. (2022) studied the effect of using water-based cutting fluids during product manufacturing. Oil-based cutting fluids are suitable for lubrication, and hence they find application in low-speed machining. This is because their cooling performance is not good. During high-speed machining, water-based cutting fluids are used. Water-based cutting fluids make use of mineral oil. Usually, water is mixed with mineral oil in a particular proportion. Also, additives such as emulsifiers and corrosion inhibitors are added to water-based cutting fluids to improve machining performance. With the introduction of cutting fluids, the process condition improves during machining, and hence there will be a reduction in power consumption and emission. Tang et al. (2022) studied the stability of water-based cutting fluids. They studied the effect of sterilization effects of various sterilization methods. They concluded that the invading microorganism could decompose molecular structure. It would result in decreased life of cutting fluid and would enhance the processing cost. Wickramasinghe et al. (2020) studied, in detail, the green cutting fluid. They observed the demerits of mineral oil-based cutting fluids. Though mineral oil-based cutting fluids can enhance dimensional accuracy, they adversely affect employee health. They recommended vegetable oil-based cutting fluids for increased performance and better health of operators. Gao et al. (2019) have studied the dispensing mechanism of vegetable oil-based cutting fluids. They also observed the interpretation of these cutting fluids during the machining of components. They did experiments by using different surfactants. They concluded that APE-10 gives the optimum performance. Guo (2017) studied the lubrication performance of a mixture of castor oil and soybean, castor oil and peanut, castor oil and sunflower, castor oil, oil and palm oil; and castor oil and rapeseed oil. The experiment was performed on a grinding machine, using high-temperature nickel-based alloy as work material. The grinding process consumes more energy per unit of material when compared to any other machining process. Observation showed that lubricating properties of a mixture of castor oil with other vegetable oils had performed better than using castor oil alone. The castor oil and soybean mixture gave an optimum performance concerning the surface quality of the component. Zhang et al. (2015) studied the effect of mixing MoS_2/CNT nanoparticles in cutting fluids on lubrication properties. MoS_2 offered good lubrication properties. At the same time, CNT proposed good thermal properties. They performed a grinding operation to study nanofluids' influence on process performance. The experimental result showed that a mixture of nanoparticles in cutting fluids performed better than pure nanoparticles. Also, the combination showed better heat transfer performance than single nanoparticles. They also concluded that the optimum mixing ratio is 2:1. Also, the optimum concentration is 6 wt.%. They have also studied the influence of nanofluids on the force generated during grinding. They studied the effect of nanofluids on the coefficient of friction. They have also studied the variation in surface quality by using nanofluids. The manufacturing and mining sectors emit 12% of global CO_2 emissions. The aerospace industry contributes to 9% of global CO_2 emissions (Monteiro et al. 2022).

10.2.1.2 Building Materials

Energy consumption in the construction sector will amount to about 40% of the total power consumption in any developed economy (Eurostat 2011). This clearly shows the significance of reducing energy consumption in the construction sector. There is a massive requirement for developing energy-efficient building materials. Salazar (2014) studied materials used to make window frame materials and concluded that future window materials should be selected based on LCA data. Here, LCA is used for the assessment of environmental impact. The selection of building materials depends upon embodied energy and carbon. Embodied energy is measured in mega-joules/kilogram (MJ/kg). Milner and Woodard (2016) studied engineered wood products. They concluded that EWP is more sustainable than competing materials.

Bamboo is one of the most environmentally friendly materials in constructing buildings, including house construction. They grow speedily, growing in many countries across the globe except in Europe and Antarctica. The main advantage of Bamboo is that it has a high strength-to-weight ratio. Experience has shown that they are perfect for making floors and cabinets, as they also last longer. The main drawback of Bamboo is that they require treatment; otherwise, it will attract insects. It will also absorb water and may swell and then crack.

Cork: They also increase, just like Bamboos. The main advantage of Cork is that it can be harvested from a living tree. Cork is resilient and hence finds application in floor making. They are also flexible and will revert to their original shape even after sustained pressure. One more advantage of these cork materials is that they absorb noise. They find applications in making insulation. They are also good at absorbing shocks, and hence they find application in floor making. The main demerit is that they require shipping from the Mediterranean, which results in increased transportation costs. Luckily, they are incredibly light, consume less energy in shipping, and have fewer emissions.

Precast concrete slabs are made at the manufacturer's site and then shipped to the construction site. They are mainly used in the construction of walls and building façades. They are also used in the construction of floors and roofs. They are made in a controlled environment and are structurally superior to their conventional concrete counterparts.

Straw Bales: They find application as a framing material. They also absorb noise and are used in insulation work. Replanting straws can be done quickly with minimal environmental impact. They can be used in filling between beams and columns. They also have excellent fire resistance qualities. When placed in walls and attic, they produce cooling effects in the summer season. Similarly, they produce warm results in winter for house dwellers.

Recycled plastic: Builders are interested in using recycled plastic as a construction material instead of conventional materials. They are being used in making concrete. The aim of recycling plastic is the researchers want to reduce the emission of carbon dioxide and other greenhouse gases into the atmosphere. A mixture of recycled and virgin plastic is used to make polymeric timbers. These plastic timbers are used in the making of fences and outdoor furniture. Recycled plastic is used to manufacture pipes, roofs, floors, etc.

Reclaimed wood: Reclaiming wood is one way to reduce deforestation and cut trees. The main advantage of reclaimed wood is that they are light in weight. Their main demerit is that they require treatment and reinforcement for protection against insects and degradation.

Recycled steel: Recycled steel would help reduce harmful greenhouse gases in the atmosphere. Recycled steel is being used as a framing material. This would make the structure durable against natural calamities, such as earthquakes. Mining raw materials, heating raw materials, and shaping consume a lot of energy. Sheep's wool has a drawback in that it is costlier as an insulating material when compared to conventional insulation materials.

Sheep's wool: It is considered one of the insulation materials in building buildings. This is eco-friendlier than chemical-based insulation. Complete home insulation can be done with Sheep's wool. This material does not consume as much energy as conventional insulation. In this way, they are better than traditional insulation.

Plant-based Polyurethane Rigid foam: Rigid foam has excellent insulation material. Surfboards usually are made out of Plant-based Polyurethane Rigid foam. Nowadays, they are used in the manufacture of turbine blades and furniture. It is also used as a sound insulation material. It offers protection against pests and mold.

Rammed earth: Rammed earth is being used to replace conventional construction materials as they are eco-friendlier than traditional materials. Rammed earth, in unstabilized conditions, is made up of soil, gravel, and sand. The main drawbacks of these materials are that they lose strength on saturation and they are eroded in the rainy season. They are generally thick walls, sometimes greater than 400 mm. They require protection against moisture. In the case of rammed earth, the stabilized state also consists of soil, gravel, and sand. Additionally, lime or cement is also added. These stabilized walls are being practiced in many countries like the USA and Europe (Verma et al. 1950; Easton 1982; Hall 2002; Houben et al. 2004; Walker et al. 2005; King 1996).

Compacted fly-ash blocks: Compacted fly-ash blocks are composed of stone crusher dust, lime, and fly ash. Lime reacts with fly ash and results in strength for the league. The reaction can be made faster by steam curing (Venkatarama Reddy and Lokras 1998; Venkatarama Reddy and Hubli 2002) or adding phosphogypsum (Bhanumathidas and Kalidas 2002).

Hempcrete: It is a concrete-like material generated from the Hemp plant. The wooden fiber is mixed with lime to produce concrete-like material. They are straight and strong. They are sound insulators and are resistant to fire. They also have good thermal properties. They are CO_2 negative. That is, they absorb CO_2. Hence, they are a more renewable resource.

Mycelium: It's a naturally occurring building material. It's a single-celled organism containing roots of mushrooms and fungi. They are made in the form of bricks. They are light in weight. The main advantage of these building materials is that they are solid. They can withstand very high temperatures.

Ferrock: They are made using steel dust obtained from the steel industry. The main advantage of these materials is that they are even more potent than concrete. It also absorbs CO_2 during drying. It is becoming an alternative to conventional cement.

Blended cement: Large volume of harmful substances like CO_2 will be emitted during cement (0.9 tonnes per ton of clinker). Adding materials like fly ash, rice husk, and granular slag will reduce CO2 emissions during cement manufacturing.

Mud blocks: Burnt clay bricks are not environmentally friendly as they consume more energy input during the brick making. On the other hand, the mud blocks are environmentally friendly as they are energy efficient. They are made by compacting soil, stabilizer, and sand. It requires a curing time of about 28 days. After which, it can be used in the construction of walls. The merit of these mud blocks is that they do not require burning. They can also be made on-site, and thus transportation of bricks can be avoided. Therefore, mud blocks are environmentally friendly. Energy in building materials is composed of following components: (1) embodied energy, that is, the energy consumed by materials and their assembly, and (2) energy required for maintaining the building during its life. The first one is like a one-time investment. The actual energy consumption depends on the materials used and the construction techniques. The amount of energy consumed depends upon the specific energy consumption. The embodied energy in buildings may vary from 5–10 GJ/m^2. The main objective in constructing buildings should be to reduce energy consumption and the magnitude of carbon emissions. The second component depends upon the region where the building is built.

10.2.2 Strategies for sustaining materials

Materials are used to construct buildings, produce vehicles, and many more products. Some of the materials used in manufacturing products are harmful to make and transport and detrimental to the environment when discarded. That is, once they are discarded, they pollute the environment. On the other hand, some materials degrade slowly and will not affect the environment. In 2002, the USA was reported to have released 4.79 billion pounds of hazardous chemicals from 24,000 industries into the environment.

Significant savings in materials will result from optimizing the design of components. Proper design will ensure a minimum number of features, thus lowering material consumption. During manufacturing a product, it was often observed that companies use more materials than required. This would result in significant energy consumption. Similarly, during process planning, a methods engineer may add unnecessary manufacturing processes, which would add up to extra energy consumption. For example, while specifying machining allowances, a methods engineer may specify additional allowance, adding to the extra time required to shape the component from the raw material. Also, more material would be wasted in the removal of chips. Thus, making the product manufacturing not an environmentally friendly process. Similarly, polychlorinated biphenyl (PCB) is used as dielectric fluid in transformers, capacitors, and heat exchangers. They were first introduced in 1929. The harmful effects of PCBs were not known at that time. Germany, Japan, and France were the pioneers in PCB production.

Reducing the yield losses will result in huge savings. During the melting of steel and aluminum, there will be a massive opportunity for controlling yield losses. By using scrap judiciously, the need for remelting scrap can be avoided. Many sheets of steel blank scrap could be used for generating small components. Melting of scrap is an energy-intensive operation. Melting consumes a lot of energy. Instead of remelting the steel scrap produced during the blanking operation, the scraps could be used to make small components. Similarly, steel beams in old buildings could be reused as it is instead of recycling them. This would reduce the energy requirement in melting the steel and protect the environment. Extending the product life can help in reducing material consumption and associated re-cycling. Thus, material and energy consumption could be reduced by extending product life. During the dismantling of old vehicles, good parts can be reused instead of sending them for recycling. They lessen the final demand for products by changing our lifestyles. For example, by using public transportation, the need for cars can be reduced. This will reduce the material requirement and reduce vehicle pollution to a large extent. Thus, the generation of greenhouse gases can be reduced to a large size. This will make a significant step in sustainability.

Reduction of final demand would also reduce the subsequent product manufacturing and the associated energy consumption and waste generation in the form of chips. For example, if each family owns a car, then we would require to produce as many cars as there are families. Instead, using public transportation by every citizen would reduce the demand for cars and thus the reduction in environmental pollution and the associated sustenance. In many industrialized nations, little attention is paid to material efficiency-related strategies. This may be due to economic and social barriers.

Today, we depend on steel and aluminum to produce many products. This would result in the need to produce aluminum and steel. The reduction in steel and aluminum production would automatically result in environmental sustenance. This demonstrates that changing our habits, such as using public transportation, would significantly reduce the production of vehicles and thereby save our environment.

Similarly, reducing over-dependency on cement would result in significant energy savings and environmental sustenance in building construction. Literature has reported using other non-conventional materials in the construction of buildings.

10.2.3 Sustainable manufacturing processes

Manufacturing can be classified into conventional, unconventional, and additive manufacturing processes.

10.2.3.1 Conventional manufacturing processes

Processes such as turning, milling, and shaping come under conventional processes. These techniques use cutting tools while shaping a component from the raw material. Cutting fluids reduce the friction between the cutting tool and the workpiece. Cutting fluids would enhance the efficiency of machining processes. The majority of the cutting fluids used in practice are made out of mineral oil. These mineral oils emit toxic fumes, which are detrimental to the environment. When inhaled by machine operators, these fumes would result in breathing complications and respiratory diseases. Literature reported that water-based oils are used whenever cooling is required during the machining process. Also, these water-based oils are less harmful to the environment.

10.2.3.2 Unconventional manufacturing processes

Manufacturing processes such as electric discharge machining (EDM) and electrochemical milling (ECM) come under unconventional manufacturing processes. In the unconventional method, material removal will not happen by shearing as in the case of conventional manufacturing. Unlike traditional machining, a cutting tool will not be used to remove extra material in the form of chips. In the EDM process, the material is removed by electrical sparks generated between the work and the electrode. In ECM, the material removal will happen by electrolysis.

EDM process makes use of dielectrics made out of hydrocarbons. During the EDM process, they generate toxic fumes, which are very harmful to the health of operators. (Dhakar et al. 2019) have explored a novel near-dry EDM process using air and EDM oil and found that the technique reduced nearly 97% of gas emissions compared to the wet EDM process.

10.2.3.3 Additive manufacturing processes

Manufacturing the components of automobiles and airplanes consumes tremendous amounts of energy. They also consume vast quantities of raw materials. The manufacturing of these components also results in enormous quantities of waste in chips and scrap. Parts made of titanium consume vast amounts of energy. Additive manufacturing has tremendous potential for reducing material consumption and waste while manufacturing a component. Thus, there is a need to develop new materials and improve the manufacturing process's efficiency. NASA and Boeing are investing in additive manufacturing (AM). They have been manufacturing components of satellites and airplanes using AM. AM is capable of producing parts having complex geometry. AM can also be used in difficult-to-machine materials.

AM processes include direct energy deposition (DED), powder bed fusion (PBF), sheet lamination, and binder jetting. DED, as well as PBF, which are widely used in aerospace industries.

Monteiro et al. (2022) made a detailed study for identifying sustainable strategies for different stages - design stage, material identification and sourcing, manufacturing process design and control, and post-sales stages. They studied in the aerospace and aeronautic sectors.

The design stage in AM is one area that offers multiple opportunities for sustenance. Literature showed that AM-designed components would consume less material than conventionally made ones – one case where an aerospace bracket reported a material reduction of 64% (Min et al. 2019). The main problem with a complex lattice structure in AM is its very high stiffness-to-weight ratio compared with bulk material. Zhang et al. (2018) proposed a bio-inspired model for achieving sustainability. The benefit of this model is that it consumes less material, generates less waste, and has the least cost. The process considered the quality of raw material, process parameters, and functionality of the component). They proposed a scheme for assessing sustainability through Life cycle assessment (LCA) and costing (LCC). The authors have used bone, diamond, and honeycomb structures in this work. Experimentally the authors have verified that bone structure resulted in the lowest cost and had less impact on the environment. The work also reported that multiple functions could be built into one component to reduce assembly time. This has resulted in reduced demand for fasteners such as nuts and bolts and the associated supply chain. This will also reduce the need for inspection and associated failures. Weight reduction is closely associated with fuel consumption in the case of airplanes. Thus, AM-made components are being used increasingly in Airplane manufacturing. Huang et al. (2016) showed the benefits of using AM-made details in the aerospace industry and reported that by 2050 consumption will be 70–170 GJ/year. Girdwood et al. (2017) demonstrated how the cost of an aircraft component could be brought down by using his framework. The framework considers 14 process factors: design, planning, surface finish, dimensional accuracy, etc. They validated the framework by making an aircraft component through AM. They came to know that the amount of material consumed by AM is much lower than a conventionally produced component, even without doing any process optimization. They demonstrated how their proposed framework could bring down significant material consumption.

Another strategy in sustainability is carrying out the AM process onsite. This would reduce the unnecessary transportation of materials from the earth to outer space. Components are AM made with locally available (extra-terrestrial) materials (Prater et al. 2017). Understanding additive manufacturing is significant for technology selection. It is also essential to understand the impact of additive manufacturing on the environment. Paris et al. (2016) studied the manufacturing of a turbine blade of titanium (Ti) material by EBM and the conventional manufacturing processes. The author has performed a sensitivity analysis and concluded that EBM performed better than traditional milling. The authors have also reported that the component weighed 1.08 kg, whereas EBM required 0.23–0.44 kg of titanium material. Authors have also observed that the environmental impact depends upon the process parameters. Many researchers have studied AM-made parts' quality by conducting mechanical tests.

Min et al. (2019) studied the environmental impact of AM processes, such as EBM, DED, and SLM, while producing Ti-rocket brackets. They also did topological optimization with various metal powders. Their study also considered recycling unused powders and the maximum number of components per print. The work also studied LPBF, EBM, and LMD processes for energy (for melting and heating) consumption, argon consumption, and powder wastage. Instead of considering the impact of a single AM product, the study evaluated the total batch size and lifetime. They ranked the process EBM, SLM, and LMD as the lowest impact per part based on production volume and lifetime. Additive manufacturing also offers other advantages that would help extend the component's life by repairing it. Also, AM would help manufacture spare parts through DED (Ford and Despeisse 2016). Researchers have also explored the benefits of combining LMD with machining that could help repair turbine blades (Gisario et al. 2019).

Production inspection time also plays a crucial role in energy conservation. Researchers have made efforts to reduce the inspection time and improve the efficiency of product inspection (Kiran 2021d). Reducing product inspection time will significantly reduce energy consumption in manufacturing. Conventional inspection techniques were suitable for sampling inspection. Modern inspection techniques using image processing will reduce the inspection time and perform 100% inspection of products.

Lean manufacturing is becoming more popular among manufacturing industries (Kiran 2021a). Lean philosophy helps in identifying waste or non-value-adding activities. It uses tools such as value stream mapping for waste identification (Kiran 2021b). By constructing the current state value stream map, the company will identify different forms of waste, such as transportation, inventory, unnecessary movement of operators, over-processing, over-production, and delays. Single-minute exchange of dies (SMED) would help reduce the number of setups and change-overs in manufacturing (Kiran 2021c). This would help in reducing energy consumption.

10.3 Conclusion and directions for future research

Changing our lifestyles would be a significant step in achieving environmental sustenance. Designing a product with optimization will lead to substantial savings in material consumption. Reusing old parts is also an essential step in reducing recycling. As recycling consumes a lot of energy, every effort should be made to lower recycling. This will lead to a reduction in energy consumption. Reducing the number of manufacturing processes during product manufacturing could significantly achieve environmental sustainability. Conventional manufacturing processes consume a lot of energy, and AM has great potential to achieve ecological sustainability. There is a great scope for increasing material efficiency and energy efficiency. Government policies alone cannot achieve environmental sustainability. It is possible only when citizens, government, and industries coordinate ecological sustainability.

References

Allwood, J. M., Cullen, J. M., Carruth, M. A., Cooper, D. R., McBrien, M., Milford, R. L., & Patel, A. C. *Sustainable Materials: With Both Eyes Open* (Vol. 2012). Cambridge: UIT Cambridge Limited, 2012.

Bhanumathidas, N., & Kalidas, N. *Fly Ash for Sustainable Development*. Ark, Communications, 2002.

Central Board for the prevention and control of water pollution (Procedure for Transaction of Business) Rules, 1975 amended 1976, Notifications.

Dhakar, K., Chaudhary, K., Dvivedi, A., & Bembalge, O. An environment-friendly, and sustainable machining method: Near-dry EDM. *Mater Manuf Process* 2019;34(12):1307–1315.

Easton, D. *The Rammed Earth Experience*, 1st edn. Blue Mountain Press, 1982.

Eurostat. *Energy, Transport, and Environment Indicators*, 2011. http://appsso.eurostat.ec.europa.eu/.

Ford, S., & Despeisse, M. Additive manufacturing and sustainability: An exploratory study of the advantages and challenges. *J Clean Prod* 2016;137:1573–1587. http://doi.org/10.1016/j.jclepro.2016.04.150.

Gao, T., Li, C., Zhang, Y., Yang, M., Jia, D., Jin, T., Hou, Y., & Li, R. Dispersing mechanism and tribological performance of vegetable oil-based CNT nanofluids with different surfactants. *Tribol Int* 2019;131:51–63.

Girdwood, R., Bezuidenhout, M., Hugo, P., Conradie, P., Oosthuizen, G., & Dimitrov, D. Investigating components affecting the resource efficiency of incorporating metal additive manufacturing in process chains. *Proc Manuf* 2017;8:52–58. http://doi.org/10.1016/j.promfg.2017.02.006.

Gisario, A., Kazarian, M., Martina, F., & Mehrpouya, M. Metal additive manufacturing in the commercial aviation industry: A review. *J Manuf Syst* 2019;53:124–149. http://doi.org/10.1016/j.jmsy.2019.08.005.

G.S.R.58(E), [27/2/1975]—The Water (Prevention and Control of Pollution) Rules, 1975 [PDF](8.14 MB).

G.S.R.378(E), [24/7/1978]—The Water (Prevention and Control of Pollution) Cess Rules, 1978.

G.S.R.830(E), [24/11/2011]—The Water (Prevention and Control of Pollution) Amendment Rules, 2011.

G.S.R.840(E), [22/11/2012]—The Central Pollution Control Board (Member-Secretary, Terms, and Conditions of Service and Recruitment) Rules, 2012.

G.S.R.860(E), [30/11/2012]—The Central Pollution Control Board (Qualifications and other Terms and Conditions of Service of Chairman) (Amendment) Rules, 2012.

Guo, S., Li, C., Zhang, Y., Wang, Y., Li, B., Yang, M., Zhang, X., & Liu, G. Experimental evaluation of the lubrication performance of mixtures of castor oil with other vegetable oils in MQL grinding of nickel-based alloy. *J Clean Prod* 2017;140(3):1060–1076.

Hall, M. Rammed earth: Traditional methods, modern techniques, sustainable future. *Building Eng* 2002;22–24.

Houben, H., & Guillaud, H. *Earth Construction—A Comprehensive Guide*. England: Intermediate Technology Publications, 2004.

Huang, R., Riddle, M., Graziano, D., Warren, J., Das, S., Nimbalkar, S., et al. Energy and emissions saving potential of additive manufacturing: The case of lightweight aircraft components. *J Clean Prod* 2016;135:1559–1570. http://doi.org/10.1016/j.jclepro.2015.04.109.

King, B. *Buildings of Earth and Straw. Structural Design for Rammed Earth and Straw Bale Architecture*. Ecological Design Press, 1996.

Kiran, M. B. Additive manufacturing of titanium alloys-a review. *Proceedings of the International Conference on Industrial Engineering and Operations Management*, 2021a, pp. 2553–2564.

Kiran, M. B. Lean transformation in electricity transmission tower manufacturing company-a case study. *Proceedings of the International Conference on Industrial Engineering and Operations Management*, 2021b, pp. 2961–2974.

Kiran, M. B. Productivity assessment studies in solenoid valve manufacturing company—a case study. *Proceedings of the International Conference on Industrial Engineering and Operations Management*, 2021c, pp. 3037–3046.

Kiran, M. B. A novel online surface roughness measuring method. *Proceedings of the International Conference on Industrial Engineering and Operations Management*, 2021d, pp. 2907–2916.

Milner, H. R., & Woodard, A. C. 8—Sustainability of engineered wood products. In Jamal M. Khatib, Eds. *Woodhead Publishing Series in Civil and Structural Engineering, Sustainability of Construction Materials* (Second Edition). Woodhead Publishing, 2016, pp. 159–180.

Min, W., Zhang, Y., Yang, S., & Zhao, Y. F. A comparative study of metal additive manufacturing processes for elevated sustainability. In ASME, Ed. *International Design Engineering Technical Conferences and Computers and Information in Engineering Conference IDETC/CIE2019*. Anaheim, CA: ASME, 2019, pp. 1–9.

Monteiro, H., Carmona-Aparicio, G., Lei, I., & Despeisse, M. Energy and material efficiency strategies enabled by metal additive manufacturing—A review for the aeronautic and aerospace sectors. *Energy Rep* 2022;8(3):298–305.

Orucu, E., & Turkmen, B. A. Evaluating the sustainability of car mat manufacturing. *Sustain Mater Technol* 2022;32:e00402.

Paris, H., Mokhtarian, H., Coatanéa, E., Museau, M., & Ituarte, I. F. Comparative environmental impacts of additive and subtractive manufacturing Technologies. *CIRP Ann—Manuf Technol* 2016;65:29–32. http://doi.org/10.1016/j.cirp.2016.04.036.

Prater, T. J., Werkheiser, N., Ledbetter, F., & Jehle, A. Nasa's in-space manufacturing project: Toward a multi-material fabrication laboratory for the international space station. *AIAA Sp Astronaut Forum Expo Sp* 2017;1–14. http://doi.org/10.2514/6.2017-5277.

Rodrigues, T. A., Duarte, V., Miranda, R. M., Santos, T. G., & Oliveira, J. P. Current status and perspectives on wire and arc additive manufacturing (WAAM). *Materials (Basel)* 2019;12. http://doi.org/10.3390/ma12071121.

Salazar, J. 21—Life Cycle Assessment (LCA) of windows and window materials. In F. Pacheco-Torgal, L. F. Cabeza, J. Labrincha, & A. de Magalhaes, Eds. *Eco-efficient Construction and Building Materials*. Woodhead Publishing, 2014, pp. 502–527.

Tang, L., Zhang, Y., Li, C., et al. Biological stability of water-based cutting fluids: Progress and application. *Chin J Mech Eng* 2022;35:3. https://doi.org/10.1186/s10033-021-00667-z.

Venkatarama Reddy, B. V., & Hubli, S. R. Properties of lime stabilized steam cured blocks for masonry. *Mater Struct (RILEM)* 2002;35:293–300.

Venkatarama Reddy, B. V., & Lokras, S. S. Steam-cured stabilized soil blocks for masonry construction. *Energy Build* 1998;29:29–33.

Verma, P. L., & Mehra, S. R. Use of soil-cement in house construction in Punjab. *Indian Concr J* 1950;24:91–96.

Walker, P., Keable, R., Martin, J., & Maniatidis, V. *Rammed Earth Design and Construction Guidelines*. BRE Bookshop, 2005.

Wickramasinghe, K. C., Sasahara, H., Rahim, E. A., & Perera, G. I. P. Green Metalworking Fluids for sustainable machining applications: A review. *J Clean Prod* 2020;257.

Zhang, H., Nagel, J. K., Al-Qas, A., Gibbons, E., & Lee, J. J.-Y. Additive manufacturing with the bio-inspired sustainable product design: A conceptual model. *Proc Manuf* 2018;26:880–891. http://doi.org/10.1016/j.promfg.2018.07.113.

Zhang, Y., Li, C., Jia, D., Zhang, D., & Zhang, X. Experimental evaluation of the lubrication performance of MoS2/CNT nanofluids for minimal quantity lubrication in Ni-based alloy grinding. *Int J Mach Tools Manuf* 2015;99:19–33.

Chapter 11

Use of additives and nanomaterials for sustainable production of biofuels

Anil Dhanola, Vijay Kumar, Arun Kumar Bambam, and Kishor Kumar Gajrani

Contents

11.1 Introduction

Energy consumption and future supplies have become a national and strategic issue for almost every country today as a result of the rising population [1]. Nowadays, conventional fossil fuels such as coal, oil, and gas, which are considered non-renewable energy resources, are predominantly used for energy consumption [2]. Non-renewable resources currently account for more than 75% of total energy production [3]. These energy resources have been used at a large scale for a long time. Due to the over-exploitation of natural resources, future consequences are overlooked [4]. As a result, difficulties with sustainability have arisen, such as the depletion of fossil fuel reserves. Many experts and environmentalists have raised various environmental issues, including forest destruction, air pollution, acid rain, global warming, ozone depletion, and global climate change [5]. More attempts are being made nowadays to limit their influence on the environment by lowering their reliance on nonrenewable energy resources like petroleum-based fossil fuels. But keep on using non-renewable energy resources indiscriminately like this, and they will surely get exhausted. Biofuels are gaining popularity as an alternative fuel around the world due to their versatility, energy security, and capacity to address global warming problems. Biofuels have been promoted as a viable alternative to nonrenewable energy sources and fossil fuels [6]. Governments of various countries have come forward to set up biodiesel industries and are offering subsidies

DOI: 10.1201/9781003291961-13

as well as waiving their taxes for the development of biodiesel. The US, for example, has set a goal of producing 36 billion gallons of biofuel by 2022. Various replacement biofuels, such as bioethanol, biodiesel, dimethyl ether, and biogas, have been widely studied in this area. These investigations demonstrate that biofuels are fuels due to their sustainable, biodegradable, oxygen-rich, and sulfur-free properties [7]. Diesel-ethanol mixture can be a permissible solution. This mixture, however, has stability issues and has lesser physicochemical qualities than diesel fuel, necessitating the use of only a few nanoadditive components to keep it stable. Potential for a reduction in greenhouse gas emissions by about 80% if diesel fuel mixed with ethanol is used in a diesel engine. While preparing the diesel-ethanol-biodiesel mixture, care should be taken to ensure that the concentration of ethanol is not less than 99%. Similarly, the amount of biodiesel should be mixed. This is not a permissible solution. Adding metal-based components, such as nanoparticles (NPs), to the base fuel is a useful strategy for enhancing the fuel's desirable properties [8–11]. NPs are readily soluble in fuels that are added to base fuel to a very small extent (100 ppm (extended up to a few thousand ppm)) [12–13]. This chapter presents recent work on the application of nanomaterials in biofuel development (microalgae biofuel). Simultaneously, this chapter demonstrates fuel properties and emission characteristics concerning the addition of various biodiesel mixtures using different nanomaterials.

11.1.1 Fuel

Fuel is a substance that reacts with other substances to release energy in the form of thermal energy, which is transformed into mechanical energy with the help of a heat engine. Energy, that is, the heat released from the combustion of fuel, has many uses, such as cooking, heating, and industrial purposes. Fuels are broadly classified in two ways: (a) based on the physical state present in nature, that is, solid, liquid, and gaseous, and (b) based on their procurement mode, that is, natural and manufactured. The classification of fuels is shown in Table 11.1. The original state of the fuel is transformed to use the energy released from the fuel more effectively and efficiently for various purposes; for example, solid state into liquid or gaseous, and liquid state into gaseous.

11.1.2 Fossil fuels

Fuels derived from the decomposition of plants and animals are high in carbon and hydrogen, both of which are abundant deep inside the earth's crust. The demand for fossil fuels is increasing rapidly, along the rate of extraction of fossil fuels has also increased [14]. Fossil

Table 11.1 Classification of Fuels

	Solid fuels	*Liquid fuels*	*Gas fuels*
Natural fuels	Coal Wood	Petroleum	Natural gas
Manufactured fuels	Charcoal Coke Straw	Oil from the distillation of petroleum Coal tar Alcohols	Coal gas Acetylene Blast furnace gas

Table 11.2 Advantages and Disadvantages of Energy From Fossil Fuels

Advantage	*Disadvantage*
High calorific value	A limited supply of fossil fuels
Technology-wise, energy production is very easy	The release of harmful gases like fly ash, CO_2, and SO_2 during combustion is the cause of the greenhouse effect and acid rain
Relatively simple manufacturing techniques for power plants powered by fossil fuels	The use of natural gas is the cause of the foul smell
Transportation is easy	Replenishment is not possible

fuels are, unfortunately, non-renewable resources that take millions of years to regenerate. The advantages and disadvantages of energy from fossil fuels are shown in Table 11.2.

According to the National Academy of Sciences, fossil fuels are responsible for nearly three-quarters of emissions from human activities over the past 20 years. Necessary action is needed to tackle the consequences of fossil fuels. Scientists and environmentalists are looking for ways to reduce pollution and make the environment healthier by reducing reliance on fossil fuels, given the harmful aspects of fossil fuels [15]. Efforts are being made to reduce emissions through post-combustion control techniques to find more sustainable and effective solutions to pollution problems [16].

11.1.3 Biofuels

Biofuels have emerged as an alternative to fossil fuels under their potential for environmental and economic benefits. They are fuels that are made from biomass material. Biofuels are divided into two groups based on the type of feedstock used: "first-generation" biofuels and "second-generation" biofuels. First-generation biofuels are generated from edible biomass (such as seeds, grains, or sugar). Second-generation biofuels are produced from non-edible lignocellulosic biomass (such as maize stalks or rice husks) or non-edible whole plant biomass (such as trees or grass planted expressly for energy). Significant commercial volumes of first-generation biofuels are being produced in many countries, while second-generation biofuels have not been produced commercially in any country to date. Biofuels are included in the category of transportation fuels. The classification of fuels is shown in Figure 11.1.

Among all available alternate energy resources, biomass resources account for about 80% of the energy produced globally. There is still much effort to be made in order to increase the use of biofuels as fuel. However, using vegetable oil directly in the engine can cause issues such as fuel system deposits, piston ring sticking, low performance, and significant NOX emissions, among other things. The increased viscosity and reduced calorific content of vegetable oils are the key reasons behind this. Various procedures (e.g., transesterification, blending, and micro-emulsification) have been utilized by several researchers to reduce the viscosity of the resulting vegetable oils, but transesterification is the most extensively used. Therefore, a mixture of diesel and biodiesel is used in CI engines. In addition, various innovative additive technologies are being worked on to improve engine performance [14]. Figure 11.2 schematically illustrates biomass growth and development, feedstock extraction, microbial hydrolysis during fermentation, and biofuel extraction and application.

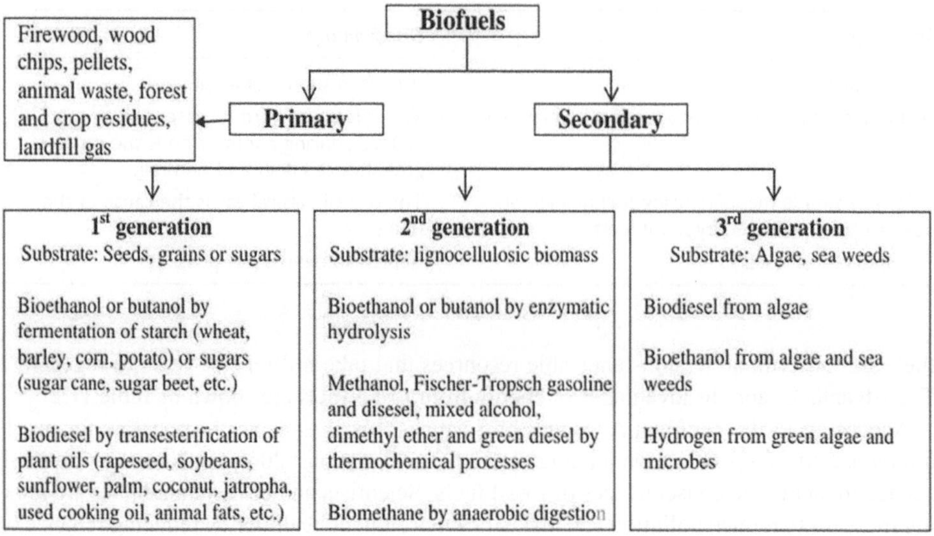

Figure 11.1 Classification of biofuels [17].

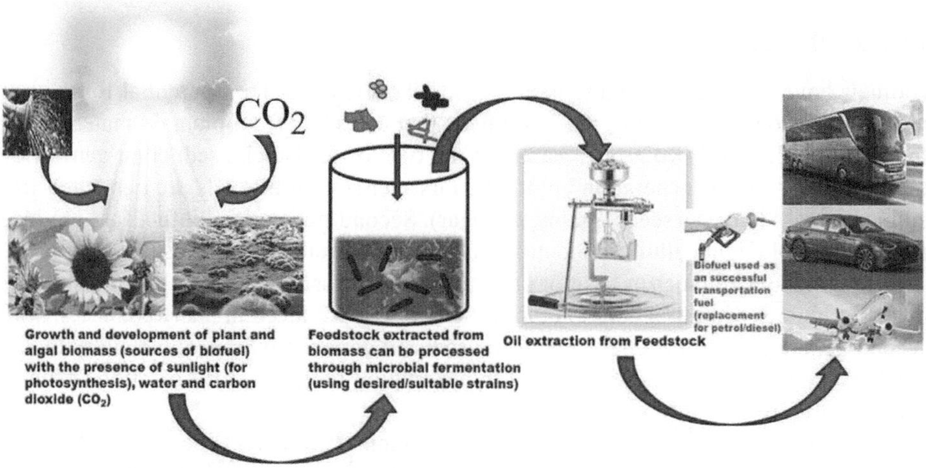

Figure 11.2 Illustration of growth and development of biomass, feedstock extraction, microbial hydrolysis through fermentation, and biofuel extraction and usage [18].

11.2 Nanotechnology

Designing a material or device at the nano-scale (i.e., 10^{-9} m) is considered under nanotechnology. As a result, nanoparticles are used in a variety of industries, including agriculture, food, cosmetics, pharmaceuticals, and electronics. The new qualities of nanoparticles, such as their nanoscale size, structure/morphology, and high reactivity, are largely responsible for the employment of nanotechnology in these various domains. In this modern age, one of the

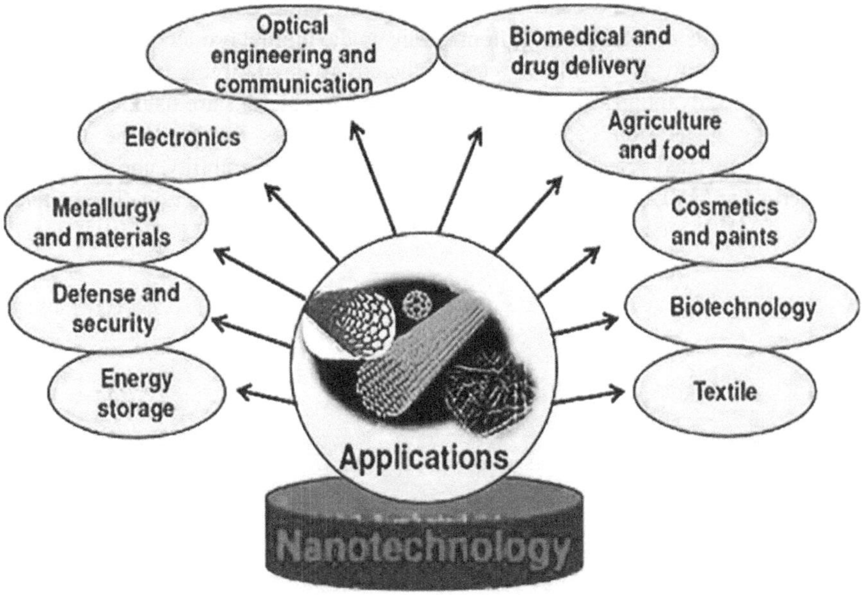

Figure 11.3 Applications of nanotechnology in daily life.

best technologies, nanotechnology, has revolutionized biofuel conversion processes and has been instrumental in improving engine performance. Nanotechnology has been introduced to improve the efficiency of biofuels through nanomaterials such as nano-fibers, nano-crystals, nano-magnets, etc. [6]. Figure 11.3 depicts the application of nanotechnology in daily life. Commercially producing biofuels has challenges related to various technologies as well as the cost of production. Biofuel production may well be achieved in an eco-friendly way with the help of nanotechnology.

11.2.1 Applications of nanomaterials in biofuel production

Due to unique structural properties, such as their small size (nanoscale size), nanomaterials are finding new applications in biofuel generation. By enhancing biosynthetic pathways, these nanomaterials improve biofuel generation. Nanomaterials can be classified as zero-dimensional (0D), one-dimensional (1D), two-dimensional (2D), or three-dimensional (3D), depending on their ultra-fine grain size (less than 50 nm). Nanomaterials can be used as substrates for lipid accumulation, extraction, and reprocessing processes in the manufacture of biofuels from biomass (edible or inedible). Some nanomaterials, such as nanoparticles, nanofibers, nanosheets, nanotubes, and other nanostructures, have been successfully researched in biofuels, such as biodiesel, biomethane, bioethanol, and others, using both direct and indirect ways [19]. Among the various types of biofuels produced from biomass, biodiesel has made a mark. From biofuel production to development, nanomaterials play an important role, such as achieving optimum operating conditions and improving product quality. The thermochemical and biological processes of conversion from biomass to biofuel can be used to produce liquid and gaseous fuels. Thermochemical processes, including gasification and direct liquefaction, are considered major conversion routes. Figure 11.4 shows products from the nano-catalytic conversion of biomass.

Despite the promising benefits of nanomaterials in the biofuel production system, its negative effects have many adversative effects on the environment, living organisms, and even the economy. The advantages and disadvantages of nanomaterials are listed in Table 11.3.

For bioenergy generation, carbon nanotubes, metal oxides, and magnetic nanoparticles (MNPs) are widely used because the high coercivity, easy recoverability, and great paramagnetic properties of MNPs also make them useful for bioenergy production. The benefits and drawbacks of MNPs are represented in Table 11.4.

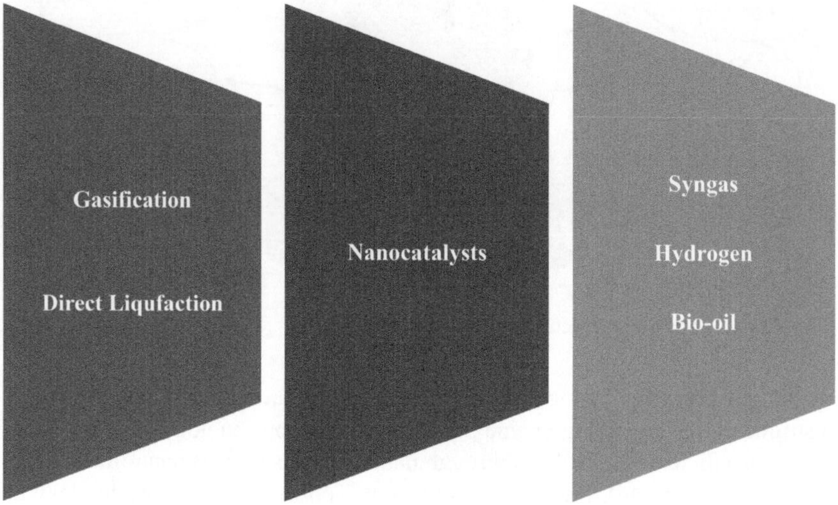

Figure 11.4 Products from nano-catalytic conversion of biomass [20].

Table 11.3 Advantages and Disadvantages of Various Nanomaterials [21]

	Advantage	*Disadvantage*
Metallic nanoparticles	Practically uniform in size and shape Uniformity in branch length	Toxicity
Carbon nanoparticles	Multiple functions Water-soluble and biocompatible Efficient loading	Toxicity
Liposomes	Biocompatible Longer duration of circulation	May trigger an immune response
Dendrimers	Practically uniform in size, shape Uniformity in branch length Enlarged surface area Improved loading targeting is attained	Complex synthetic route
Polymer micelles	Biodegradable, self-assembling, and biocompatible Potential targeting	Need for surface modification

Table 11.4 Advantages and Disadvantages of Magnetic Nanoparticles [22]

Advantages	*Disadvantages*
Excellent biodegradability	Poor dispersion abilities
Readily to be customized	High cost of synthesis material
Ease of separation	Limitations in scale-up production processes
Low cytotoxicity to biomass cell	Mobility dependent on environment compatibilities
Ease of synthesis	Maintain stability after mechanical, physical, and chemical modification
Ability to bind multiple targeted compounds	
Large surface-to-volume ratio	

11.3 Biofuel feedstock and production

The two most commonly used biofuels are biodiesel and bioethanol, which are mainly obtained from vegetable oils (i.e., edible or non-edible) and sugar seeds. They are considered petroleum diesel substitutes and petroleum gasoline substitutes, respectively. Biodiesel feedstocks are obtained from edible oils (i.e., palm, soybean, rapeseed, etc.) and non-edible oils (i.e., Jatropha oil, Karanja, Neem, etc.), whereas bioethanol feedstocks are obtained from the fermentation of sugarcane and seeds. Another biofuel feedstock microalgae are considered to be the third generation of biofuels [17, 23]. Reducing the extraction of fossil fuels, reducing carbon emissions, etc., are the positive effects of using biodiesel or bioethanol. In most research studies, the transformation of biomass to biodiesel is done with the help of nanomaterials using a variety of feedstocks.

11.3.1 Biofuels from palm

With the increasing use of renewable feedstocks and global trends, the palm oil crop has become more important as a versatile oil crop due to its ability to produce bioenergy, or biofuel, from the fruit of the palm oil tree to oil palm residues. India is the primary global consumer of palm oil, as approximately 9.3 million tons of palm oil were consumed between 2016 and 2017, and palm oil consumption is expected to be twofold by 2030.

Palm oil has established itself as a potential biofuel source with many advantages such as higher yield than other oils (3.93 tons per hectare per year), cheaper than other oils, a perennial crop, very low pour point, produces less carbon residue and sulfur content is much lower than petroleum-based diesel, etc. [24, 25]. A comparison of palm oil yields per hectare with other vegetable oils is shown in Figure 11.5.

11.3.2 Biofuels from sugarcane

A tropical perennial grass used for the production of sugar is known as sugarcane. In today's time, everyone is using traditional petroleum fuel in abundance, which is not at all good for the environment. In the meantime, there is a ray of hope – sugarcane. Sugars extracted from sugarcane can be easily fermented to produce biofuels such as ethanol. Additionally, sugar mills are used to generate steam and electricity [26, 27]. The overall global production of

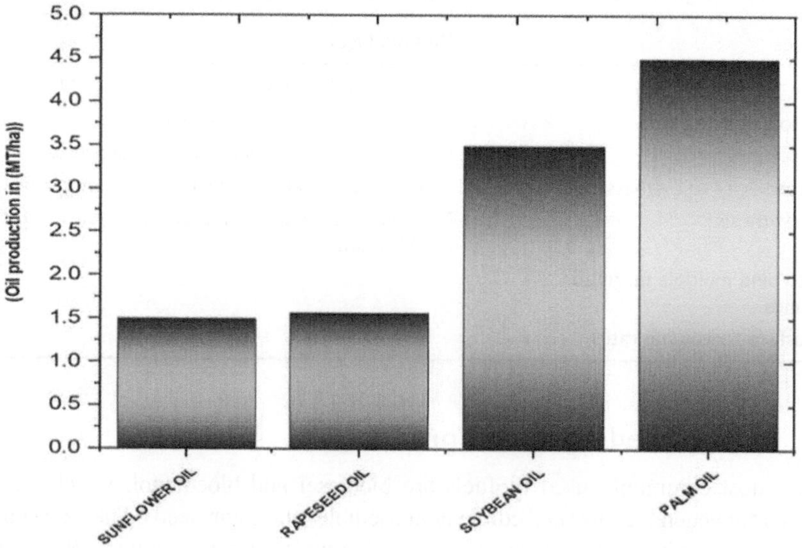

Figure 11.5 Illustrated significant plant-based oil yields per hectare [24].

renewable fuels is currently 50 billion L/year, with sugarcane accounting for almost 40% of that, making it a key source of biofuel production [28]. Figure 11.6 represents the block flow diagram of the bioethanol production steps from sugarcane. Pretreatment, enzymatic hydrolysis, and fermentation are all steps in the bioethanol production process. One ton of sugarcane bagasse might theoretically yield up to 300 L of ethanol [29]. However, various factors, like the quality of bagasse and the ethanol production process, have a direct impact on ethanol yield [29].

11.3.3 Biofuels from Jatropha

Due to the increasing attraction of biofuels, the concept of cultivation of Jatropha for the manufacture of biofuels attracted its attention. On this eve of the discovery of suitable biomass for biofuels, the concept of Jatropha farming came into the limelight for liquid biofuel production due to several characteristics such as (i) ease of propagation, (ii) longevity of plantations, (iii) high oil content in seed, (iv) disease tolerant properties, (v) resistant to pests, and (vi) low management requirement [31]. Because of its high oil content, Jatropha, a non-edible oil plant native to Central and South America, is regarded as a reliable source for the generation of biodiesel.

11.3.4 Biofuels from microalgae

Due to the increasing attraction of biofuels, the conception of microalgae cultivation for the manufacture of biofuels came into the limelight with many positive aspects, such as (i) plentiful oil content, carbohydrate, and protein; (ii) can grow in an aquatic medium, such as freshwater, brackish water, demands less water, however, can also assimilate nutrients from

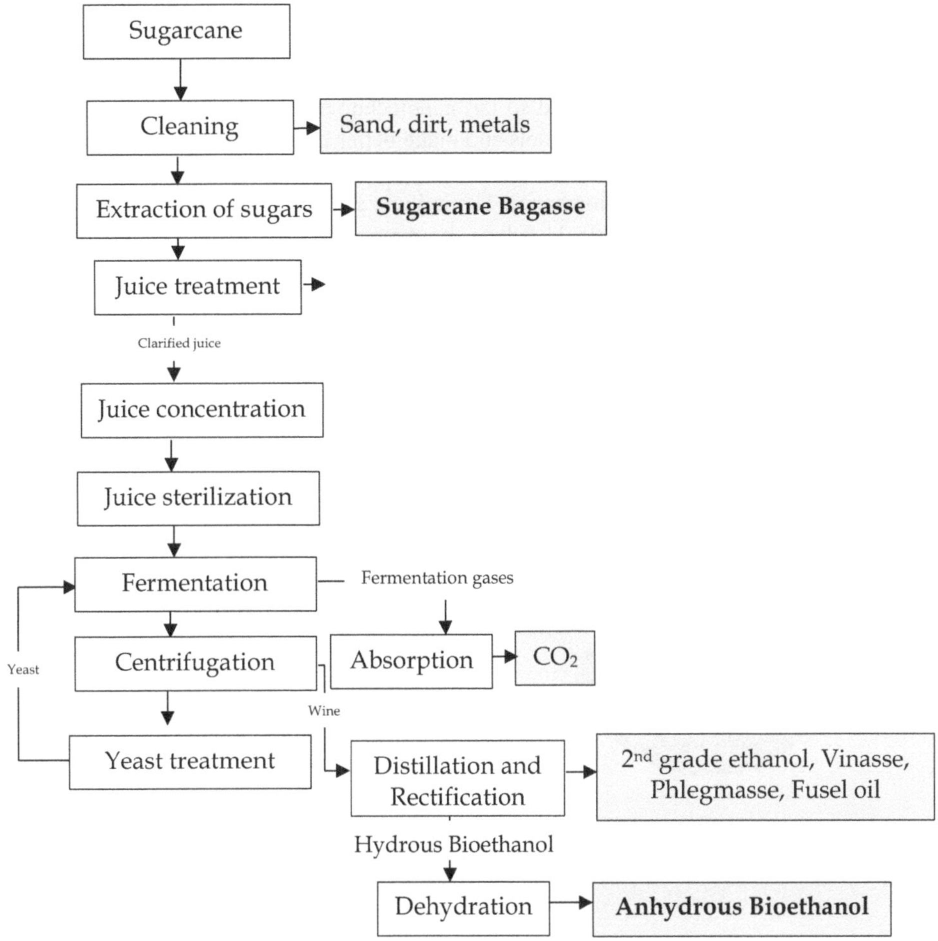

Figure 11.6 Block-flow diagram of the bioethanol production process from sugarcane in an autonomous distillery [30].

excessive water; (iii) do not collide with human and animal food chains; (iv) ability to grow naturally throughout the year and in the presence of sunlight; (v) can be cultivated in rivers, ponds, seas, wet bare land, and waste dump areas; (vi) significant contribution in developing sustainable oxygen production system; and (vii) play an important role in reducing carbon dioxide by the process of photosynthesis respiration [32–36]. Furthermore, the life cycle of microalgae harvesting is very short, yet it produces high biofuel productivity as nascent biomass continues to be generated [34, 37].

Biofuel industries have implemented nanotechnology applications for biofuel production. There are many controversial challenges, from microalgae production to the cost of harvesting and even the energy consumption of microalgae for biofuel production. There are the following benefits of nanomaterials, such as the prevention of microalgae from dying, and with the help of this extracted microalgae, the process of re-cultivation is carried forward.

According to the studies conducted so far, nanotechnology applications have been used to maximize the yield of many microalgae biofuels as well as make better use of microalgae-biofuels in petrol and diesel engines. Figure 11.7 depicts the approach of nano-additives from microalgae cultivation to microalgae-biofuel implementation [6, 38–41].

According to the studies conducted so far, nanotechnology applications have been used to maximize the yield of various biofuels and make better use of biofuels in diesel or petrol engines. Various nanoparticles with different ranges are blended with biofuels to improve the efficiency of biodiesel and bioethanol. The influence of nanomaterials in biofuels is represented in Table 11.5, and Table 11.6 lists the nanocatalysts utilized in the biomass-to-biodiesel conversion process.

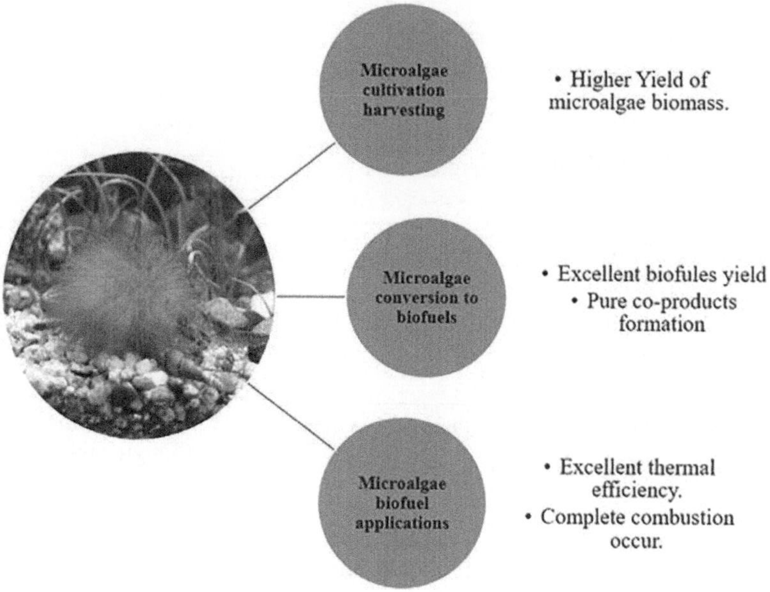

Figure 11.7 Applications of nano-additive to enhance the cultivation of microalgae for biofuel operation [6].

Table 11.5 Applications of Various Amounts of Nanomaterials on Biofuels and Arrangements of Biofuels Nano-Particles Blending and Output Application [6]

Biofuel	*Source of biofuel extraction*	*Nano-particles*	*Amount*	*Combination of nanoparticles in biofuels*	*Output*	*References*
Biodiesel	*Jatropha curcas*	Al_2O_3 CeO_2	30 ppm 30 ppm	B100A30C30 ppm	Enhanced thermal efficiency Reduction in CO and NO_2	[42]
Biodiesel	*Pongamia pinnata*	Rh_2O_3	100 nm	$B100Rh_2O_3$	Reduction in CO 37% NOx Improvement of thermal efficiency	[43]

Biofuel	*Source of biofuel extraction*	*Nano-particles*	*Amount*	*Combination of nanoparticles in biofuels*	*Output*	*References*
Biodiesel	*Azadirachta indica*	Ag_2O	05 ppm	B100Ag_2O 05 ppm	Reduction in HC (10.89%), NOx (4.24%), CO (12.22%), and smoke (6.61%). Improvement in brake thermal efficiency Reduction in brake-specific fuel consumption	[44]
Biodiesel and bioethanol	Vegetable oil and alcohol	Fe_2O_3	150 ppm	BBFe_2O_3 150 ppm	Increase in thermal efficiency by 1% Decrease in emissions by 60% Decrease in the amounts of HC, NOx, CO, and smoke Improvement in mixing appearance of secondary atomization	[45]
Biodiesel, castor oil, diesel, and bioethanol	Vegetable oil and *Ricinus communis* oil Vegetable oil and alcohol	CeO_2-CNT	25 ppm 50 ppm 100 ppm	–	Reduction in CO, CO_2, and smoke Increase in calorific value Improvement in thermal efficiency of the brake	[46]

Table 11.6 Nanocatalysts Used for Biodiesel Production Along With Their Operating Conditions [20]

Nanocatalysts	*Size*	*Feedstock*	*Operating conditions*					*References*
	(nm)		Temp. (°Celsius)	Alcohol-oil ratio	Catalysts (wt.%)	Reaction time (minutes)	Biodiesel yield (percentages)	
TiO_2-ZnO	34.20	Palm oil	60	12:1	_	300	92.2%	[47]
ZnO	28.4						83.2%	
(Li-Cao)	40	Jatropha oil	65	12:1	05	120	99%	[48]
		Karanja oil				60		
CaO/Fe_3O_4	49	Jatropha oil	70	15:1	02	80	95%	[49]
Mg/Al	7.3	Jatropha oil	45	4:1	01	90	95.2%	[50]
CaO/MgO	_	Jatropha oil	64.50	18:1	02	210	92%	[51]
Ca $(OH)_2$-Fe_3O_4 (Ca^{2+}: Fe_3O_4=7)	_	Jatropha oil	70	15:1	02	240	99%	[52]
CaO	20	Soybean oil	23–25	27:1	_	720	99%	[53]
MgO	60	Soybean oil	200–260	6:1	0.5–03	12	99.04%	[54]
KF/Al_2O_3	50	Canola oil	65	6:1	03	480	97.7%	[55]

11.4 Conclusion

The scarcity and high energy demand for fossil fuels have prompted researchers to think about alternative renewable energy resources. Among the existing alternatives, biomass is the best choice because renewable energy generated from biomass is more environmentally friendly. Nano-additive applications for biofuel manufacturing are classified into numerous stages, from raw material production to final product implications. These steps are as follows: (i) nanomaterials for the cultivation process, (ii) nanomaterials for converting biomass into biofuel, and (iii) nanomaterials for biofuel applications. Biofuels with rock-solid nano-particles, nano-liquids, or nano-droplets have been proven to improve lubricity, catalytic performance, cetane number, burn rate, chemical reactivity, fire point, flashpoint, etc. [50–51]. As documented in this chapter, it is quite clear that nanoparticles can play a significant role in everything from biofuel production processes to enhancing engine performance. However, some technical problems arise during the production of biofuels. To accelerate the production of biofuels, we have to address these problems, such as (i) using less expensive nanoparticles, (ii) synthesis of non-toxic nanoparticles for the benefit of micro-organisms, and (iii) using environmentally friendly nanoparticles. This chapter will be helpful in determining future research work, which is as follows:

- Safety assessment on integrating nanomaterials for biofuel production processes is essential to provide insights for future research.
- Study of a wide variety of concentrations of nanomaterials to establish optimal process conditions.
- Application of nanomaterials for commercial approaches in terms of various shapes and sizes during biofuel production.
- Before using it on a large scale, a comprehensive study of the good characterization of nanoparticles should have to be done.
- The impact of nano-catalyst implementation on engine performance, combustion quality of biofuel, and gas emissions should be thoroughly studied and understood before implementation.

References

1. Waqas M., Aburiazaiza A.S., Minadad R., Rehan M., Barakat M.A. and Nizami A.S. "Development of biochar as fuel and catalyst in energy recovery technologies", *J. Clean. Prod.*, volume no. 188, pp. 477–488, 2018.
2. Senjyu T. and Howlader A.M. "Chapter 3—Operational aspects of distribution systems with massive DER penetrations", In *Integration of Distributed Energy Resources in Power Systems*, Funabashi, T., Ed. Academic Press, Cambridge, MA, volume no. 12, pp. 51–76, 2016.
3. Nizami A.S., Rehan M., Waqas M., Naqvi M., Ouda O.K.M., Shahzad K., Miandad R., Khan M.Z., Syamsiro M., Ismail I.M.I. and Pant D. "Waste Bio refineries: Enabling circular economies in developing countries", *Biores. Tech.*, volume no. 251, pp. 1101–1117, 2017.
4. Nizami A. and Rehan M. "Towards nanotechnology-based biofuel industry", *Biofuel. Res. J.*, volume no. 5, pp. 798–799, 2018.
5. Karthikeyan S. and Prathama A. "Microalgae biofuel with CeO_2 Nano additives as an eco-friendly fuel for CI engine", *Energy Source Part A: Rec. Utilize. Environ. Eff.*, volume no. 39, pp. 1332–1338, 2017.

6. Hossain N., Mahlia T.M.I. and Saidur R. "Latest development in microalgae-biofuel production with nano-additives", *Biotech. Biofuel.*, volume no. 12, pp. 1–16, 2019.
7. Bidir M.G., Millerjothi N.K., Adaramola M.S. and Hagos F.Y. "The role of nanoparticles on biofuel production and as an additive in ternary blend fuelled diesel engine: A review", *Energy Rep.*, volume no. 7, pp. 3614–3627, 2021.
8. Mofijur M., Rasul M.G., Hyde J. and Bhuyia M.M.K. "Role of biofuels on IC engines emission reduction", *Energy Proced.*, volume no. 75, pp. 886–892, 2015.
9. Shahir S.A., Masjuki H.H., Kalam M.A., Imran A. and Ashraful A.M. "Performance and emission assessment of diesel–biodiesel–ethanol/bioethanol blend as a fuel in diesel engines: A review. *Renew. Sustain. Energy Rev.*, volume no. 48, pp. 62–78, 2015.
10. Chandrasekaran V.K., Arthanarisamy M., Nachiappan P., Dhanakotti S. and Moorthy B. "The role of nano additives for biodiesel and diesel blended transportation fuels", *Trans. Res. Part D.*, volume no. 46, pp. 145–156, 2016.
11. Tamilselvan P., Nallusamy N. and Rajkumar S. "A comprehensive review on performance, combustion and emission characteristics of biodiesel fuelled diesel engines", *Renew. Sustain. Energy Rev.*, volume no. 79, pp. 1134–1159, 2017.
12. Basha J.S. and Anand R.B. "The influence of nano additive blended biodiesel fuels on the working characteristics of a diesel engine", *J. Braz. Soc. Mech. Sci. Eng.*, volume no. 35, pp. 257–264, 2013.
13. Wu Q., Xie X., Wang Y. and Roskilly T. "Effect of carbon-coated aluminium nanoparticles as an additive to biodiesel-diesel blends on performance and emission characteristics of a diesel engine", *Appl. Energy.*, volume no. 221, pp. 597–604, 2018.
14. Shalin M. and Sinha V.K. "A review approach on exhaust emission reduction by nanoparticle additives in diesel engine", *Int. J. Innov. Res. Sci. Eng. Technol.*, volume no. 54, p. 160, 2000.
15. Nasrollahzadeh M., Sajadi S.M., Sajjadi M. and Issaabadi Z. "Applications of nanotechnology in daily life", *Interface Sci. Technol.*, volume no. 28, pp. 113–143, 2019.
16. Panahi H.K.S., Dehhaghi M., Kinder J.E. and Ezeji T.C. "A review on green liquid fuels for the transportation sector: a prospect of microbial solutions to climate change", *Biofuel Res. Journ.*, volume no. 06, pp. 995–1024, 2019.
17. Nigam P.S. and Singh A. "Production of liquid biofuels from renewable resources", *Prog. Energy Combust. Sci.*, volume no. 37, pp. 52–68, 2011.
18. Manikandan S., Subbaiya R., Biruntha M., Krishnan R.Y., Muthusamy G. and Karmegam N. "Recent development patterns, utilization and prospective of biofuel production: Emerging nano-technological intervention for environmental sustainability-A review", *Fuel*, volume no. 314, pp. 1–12, 2022.
19. Tiwari J.N., Tiwari R.N. and Kim K.S. "Zero-dimensional, one-dimensional, two-dimensional and three-dimensional nanostructured materials for advanced electrochemical energy devices", *Prog. Mater. Sci.*, volume no. 57, pp. 724–803, 2012.
20. Akia M., Yazdani F., Motaee E., Han D. and Arandiyan H. "A review on conversion of biomass to biofuel by nanocatalysts", *Biofuel Res. J.*, volume no. 01, pp. 16–25, 2014.
21. Lamberti M., Zappavigna S., Sannolo N., Porto S. and Caraglia M. "Advantages and risks of nanotechnologies in cancer patients and occupationally exposed workers", *Expert Opin. Drug. Deliv.*, volume no. 11, pp. 1087–1101, 2014.
22. Khoo K.S., Chia W.Y., Tang D.Y.Y., Show P.L., Chew K.W. and Chen W.H. "Nanomaterials utilization in biomass for biofuel and bioenergy production", *Energies*, volume no. 13, pp. 1–13, 2020.
23. Hassan M.H. and Kalam M.A. "An overview of biofuel as a renewable energy source: Development and challenges", *Proce. Eng.*, volume no. 56, pp. 39–53, 2013.
24. Kaniapan S., Hassan S., Ya H., Nesan K.P and Azeem M. "The utilisation of palm oil and oil palm residues and the related challenges as a sustainable alternative in biofuel, bioenergy, and transportation sector: A review", *Sustain (Switzerland)*, volume no. 13, 2021.

25. Kurnia J.C., Jangram S.V., Akhtar S., Sasmito A.P. and Mujumdar A.S. "Advances in biofuel production from oil palm and palm oil processing wastes: A review", *Biofuel Res. J.*, volume no. 3, pp. 332–346, 2016.
26. Talukdar D., Verma D.K., Malik K., Mohapatra B. and Yulianto R. "Sugarcane as a potential biofuel crop", *Sugar. Biotech. Challenges and Pro.*, pp. 1–176, 2017.
27. Mandegari M.A., Farzad S. and Görgens J.F. "Recent trends on techno-economic assessment (TEA) of sugarcane bio refineries", *Biofuel Res. J.*, volume no. 4, pp. 704–712, 2017.
28. Talukdar D., Verma D.K., Malik K., Mohapatra B. and Yulianto R. "Sugarcane as a potential biofuel crop", In *Sugarcane Biotechnology: Challenges and Prospects*, Springer, Cham, pp. 123–137.
29. Griffin W.M. and Scandiffio M.I.G. "Can Brazil replace 5% of the 2025 gasoline world demand with ethanol?", *Energy*, volume no. 34, pp. 655–661, 2009.
30. Dias M.O., Modesto M., Ensinas A.V., Nebra S.A., Maciel Filho R. and Rossell, C.E. "Improving bioethanol production from sugarcane: Evaluation of distillation, thermal integration and cogeneration systems", *Energy*, volume no. 36, pp. 3691–3703, 2011.
31. Jingura R.M. and Kamusoko R. "A multi-factor evaluation of Jatropha as a feedstock for biofuels: The case of sub-Saharan Africa", *Biofuel Res. J.*, volume no. 02, pp. 254–257, 2015.
32. Laurens L., Chen-Glasser M. and McMillan J. "A perspective on renewable bioenergy from photosynthetic algae as feedstock for biofuels and bioproducts", *Alga. Res.*, volume no. 24, pp. 261–264, 2017.
33. Collotta M., Champagne P., Mabee W. and Tomasoni G. "Wastewater and waste CO_2 for sustainable biofuels from microalgae", *Alga. Res.*, volume no. 29, pp. 12–21, 2018.
34. Hossain N. and Jalil R. "Analysis of bio energy from Malaysian local plants sentang and sesendok", *Asia J. Pac. Energy Environ.*, volume no. 02, pp. 141–144, 2015.
35. Goh B.H.H., Ong H.C., Cheah M.Y., Chen W.H., Yu K.L. and Mahlia T.M.I. "Sustainability of direct biodiesel synthesis from microalgae biomass: A critical review", *Renew. Sust. Energy Rev.*, volume no. 107, pp. 59–74, 2019.
36. Hossain N., Zaini J. and Mahlia T.M.I. "Experimental investigation of energy properties for *Stigonematales* sp. microalgae as potential biofuel feedstock", *Int. J. Sustain Eng.*, volume no. 12, pp. 123–130, 2019.
37. Shahir S.A., Masjuki H.H., Kalam M.A., Imran A., Fattah I.M.R. and Sanjid A. "Feasibility of diesel-biodiesel-ethanol/bioethanol blend as existing CI engine fuel: An assessment of properties, material compatibility, safety and combustion", *Renew. Sustain. Energy Rev.*, volume no. 32, pp. 379–395, 2014.
38. Hasannuddin A.K., Yahya W.J., Sarah S., Ithnin A.M., Syahrullail S., Sidik N.A.C., Kasim K.A., Hirofumi N., Ahmad M.A., Sugeng D.A., Zuber M.A. and Ramlan N.A. "Nano-additives incorporated water in diesel emulsion fuel: Fuel properties, performance and emission characteristics assessment", *Energy Convers. Man.*, volume no. 169, pp. 291–314, 2018.
39. Kim J., Jia H. and Wang P. "Challenges in biocatalysis for enzyme-based biofuel cells", *Biotechnol. Adv.*, volume no. 24, pp. 296–308, 2006.
40. Strzalka R., Schneider D. and Eicker U. "Current status of bioenergy technologies in Germany: review", *Sustain. Energy Rev.*, volume no. 72, pp. 801–820, 2017.
41. Bharathiraja B., Selvakumari A.E. and Kumar P. "Future prospects, opportunities, and challenges in the application of nanomaterials in biofuel production systems", In *Application in Biofuels and Bioenergy Production Systems*, Elsevier, volume no. 39, pp. 797–806, 2021.
42. Prabu A. "Nanoparticles as additive in biodiesel on the working characteristics of a DI diesel engine", *Ain Shams Eng. J.*, volume no. 09, pp. 2343–2349, 2018.
43. Karthikeyan S. and Prathima A. "Analysis of emissions from use of an algae biofuel with nano-ZrO_2", *Energy Source Part A.*, volume no. 39, pp. 473–479, 2017.
44. Devarajan Y., Munuswamy D.B. and Mahalingam A. "Influence of nanoadditive on performance and emission characteristics of a diesel engine running on neat neem oil biodiesel", *Environ. Sci. Pollut. Res. Int.*, volume no. 25, pp. 26167–26172, 2018.

45. Sabarish R., kumar D.M., Kumar D.M.P.J. and Manavalan R. "Experimental study of nanoadditive with biodiesel and its blends for diesel engine", *Int. J. Pure Appl. Math.*, volume no. 118, pp. 967–979, 2018.
46. Shaafi T., Sairam K., Gopinath A., Kumaresan G. and Velraj R. "Effect of dispersion of various nanoadditives on the performance and emission characteristics of a CI engine fuelled with diesel, biodiesel and blends: A review", *Renew. Sust. Energ Rev.*, volume no. 49, pp. 563–573, 2015.
47. Madhuvilakku R. and Piraman K. "Biodiesel synthesis by TiO_2–ZnO mixed oxide nanocatalysts catalyzed palm oil transesterification process", *Bioresour. Technol.*, volume no. 150, pp. 55–59, 2013.
48. Kaur M. and Ali A. "Lithium ion impregnated calcium oxide as nano catalyst for the biodiesel production from Karanja and Jatropha oils", *Renew. Energy*, volume no. 36, pp. 2866–2871, 2011.
49. Chang A.C., Louh R.F., Wong D., Tseng J. and Lee Y.S. "Hydrogen production by aqueous-phase biomass reforming over carbon textile supported Pt-Ru bimetallic catalysts", *Int. J. Hydrog. Energy*, volume no. 36, pp. 8794–8799, 2011.
50. Deng X., Fang Z., Liu Y.H. and Yu C.L. "Production of biodiesel from Jatropha oil catalyzed by nanosized solid basic catalyst", *Energy*, volume no. 36, pp. 777–784, 2011.
51. Reddy C.R.V., Oshel R. and Verkade J.G. "Room-temperature conversion of soybean oil and poultry fat to biodiesel catalyzed by nanocrystalline calcium oxides", *Energy & Fuels*, volume no. 20, pp. 1310–1314, 2006.
52. Wang L. and Yang J. "Transesterification of soybean oil with nano-MgO or not in supercritical and subcritical methanol", *Fuel*, volume no. 86, pp. 328–333, 2007.
53. Boz N., Degirmenbasi N. and Kalyon M.D. "Conversion of biomass to fuel: Transesterification of vegetable oil to biodiesel using KF loaded nano-γ-Al_2O_3 as catalyst", *Appl. Catal.*, volume no. 89, pp. 590–596, 2009.
54. Silitonga A.S., Masjuki H.H., Ong H.C., Sebayang A.H., Dharma S., Kusumo et al. "Evaluation of the engine performance and exhaust emissions of biodiesel—Bioethanol—diesel blends using kernel-based extreme learning Machine", *Energy*, volume no. 159, pp. 1075–87, 2018.
55. Morone P. and Cottoni L. "Biofuels: Technology, economics, and policy issues", In *Handbook of Biofuels Production: Processes and Technologies*, Elsevier, Second Edition, pp. 61–83, 2016.

Chapter 12

Modification of SS316 steel with the assistance of high velocity oxy fuel (HVOF) process to upsurge its sustainability

Vikrant Singh, Anuj Bansal, Anil Kumar Singla, Deepak Kumar Goyal, Rampal, and Navneet Khanna

Contents

Abbreviations

CE: Cavitation Erosion
VC: Vanadium Carbide
CuNi-Cr: Cupronickel-Chromium
HVOF: High Velocity Oxy Fuel
C: Carbon
Cr: Chromium
P: Phosphorous
Si: Silicon
Mn: Manganese
S: Sulfur
Ni: Nickel
Mb: Molybdenum
SLPM: Standard Liter Per Minute

DOI: 10.1201/9781003291961-14

12.1 Introduction

India, a tremendously developing nation, requires a large amount of electricity production. Presently, India produces approximately 20.76% share of its total produced electricity from hydropotential electricity, and it can be increased by a large extent as hydroelectricity is very promising in all the available sources of renewable energy. And also, India is blessed with a number of rivers, which increases the probability of extending the hydroelectricity share mentioned earlier. For this, our hydropower plants must be up to date with the technology available and with the knowledge of research studies [1–3]. In a hydropower plant, water flows at several velocities at different parts. Due to this flowing water, the materials of hydropower plant machinery components are subjected to abrasive slit erosion, corrosion erosion, and cavitation erosion. The detailed classification of different types of wear, along with the symptoms and appearance of worn-out surfaces, is given in Table 12.1. Out of the said erosion that occurs in hydromachinery, cavitation erosion is mainly focused on in this chapter. Cavitation erosion is usually heard for hydraulically operated systems, such as propellers, valves, and hydraulic pumps [4]. Cavitation happens because of the rapid variations of pressure in a liquid that lead to the creation of minor vapor-filled bubbles or cavities [5–6]. The force of collapsing of these vapor bubbles is great enough to remove material from the hydraulic components, making the hydraulic equipment weak and unsustainable.

Due to the ease of availability of iron, its abundance in the earth's crust (5%), and its economic extraction, it is widely used to cooperating with the demands. The iron is alloyed with several different elements according to the requirements. One of the major alloys of iron is steel, when the iron is alloyed with carbon and sometimes with other elements, to

Table 12.1 Classification of Different Types of Wear

Types of Wear	*Symptoms*	*The appearance of the degraded surface*
Abrasive	Occurrence of spotless furrows cut out by coarse abrasive particles	Groves
Corrosion	Existence of corroded metal constituents	Depressions or rough pits
Adhesive	Prime symptom is metal allocation from the surface	Catering, seizure surfaces, and torn out
Erosion	Existence of solid abrasive particles in the rapid moving slurry/fluid and short abrasion grooves	Waves and troughs
Fatigue	Existence of surface cracks with some small pits and spalls	Sharp and short angular edges around pits
Delamination	Cracks running parallel to the surface with partially dislodged or movable flakes	Long, loose, and thin particles
Fretting	Production of a voluminous amount of loose debris	Seizure, roughening, and development of oxides ridges
Cavitation	Vibration and noise due to the sudden collapsing of bubbles	Ploughing, overlapped cavities, micropits, etc.

control the physical as well as chemical properties of steel. Stainless steel is such steel that has corrosion resistance, rust resistance, and water stain resistance, but ordinary steel does not have such properties. However, stainless steel is not fully stain-proof in poor air-circulation environments or low-oxygen and in high-salinity. Austenitic stainless steel is commonly used mainly for injectors due to its good mechanical assets, such as high fracture toughness, high impact energy, and weldability, among other beneficial characteristics in the corrosive environment as well. Austenitic stainless steels, however, may be sufficient to withstand cavitation attacks in harsh environments due to their exceptional properties. But still, the equipment is not sustainable enough to protect the hydraulic equipment from hazardous cavitation bubbles. Now, to overcome this dangerous hurdle in the field of hydraulic components and make the component sustainable for the environment and the industries as well, appropriate protective coating of harder metals/alloys is majorly suggested as a good solution in the present literature [7, 8].

A protective coating can be characterized as a film of material framed normally or artificially or stored intentionally on the exterior of a component made of material other than the coated film with the point of getting needed specialized or enriching properties. It is a coating that can be imposed on an article's surface, frequently called base material. The inspiration behind the application of covering is the upgrade of the base material by improving its physical characteristics, erosion resistance, wear resistance, and many more. The covering technique includes using a thin covering of protective material on a substrate. The protective material can be metallic or non-metallic, natural or inorganic, solid, fluid, or gas in a wide range of utilization. The material must be worked under severe conditions, for example, cavitation, disintegration, erosion, and oxidation, at a higher temperature in aggressive chemical situations. Hence, surface modification of these segments is important to secure them against different sorts of surface disintegration. The base material gives vital mechanical properties, and coatings give a method for broadening the breaking points of the utilization of materials at the upper end of their execution capacities, by allowing the mechanical possessions of the substrate material to be kept up. Protective coatings might be applied to change the surface properties of the substrate, for example, wettability, adhesion, erosion wear, or wear resistance. Various researchers have already deposited a lot of protective layers over the steel surface to make it sustainable, but even though some of the materials are still left over, some work can still be performed, increasing the life of hydromachinery steel. Some researchers have observed that coating of WC can prevent the hydraulic components from cavitation and corrosion erosion [9, 10]. They have observed that the harder WC particles are capable enough to provide the desired wear-resistive property to the substrate [11, 12]. Further, binders like Co, Cr, and Ni provide enough stiffness, leading to the effect of shockwaves produced by the bursting of bubbles. In one of his investigations, Zhang et al. [13] revealed the corrosion and erosion resistance of AISI 420 cladded with VC weight percentages (0 wt.%–40 wt.%) increased when the VC fraction was raised. Other researchers also observed that VC could be a hard coating material in order and sustainable coating material to increase the wear resistive property of the surface [14, 15].

Of the various coating techniques, the thermal spraying technique is one of the most important surface engineering techniques. It was observed as an effective solution for various applications that requires resistance to cavitation, heat, and corrosion erosion [14–16]. The major thermal spraying processes are HVOF, arc wire spraying, flame spraying, plasma spraying, etc. [17–19]. The detailed classification of different thermal spraying processes and their energy requirement is given in Table 12.2. Among these thermal spray coating

Table 12.2 Detailed Classification of Different Thermal Spraying Processes Along With Their Energy Requirement

Processes		*Energy sources*	*Different nomenclature*
High-energy processes	Detonation flame spraying	Chemical	D-gun
	Rodojet spraying	Chemical	Rodojet
	HVOF spraying	Chemical	HVOF spraying
			High-velocity flame spraying (HVFS)
			High-velocity air fuel
	Plasma spraying	Electrical	Vacuum plasma spraying (VPS)
			Air plasma spraying (APS)
			Water-stabilized plasma spraying (UWS)
			Inductive plasma spraying
			Low-pressure plasma spraying (LPPS)
Low-energy processes	Flame spraying	Chemical	Metallizing
			Oxyfuel gas-powder spraying
			Oxyfuel gas-wire spraying
	Arc spraying	Electrical	Twin-wire arc spraying
			Electric arc spraying
			Twin-wire arc spraying

techniques, HVOF process technology has attracted much attention in the past few years because of its supreme characteristics of providing a superior coating of a high adhesion strength [20–22]. In one of their investigations, Bansal et al. [23] found that HVOF spray (Tungsten carbide + cobalt-chromium)-based coating has successfully reduced the slurry erosion wear of SS316 steel owing to its improved strength. Santa et al. [24] analyzed that the slurry erosion resistance of specimens was improved up to 16 times when martensitic steel was coated using the oxy fuel powder (OFP) process and the HVOF process. Further, Amarendra et al. developed 70Ni30Cr coatings on SS410 steel substrate using the HVOF process and further performed tests to investigate the slurry and cavitation erosion behavior of the above-mentioned coatings. Specimens with deposited coatings better resisted erosion than uncoated specimens. Moreover, Babu et al. produced WC-10Co-4Cr and Ni coatings on the surface of SS316 steel by employing Detonation Gun thermal spray process. The authors further processed the prepared specimens using microwave irradiations and probed the behavior of prepared specimens against slurry and cavitation erosion. Post-processed specimens were found to be nine times more resistant to cavitation erosion than the specimens without post-processing.

From the literature survey, it can be easily concluded that high-velocity spraying system has produced some good coatings, which have unbelievable strength in the field of hydro-machinery equipment [25–27]. And the same can also be explained with the help of a schematic diagram, as shown in Figure 12.1. In the present work, 75% VC blended with a 25 wt.% binder CuNi-Cr has been deposited on hydromachinery SS316 steel using the HVOF spraying process with an aim to prepare a sustainable coated surface that can increase the life of hydromachinery steel. The basic impression behind the preparation of a

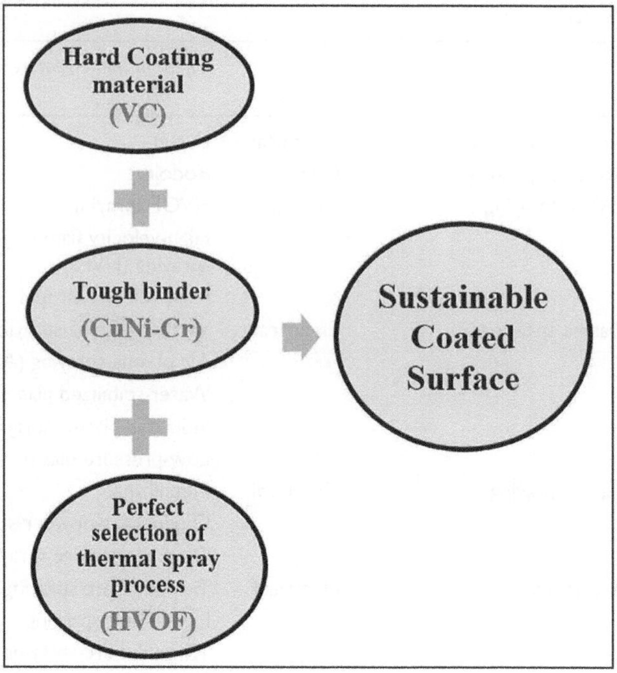

Figure 12.1 Sustainable coating surface development idea.

sustainable prepared surface is given in Figure 12.1. Further, cavitation erosion testing was accomplished with the assistance of an in-house cavitation erosion test rig. As per ASTM G134 standard with variability in impingement parameters comprising of jet velocity (m/s) and stand-off distance (cm) [28, 29]. Furthermore, the microstructures and phase composition of the coating are acknowledged by scanning electron microscopy (SEM) and energy dispersive spectroscopy (EDS). Moreover, the possible mechanism behind the eroded surfaces is also examined by means of SEM.

12.2 Experimental procedure

12.2.1 Materials and HVOF thermal spray process

For the present study, the metallic powders were procured from Chemi Enterprises LLP, Mumbai, Maharashtra, India. The major powder (VC) used for the coating substrate steel has bigger and dense morphology with the size of particles varying from 10 to 40 μm, as shown in Figure 12.2 (a). Contrarily, CuNi powder was reported to have dense and globular morphology with an average size of particle as 30 μm (refer to Figure 12.2 (b)). Further, it was also observed that Cr attained bulky, irregular, and solid morphology (refer to Figure 12.2 (c)). It was also observed from Figures 12.2 (a), (b), and (c) that the powders were also composed of many small-sized particles of 1–10 μm in non-agglomerated and a bit of sintered state. The morphology of the powders was deep-rooted by a FESEM (JSM-7610Fplus, JEOL,

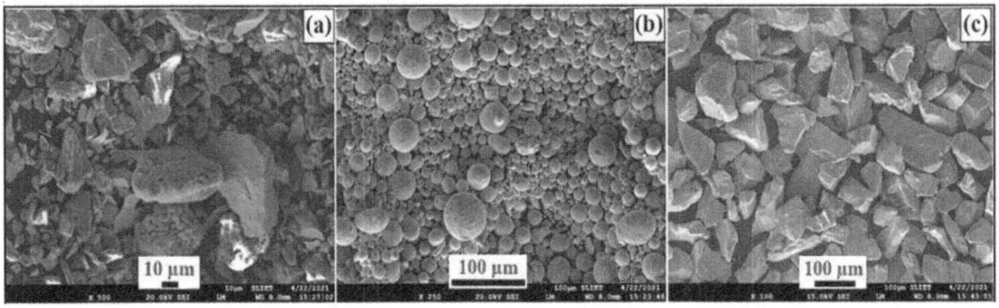

Figure 12.2 SEM micrograph of (a) vanadium carbide (VC) X500, (b) cupronickel (CuNi) X250, and (c) chromium (Cr) powder X100.

Table 12.3 As-Received Composition (wt.%) of SS316 Compared With ASTM Standard

SS316 steel	*C*	*Cr*	*P*	*Si*	*Mn*	*S*	*Ni*	*Mb*	*Fe*
ASTM standard	<0.08	16–18	<0.045	<1	<2	<0.03	8–12	2–3	Balance
As received	0.0739	17.01	0.043	0.96	1.98	0.023	9.86	2.76	Balance

Table 12.4 HVOF Process Parameters Acquired for Coating at MECPL, Jodhpur, India

Parameters	*Oxygen flow rate (slpm)*	*Oxygen pressure (kg/cm²)*	*Fuel (LPG) flow rate (slpm)*	*Fuel pressure (kg/cm²)*	*Air flow rate (slpm)*	*Air pressure (kg/cm²)*	*Spray distance (cm)*	*Powder feed rate (g/min)*
Value	250–260	11–12	65–75	6.5–7.0	650–750	5.5–6.5	15 cm	40

Japan) and is mentioned in Figures 12.2 (a), (b), and (c). Further, the substrate steel used for the present work was SS316 steel, and it was procured from Micro Engineering Company, Ludhiana, Punjab. In order to confirm the purity of the substrate, the chemical composition of the substrate is also done at MECPL, Jodhpur, India. The result and typical composition are shown in Table 12.3.

Furthermore, the specimens were prepared for coating by sandblasting their surfaces with a pressure blaster (PB-9182) machine before coating. Air, serving as a carrier gas, struck Al_2O_3 particles (16 mesh size) across the surface of the substrate at a pressure of 5 bar, resulting in average surface roughness of 12–14 µm. Then, the substrate was coated by VC+CuNi-Cr based coating using a Hipojet-2700 gun spraying system. The HVOF spraying parameters acquired while the coating is mentioned in Table 12.4. After 12–13 passes of the spraying gun, a coating of an average thickness of 250 µm is deposited over the substrate. Further, the self-explanatory schematic diagram for the experimental coating setup has been manifested in Figure 12.3. From Figure 12.3, it can be explained that the SS316 material was first grit blasted with the aid of grits, and a rough surface was obtained. After that, coating powders were fed to the mixing chamber, where these powders were mixed and transferred in a Hipojet-2700 gun, which was operated with a gas control unit. Owing to the high-temperature characteristic of the HVOF spraying system, the powders got evacuated out of

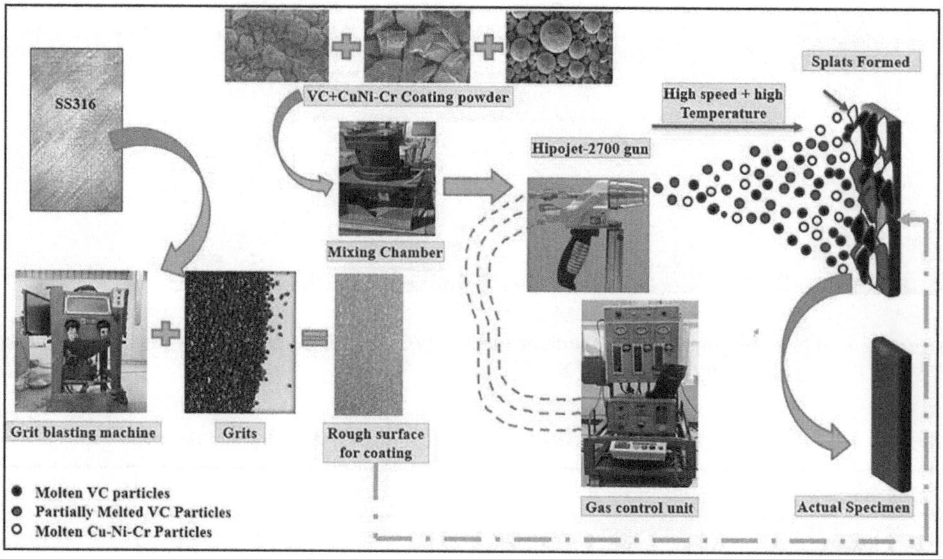

Figure 12.3 Schematic representation of coating – experimental setup.

the gun in the form of splats and got wedged over the base surface owing to the supersonic velocity characteristic of the HVOF spraying technique.

12.2.2 Cavitation erosion tests

Coated and uncoated specimens were evaluated for their cavitation erosion behavior by using high velocity cavitation erosion test rig configurations per ASTM G134 standards. The detailed setup of the test apparatus can be found in the existing literature [14]. In essence, the test rig has such a design that it permits the impingement of a high-speed jet of water on the submerged specimen. Thus, the effect of the straight impact of the jet becomes negligible owing to the underwater placed specimen. Further, the rapid velocity at the nozzle's tip can lower the pressure much beneath the saturated vapor pressure, which results in the creation of underwater bubbles and cavitation clouds [23]. In this course of work, both specimens were tested with the variability in impingement parameters, comprising jet velocity (m/s) and stand-off distance (cm) with a fixed impingement angle (degree) to normal. Furthermore, both the specimens (coated and uncoated) were imperiled to CE for 10 h with the effective parameter combination corresponding to the maximum erosion from the different combinations of the full factorial method, that is, velocity 40 m/s and stand-off distance 12 cm, while the nozzle diameter and angle of attack as 3 mm and 90°, respectively. Furthermore, the samples were cleaned and then weighed using Shimadzu AUW-120D weighing balance having a least count of 0.0001 g.

12.2.3 Microstructure characterization

The cross-sectional SEM images of coated specimens have been captured and apprehended using FESEM (JSM-7610Fplus, JEOL, Japan). Further, with assistance from the ImageJ software, these SEM images of cross-sections have been used to determine the porosity and average thickness of the coating. Further, the surface irregularities of the coated specimen are were measured by using (Surtronic, Taylor Hobson: SE1200) apparatus. Furthermore, Vickers hardness tests of the coated and uncoated specimens were measured at a load of 500 g for an average time of 20-s with the assistance of Vickers micro-hardness tester (RMHT-201 by Radical Scientific Equipment Private Limited, India). Moreover, to ensure the repeatability of results, the reading was taken at 15 different locations.

12.3 Results and discussions

12.3.1 Characterization of developed coating

After the deposition, the above-mentioned coatings were characterized for various attributes, namely percentage porosity, surface roughness, and microhardness. In that context, Figure 12.4 shows the cross-sectional view of 75%VC+25%CuNi-Cr coating deposited over SS316 Steel. It is very clear from the cross-sectional image that the coatings have splats-like structures and strong mechanical interlocking at the interface, which might be attributed to the high-velocity behavior of the HVOF spraying technique. Further, variable distribution of coating splats can be observed in asperities of the SS316 surface. Coating thickness values were measured at 10 different locations using the ImageJ software, and an average coating thickness of 229.5 μm was recorded. Some part of the interface is enclosed in a shape, and its enlarged view is shown adjacent to it. The enlarged view verifies the existence of micro-pores and cracks in the coating, which might have deposited during the deposition. High-melting point VC particles might have caused the generation of pores in the above-mentioned coatings. Some of the VC particles could not melt properly and hence remained in partially solid form. These partially melted particles cause the non-uniform distribution of further depositing splats. It generates some pores as well as surface irregularities in the developed coatings. Therefore, porosity in coatings enhances coating thickness, which, in turn, provides cushioning effect while the bursting of bubbles during CE tests.

Further, Table 12.5 clarifies the average surface roughness and microhardness values of SS316 steel and 75%VC+25%CuNi-Cr coated specimen. It is clear from Table 12.5 that the coated specimen has very high microhardness as compared with stainless steel. Moreover,

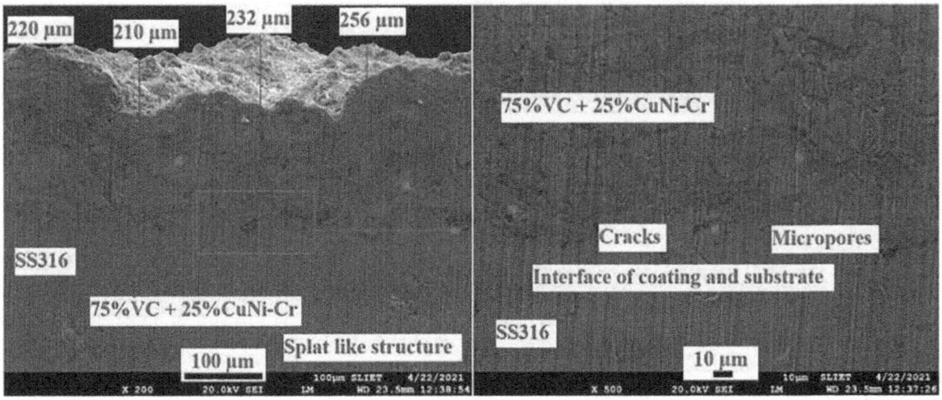

Figure 12.4 Cross-sectional SEM image of HVOF sprayed (75%VC+25%CuNi-Cr).

Table 12.5 Average Surface Roughness and Microhardness Values of SS316 steel and 75%VC+25% CuNi-Cr

Specimens	*SS316*	*75%VC+25%CuNi-Cr*
Average surface roughness Ra (μm)	1.785 ± 0.13	3.672 ± 0.26
Microhardness (HV9.81N)	156 ± 8	974 ± 26

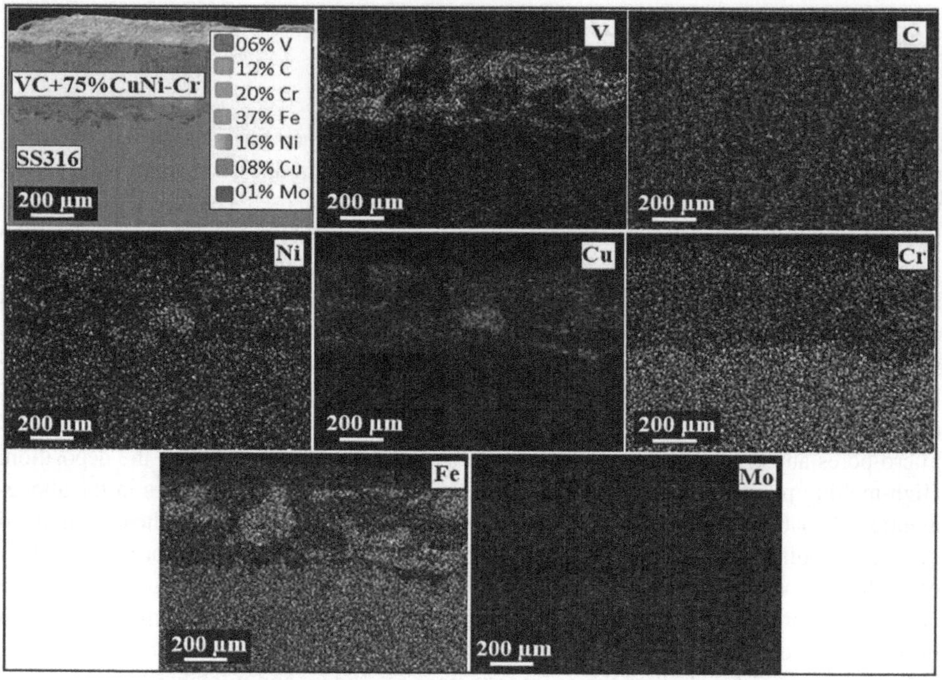

Figure 12.5 Cross-sectional mapping of HVOF sprayed (75%VC+25%CuNi-Cr).

ten values of microhardness values were taken at the spacing of 100 μm corresponding to a particular specimen, and then, the average value was considered to get the precise result.

Moreover, to find the homogeneity of the sprayed coating particles in the coating matrix, the EDAX mapping of the cross-section was also evaluated, as shown in Figure 12.5. From the matrix, it can be evaluated that particles are homogenous and spread all over the coating surface, which leads to a better and more sustainable surface owing to the adequate proportion of hardness and toughness present within.

12.3.2 Cavitation erosion characteristics

Figure 12.6 depicts the cumulative cavitation erosion mass loss curves of HVOF sprayed 75%VC+25%CuNi-Cr coating and SS316 steel. Further, it has been conspicuously observed that HVOF sprayed 75%VC+25%CuNi-Cr coating has an excellent cavitation erosion resistance than SS316 steel. The mass loss of the coated and uncoated specimens was evaluated

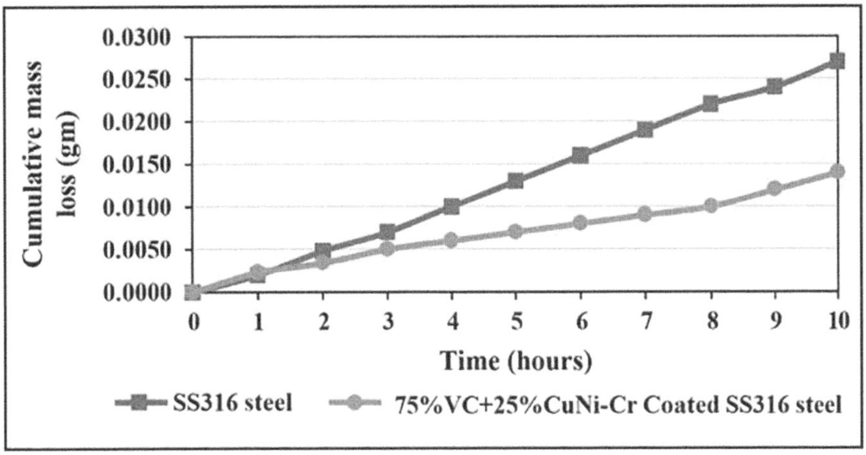

Figure 12.6 Cumulative mass loss SS316 steel and 75%VC+25%CuNi-Cr with time.

for 10 h at a regular time interval of 1 h. It was found that the cumulative mass loss of HVOF sprayed 75%VC+25%CuNi-Cr coating was 14.1 mg, which was only 51% of the mass loss of SS316 steel (27.4 mg).

12.3.3 Failure and mass loss due to impingement variables

Various researchers, including Pandey et al. [14], concluded that cavitation erosion was mainly affected by variability in impingement parameters comprising jet velocity (m/s), stand-off distance (cm), etc. Keeping this in mind, the erosion tests were performed corresponding to the above-mentioned operating parameters. The full factorial design method was used to get the optimized result which comprises two factors and three levels. Thus, a total of nine sets of experiments were performed in each case of coated and uncoated specimens. Further, Figure 12.6 shows the effect of jet velocity and SOD at a normal impingement angle (90°). From the readings obtained, it was perceived that mass loss increased with the velocity of the jet. The reason is quite obvious. With the increase in jet velocity, the pressure loss in flowing fluid is also increased; thus, the more the velocity of jet, the more the interruptions in water flow, and therefore, the higher the pressure loss. Furthermore, it also proliferates the establishment of bubbles, which causes extra erosion due to cavitation.

SOD distance also proved to be a major parameter while obtaining the mass loss for both coated and uncoated specimens. From Figure 12.7, it can be easily concluded that the mass loss is highest at intermediate SOD of 8 cm, lower at 12 cm, and minimum at 4 cm. That might be because at a lower stand-off distance, although bubbles are formed, they could not persist on the surface of specimens and are taken away by fluid flowing at high velocity. Further, at higher SOD, the bubbles can grow properly and gets burst in the fluid itself before striking the targeted area. Therefore, the mass loss is maximum at intermediate SOD because at intermediate SOD, the generated bubbles have enough time to stabilize and target the surface of specimens. Moreover, the coated specimen provides good resistance as compared to

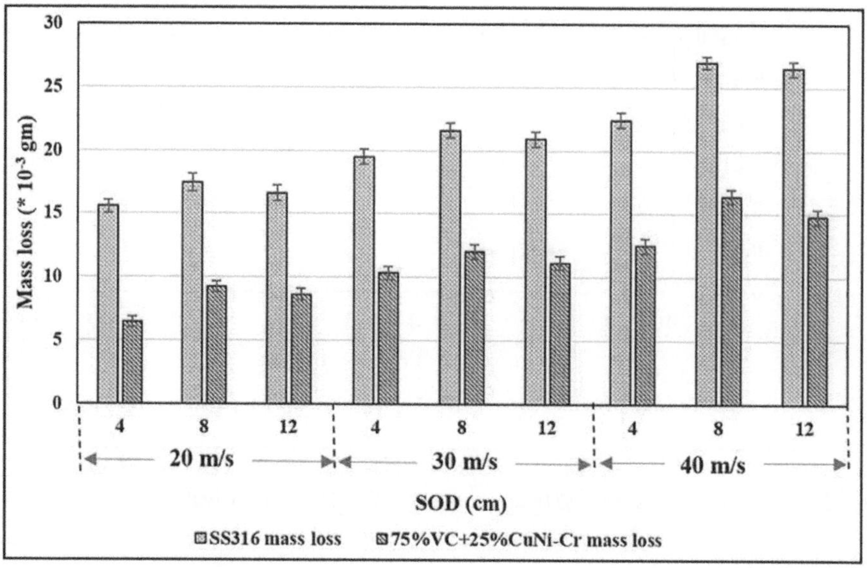

Figure 12.7 Mass loss variations of SS316 steel and HVOF sprayed 75%VC+25%CuNi-Cr coated SS316 steel with a velocity of jet and stand-off distance (SOD) at normal impact angle (90°).

the substrate. The probable reason might be the increased hardness of the coating specimen as compared to the base substrate.

12.3.4 Cavitation failure mechanism

With the aim of finding the mechanism of erosion behind the coated and uncoated specimens, the SEM images of both specimens were deeply interrogated. Figure 12.8 shows the eroded SEM images of 75%VC+25%CuNi-Cr coated SS316 steel during cavitation resistance experimentation. Coating splats were observed in the SEM image, which may correlate with the major attribute of the HVOF spraying technique. The low melting point of CuNi-Cr particles allows their complete melting, which, in turn, deposits the material in the form of splats. A deep enlarged cavity is enclosed in an area, and its magnified view is presented in an adjacent SEM image. Desquamation of material was also observed, which might be because of the presence of cracks and pores on the coating surface. A plastically deformed zone or a deep enlarged cavity is also visible in Figure 12.8, which might be because of the bursting of multiples bubbles at a softer region produced by binders (CuNi-Cr). Micro pits and pores can easily be observed close to the plastically deformed zone. In crux, this type of coating acquires moderate hardness and adequate stiffness with the combined effect of VC and CuNi particles. Further, this combined effect cushioned the dangerous shock waves and increased the resistance to cavitation erosion. Contrarily, if we talk about the mechanism behind the erosion of SS316 then it is observed that the substrate is majorly eroded by ductile failure mode. The major cause behind the erosion of SS316 steel might be a large number of short-term impacts of the bursting of bubbles. The same type of mechanism for SS316 steel is also reported in the existing literature.

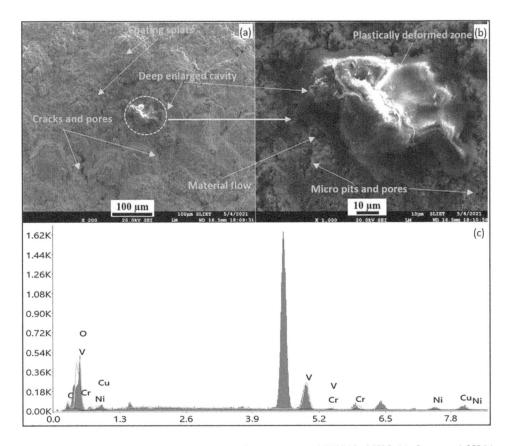

Figure 12.8 Eroded SEM images along with EDS attachment of 75%VC+25%CuNi-Cr coated SS316 steel.

Moreover, the EDS plot also verifies the presence of oxygen on the coated surface. Therefore, it can be concluded that the specimen surface is coated with desired V, C, Cu, Ni, and Cr elements, along with some oxides that are developed during the travel of powder particles to the coating surface.

12.4 Conclusions

This chapter emphasized on preparing sustainable coated material that can be used as a replacement for hydromachinery equipment. Based on the experimental results and erosion mechanism, the following conclusions were made:

- 75%VC+25%CuNi-Cr coating was magnificently prepared over SS316 steel by the HVOF spraying process. The coating was denser in nature, with tight mechanical interlocking in between the splats.
- Vickers microhardness of 75%VC+25%CuNi-Cr coating was found to be 974 HV9.81N, which is much higher than that of SS316 steel (156 HV9.81N).

- From the results of the cavitation erosion test-coated specimen, it was concluded that the specimen with the highest microhardness corresponded to the minimum cavitation erosion resistance. The conclusion drawn was in accordance with the previous studies [26].
- Resistance to cavitation erosion was found to be much higher (about 52%) for 75%VC+25%CuNi-Cr coated SS316 steel as compared to bare SS316 steel. That clarifies that 75%VC+25%CuNi-Cr coated surface can be used as a sustainable coated surface in place of bare SS316 steel.
- Both the coated and uncoated specimens were found to have different mechanisms of failure. In the case of SS316 steel, a ductile manner of failure was identified, while in the case of 75%VC+25%CuNi-Cr coated SS316, mix manner of failure was identified (i.e., brittle and ductile).

Acknowledgments

Authors thankfully acknowledge the Central Research Facility, SLIET Longowal, for carrying out SEM and EDS analysis and Research Grant under TEQIP-III (Subcomponent 1.3.2.5), SLIET, Longowal, India for carrying out this R&D work on "Design and Development of Cavitation Test Rig to Analyze the Cavitation Erosion of Hydro-Machinery Steels under Surface Modifications" vide reference number SLIET/Dean (R&C)/2019/693, dated 26/03/2019.

References

1) Hong, S., Wu, Y., Wu, J., Zhang, Y., Zheng, Y., Li, J., & Lin, J. (2021). Microstructure and cavitation erosion behavior of HVOF sprayed ceramic-metal composite coatings for application in hydro-turbines. *Renewable Energy*, 164, 1089–1099.
2) Hong, S., Wu, Y., Wu, J., Zheng, Y., Zhang, Y., Cheng, J., . . . & Lin, J. (2020). Effect of flow velocity on cavitation erosion behavior of HVOF sprayed WC-10Ni and WC-20Cr3C2–7Ni coatings. *International Journal of Refractory Metals and Hard Materials*, 92, 105330.
3) Zhang, H., Chen, X., Gong, Y., Tian, Y., McDonald, A., & Li, H. (2020). In-situ SEM observations of ultrasonic cavitation erosion behavior of HVOF-sprayed coatings. *Ultrasonics Sonochemistry*, 60, 104760.
4) Bansal, A., Singh, J., Singh, H., & Goyal, D. K. (2021). Influence of thickness of hydrophobic polytetrafluoroethylene (PTFE) coatings on cavitation erosion of hydro-machinery steel SS410. *Wear*, 477, 203886.
5) Karaoglanli, A. C., Oge, M., Doleker, K. M., & Hotamis, M. (2017). Comparison of tribological properties of HVOF sprayed coatings with different composition. *Surface and Coatings Technology*, 318, 299–308.
6) Bansal, A., Goyal, D. K., Singh, P., Singla, A. K., Gupta, M. K., Bala, N., . . . & Setia, G. (2020). Erosive wear behaviour of HVOF-sprayed Ni-20Cr2O3 coating on pipeline materials. *International Journal of Refractory Metals and Hard Materials*, 92, 105332.
7) Wang, J., Chen, H., Qin, L., Li, Y., & Chen, D. (2008). Key roles of micro-particles in water on occurrence of cavitation-erosion of hydro-machinery. *Chinese Science Bulletin*, 53(10), 1603–1607.
8) Petkovšek, M., & Dular, M. (2013). Simultaneous observation of cavitation structures and cavitation erosion. *Wear*, 300(1–2), 55–64.
9) Santa, J. F., Blanco, J. A., Giraldo, J. E., & Toro, A. (2011). Cavitation erosion of martensitic and austenitic stainless-steel welded coatings. *Wear*, 271, 1445–1453.

10) Singh, S., Kumar, P., Goyal, D. K., & Bansal, A. (2021). Erosion behavior of laser cladded Colmonoy-6+ 50% WC on SS410 steel under accelerated slurry erosion testing. *International Journal of Refractory Metals and Hard Materials*, 98, 105573.
11) Goyal, D. K., Singh, H., Kumar, H., & Sahni, V. (2012). Slurry erosion behaviour of HVOF sprayed WC–10Co–4Cr and Al_2O_3 + 13TiO_2 coatings on a turbine steel. *Wear*, 289, 46–57.
12) Krelling, A. P. (2018). HVOF-sprayed coating over AISI 4140 steel for hard chromium replacement. *Materials Research*, 21(4), e20180138.
13) Zhang, H., Chen, X., Gong, Y., Tian, Y., McDonald, A., & Li, H. (2020). In-situ SEM observations of ultrasonic cavitation erosion behavior of HVOF-sprayed coatings. *Ultrasonics Sonochemistry*, 60, 104760.
14) Pandey, S., Bansal, A., Omer, A., Singla, A. K., Goyal, D. K., Singh, J., & Gupta, M. K. (2021). Effect of fuel pressure, feed rate, and spray distance on cavitation erosion of Rodojet sprayed Al_2O_3+50%TiO_2 coated AISI410 steel. *Surface and Coatings Technology*, 410, 126961.
15) Bansal, A., Singla, J., Pandey, S., & Raj, P. (2020). Design and development of high-velocity submerged water jet cavitation erosion test rig. In *Manufacturing Engineering* (pp. 85–93). Springer, Singapore.
16) Stachowiak, G. B., & Stachowiak, G. W. (2010). Tribological characteristics of WC-based claddings using a ball-cratering method. *International Journal of Refractory Metals and Hard Materials*, 28(1), 95–105.
17) Murthy, J. K. N., & Venkataraman, B. (2006). Abrasive wear behaviour of WC–CoCr and Cr3C2–20 (NiCr) deposited by HVOF and detonation spray processes. *Surface and Coatings Technology*, 200(8), 2642–2652.
18) Barletta, M., Bolelli, G., Bonferroni, B., & Lusvarghi, L. (2010). Wear and corrosion behavior of HVOF- sprayed WC-CoCr coatings on Al alloys. *Journal of Thermal Spray Technology*, 19(1), 358–367.
19) Zhang, Z., Yu, T., & Kovacevic, R. (2017). Erosion and corrosion resistance of laser cladded AISI 420 stainless steel reinforced with VC. *Applied Surface Science*, 410, 225–240.
20) Herman, H., Sampath, S., & McCune, R. (2000). Thermal spray: Current status and future trends. *MRS Bulletin*, 25(7), 17–25.
21) Brandt, O. C. (1995). Mechanical properties of HVOF coatings. *Journal of Thermal Spray Technology*, 4(2), 147–152.
22) Dongmo, E., Wenzelburger, M., & Gadow, R. (2008). Analysis and optimization of the HVOF process by combined experimental and numerical approaches. *Surface and Coatings Technology*, 202(18), 4470–4478.
23) Bansal, A., Singh, J., & Singh, H. (2019). Slurry erosion behavior of HVOF-sprayed WC-10Co-4Cr coated SS 316 steel with and without PTFE modification. *Journal of Thermal Spray Technology*, 28(7), 1448–1465.
24) Santa, J. F., Blanco, J. A., Giraldo, J. E., & Toro, A. (2011). Cavitation erosion of martensitic and austenitic stainless steel welded coatings. *Wear*, 271(9–10), 1445–1453.
25) Bansal, A., Singh, J., & Singh, H. (2020). Erosion behavior of hydrophobic polytetrafluoroethylene (PTFE) coatings with different thicknesses. *Wear*, 456, 203340.
26) Singh, V., Singh, I., Bansal, A., Omer, A., Singla, A. K., & Goyal, D. K. (2022). Cavitation erosion behavior of high velocity oxy fuel (HVOF) sprayed (VC+ CuNi-Cr) based novel coatings on SS316 steel. *Surface and Coatings Technology*, 128052.
27) Cheng, F., Jiang, S., & Liang, J. (2013). Cavitation erosion resistance of microarc oxidation coating on aluminium alloy. *Applied Surface Science*, 280, 287–296.
28) Qiu, N., Wang, L., Wu, S., & Likhachev, D. S. (2015). Research on cavitation erosion and wear resistance performance of coatings. *Engineering Failure Analysis*, 55, 208–223.
29) Hutli, E., Nedeljkovic, M. S., Bonyár, A., & Légrády, D. (2017). Experimental study on the influence of geometrical parameters on the cavitation erosion characteristics of high speed submerged jets. *Experimental Thermal and Fluid Science*, 80, 281–292.

Chapter 13

Effect of HVOF sprayed TiC+25%CuNi-Cr coatings on sustainability and cavitation erosion resistance of SS316 steel

Rampal, Anuj Bansal, Anil Kumar Singla, Deepak Kumar Goyal, Jonny Singla, and Vikrant Singh

Contents

Abbreviations

ASTM: American Society for Testing and Materials
CER: Cavitation Erosion Resistance
Cr: Chromium
CuNi: Copper Nickel/Cupronickel
HV: Vickers Hardness
HVAF: High Velocity Air Fuel
HVOF: High Velocity Oxy Fuel
SCHF: Standard Cubic Feet per Hour
SEM: Scanning Electron Microscopy
SLPM: Standard Liter per Minute
TiC: Titanium Carbide
SOD: Stand-Off Distance

DOI: 10.1201/9781003291961-15

13.1 Introduction

The utilization of hydromachinery components is increasing day by day in almost all countries as they take part in the conversion of water energy into some other sort of energy such as electrical and mechanical. The shape and size of these water-handling components, such as turbine water buckets, impellers, and guide vanes, are a matter of great concern, especially in countries where hydroelectricity is the prominent source of energy [1, 2].

These components confront the problem of expunging of material from the surface through a well-known phenomenon called cavitation [2, 3]. In this phenomenon, the surface of the component is worn out by pressure waves induced by the bursting of bubbles over it. Bubbles are generated when the flow of the water is restricted, and its pressure is suddenly alleviated even lower than its vapor pressure. These bubbles are then carried to a high-pressure area by the flow of water, where they collapse [4–6]. High-frequency bursting of bubbles induces high-intensity shock waves, which further results in the generation of fatigue or cyclic stresses in the component and thus cause erosion of the surface [7]. This erosion occurs in the form of pits over the surface, which further hinders the flow of water and is followed by a lower efficiency of hydromachinery components [8].

To overcome the problem of cavitation erosion, one of the surface engineering techniques, called "coatings", have been explored in this work. Coatings have proved to be one of the best alternatives for the enhancement of surface properties [9]. Researchers are continuously working on the materials that can be coated on substrates by employing various techniques of coatings. Chiu et al. [10] coated NiTi powder over SS316 substrate using the laser technique of coating and concluded that the CE resistance of the specimens was increased by 29 times when compared to that of uncoated SS316 steel. Cheng et al. deployed NiTi coatings on SS316 steel using tungsten inert gas (TIG) surfacing process with an ambition to increase the resistance to cavitation erosion. The authors found that the cavitation erosion resistance of the above-mentioned coatings increased nine times that of SS316 steel specimens without coatings. It was attributed to the ability of the coatings to persist the super elasticity and retain high hardness during the application. Therefore, it can be deduced that coatings can provide desirable properties over the surface and can make the overall components more sustainable.

In this course of work, the thermal spray process was employed to deposit the required materials over the substrate in order to enhance the sustainability of the parent component by improving its surface. In thermal spray deposition processes, materials in melted or partially melted form are sprayed over the substrate to produce the coating using both thermal and kinetic energies of the particles to be coated. The researchers have employed a number of thermal spray processes so far to produce surface protective coatings.

Babu et al. produced WC-10Co-4Cr and Ni coatings on the surface of SS316 steel by employing the detonation gun thermal spray process [11]. The authors further processed the prepared specimens using microwave irradiations and probed the behavior of prepared specimens against slurry and cavitation erosion. Post-processed specimens were found to be nine times more resistant to cavitation erosion than the specimens without post-processing. Wang et al. produced CoMoCrSi coatings on steel substrate using atmospheric plasma spraying (APS) process and subsequent heating of the specimens [12]. The authors conducted ultrasonic tests to investigate the cavitation erosion behavior of prepared coatings.

In the results, it was found that heat treatment significantly reduced the depth of erosion in deposited coatings. Moreover, weak adhesion of splashes produced during the plasma spray process and delamination of fragments resulting from the bursting of bubbles were found to be the two major reasons for the damage over deposited coatings. Cheng et al. deposited FeBSiCrNbMnY alloy coatings on stainless steel substrate using the wire arc spraying process [13]. The authors analyzed the microstructure and mechanical properties of the developed coatings. The above-mentioned coatings recorded high microhardness and excellent wear resistance. Zeng et al. sprayed 316L + stainless steel coatings on carbon steel substrate using high-velocity air fuel (HVAF) process and studied the effect of powder feed rate and particle size on the microstructure and oxide content of the coatings [14]. The authors found a decrease in oxide content in the coatings with an increase in powder feed rate and particle size. The corrosion resistance of the coatings was also evaluated. Amarendra et al. developed 70Ni30Cr coatings on SS410 steel substrate using the HVOF process and further performed tests to investigate the slurry and cavitation erosion behavior of the above-mentioned coatings [15]. Specimens with deposited coatings better resisted the erosion than uncoated specimens.

In the present work, high velocity oxy fuel (HVOF) thermal spray process of coating has been explored. In this process, fuel gas and oxygen, after mixing with each other, are fed to a combustion chamber to ignite. The resultant high-velocity gas stream is allowed to egress through a nozzle at a very high speed, which is further amalgamated with powder to be coated. Fully or partially melted powder, carried by a high-velocity gas stream, is then struck to the substrate and thus gets deposited [16]. HVOF is selected as the process of coating owing to its benefits over other thermal spray processes, such as comparatively less porosity and dense coatings. These advantages are attributed to the high velocity of coating particles with which they impinge over the substrate. Further, substrate and coating materials should have a high modulus of elasticity which cause high stiffness in the material. More the stiffness, more the resistance for deformation due to shock waves. Materials should be tough in nature and should have moderate hardness [17].

SS316 is selected as the substrate in the present work for having a good proportion of chromium (Cr) and nickel (Ni) in it [18], which provides great stiffness and also protects it from corrosion [19]. The high hardness of titanium carbide (TiC) and adequate amount of stiffness provided by cupronickel (CuNi) make them appropriate to be chosen as coating materials as more the stiffness, more the resistance for shock waves [20], and less the CE.

Therefore, in this work, TiC was deposited along with 25 wt.% of CuNi and Cr as binding materials over stainless steel (SS316), acting as the substrate, using the High Velocity Oxy Fuel (HVOF) thermal spray process of coating.

13.2 Experimental procedure

13.2.1 Materials and method of coating deposition

Fine particles of TiC and CuNi-Cr were deposited on the SS316 steel specimen by employing the HVOF process of coating. With a very high Mohs hardness value (9–9.5) [21], TiC provides great resistance to surface erosion by flowing water bubbles. Moreover, the elastic modulus (E) of about 400 GPa is adequate for TiC to provide high resistance to cavitation erosion as it gives cushioning effect to the material during high-intensity pressure waves.

CuNi and Cr are selected as the binders in the above-mentioned coatings. These binders, having a low melting point, melt earlier than TiC particles and bind the coatings together during the deposition. Stainless steel (SS316) is selected as the substrate as it consists of a great proportion of Cr (16.74%) and Ni (10.55%) in it. Cr provides high resistance to corrosion and Ni adds to the stiffness, which helps the specimens bear pressure waves during cavitation.

Stainless steel was procured in the form of a plate having dimensions 600 mm × 150 mm × 8 mm. It was later cut into pieces (70 mm × 30 mm × 8 mm) so as to act as specimens for HVOF coatings. Table 13.1 shows the chemical composition of procured SS316 steel, which was acquired by conducting its spectroscopy. Table 13.1 further compares the acquired chemical composition of SS316 steel with that as per ASTM standards. The chemical composition of procured SS316 steel was clearly within the range of ASTM standards.

Coating materials were procured in powdered form from Parshwamani Metals, an industrial unit in Mumbai city of Maharashtra in India. Powders were then analyzed through SEM technique in order to cognize their morphology. Figure 13.1 (a, b, and c) represents the images of TiC, CuNi, and Cr powders, respectively, acquired from SEM technology. TiC particles, as shown in Figure 13.1 (a), were found to have irregular shapes with sizes varying from 5 to 45 μm, constituting an average particle size of 35 μm. Cr was also recorded to have jagged powder particles with sizes varying from 60 to 110 μm, resulting in average particle size of 75 μm (with reference to Figure 13.1 (c)). CuNi powder particles (refer to Figure 13.1 (b)) were reported to have non-uniform spherical shapes with their sizes varying from 10–50 μm and an average particle size of 30 μm.

Table 13.1 Composition of SS316 as Per Spectroscopy and Its Comparison With Standard Composition

SS316 steel	*C*	*Cr*	*P*	*Si*	*Mn*	*S*	*Ni*	*Mb*	*Fe*
ASTM standard	<0.08	16–18	<0.045	<1	<2	<0.03	8–12	2–3	Balance
As received	0.0759	17.08	0.043	0.991	1.91	0.024	9.86	2.74	Balance

Figure 13.1 SEM images of (a) titanium carbide (TiC), (b) cupronickel (CuNi), and (c) chromium (Cr) powder particles.

13.2.2 Fabrication of coating

Coatings are deposited by employing HVOF thermal spray process because of their numerous advantages, namely high density of coatings, low porosity, and less chance of metallurgical transformations amid the travel of particles from the nozzle to substrate owing to their high velocities. Specimens were coated with the above-mentioned materials in MECPL (Metalizing Equipment Co. Pvt. Ltd), one of the prestigious organizations known for HVOF coatings and situated in Jodhpur district of Rajasthan in India. Before coatings, specimens were prepared by sand blasting their surfaces using a pressure blaster (PB-9182) machine. Air acts as a carrier gas for sand blasting the specimens to be coated. With 5 bars of pressure, the air is struck over the surface along with Al_2O_3 particles (16 mesh size) contained in it. Average surface roughness of 6–8 μm is thus obtained. Sand blasting the specimens resulted in the formation of asperities over the surface, which further proves beneficial for good mechanical interlocking of coatings and substrate.

Sand-blasted specimens were taken to the HVOF coatings deposition setup. The setup comprised a programmable controller, gas supply controller, powder feed unit, and HVOF torch setup (Figure 13.2). A set of instructions in the form of a program were given to the programmable controller that further sent the controlling signals to the gas supply unit and powder feed unit. A controlled amount of gasses and coating powder were fed to the HVOF torch. Products of combustion partially or fully melted the powder, which was further deposited in the form of coatings.

Prepared specimens were coated by imitating above stated parameters in Table 13.2 during the HVOF process. Oxygen, fuel, and air were provided with flow rates of 260, 60, and 450 SLPM (Standard Liter Per Minute). Nitrogen as carrier gas was provided with a flow rate of 20 SCFH (Standard Cubic Feet per Hour). Particles were sprayed over the specimen surface at a distance of 6 inches with a spraying velocity of 700 m/s (refer to Table 13.2). High-velocity coating particles in molten or semi-molten form were taken to the substrate and were impinged over the surface of the substrate in the form of splats. Asperities obtained from sand blasting the surface were filled first by high-velocity coating particles. There occurred mechanical interlocking between the substrate and the deposited splats. A continuous spray of coating particles makes the molten or semi-molten splats deposit over one another once the asperities are filled. This deposition, up to a certain extent, comprises the thickness of coatings.

13.2.3 Characterization of developed coatings

For investigating the coating morphologies, SEM images of the surface as well as of the cross-section of coatings were taken and analyzed. Corresponding to each magnification of 100×, 500×, and 1500×, five SEM images of the surface and the cross-section were taken. Various attributes of coatings, namely thickness and microhardness were evaluated by referring to acquired SEM images. ImageJ software was utilized to evaluate the extent of porosity and thickness of developed coatings with the assistance of their cross-sectional SEM images taken at various magnifications.

Micro hardness of specimens was investigated by employing a standard Vickers micro hardness testing machine (RMHT-201). Loads of 1000 and 500 g were utilized for a dwell time of 30 seconds during the evaluation of micro hardness values of uncoated and coated specimens, respectively. Moreover, the Surtronic device (Taylor Hobson: SE1200) was used for obtaining a surface roughness value of sustainable coated substrates.

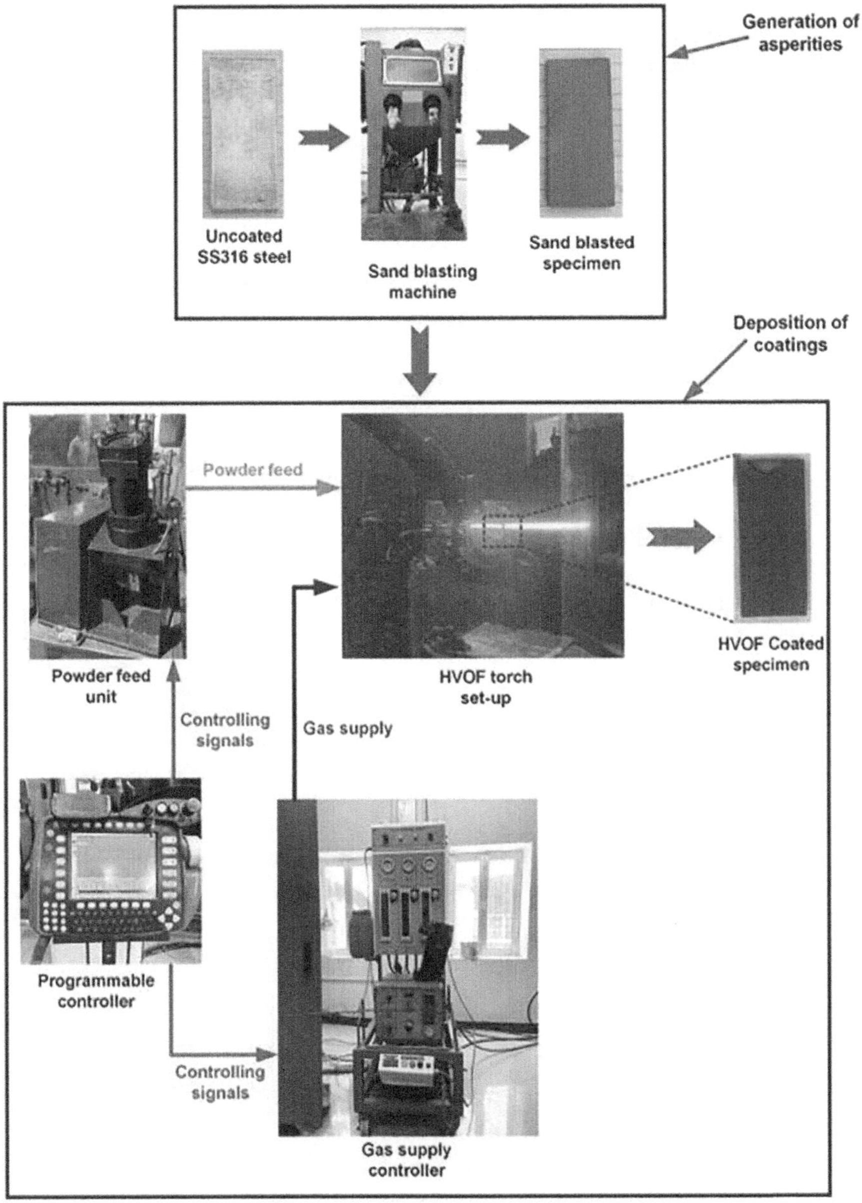

Figure 13.2 Setup for the deposition of high velocity oxy fuel (HVOF) sprayed coatings.

Table 13.2 HVOF Process Parameters Acquired for Coating at MECPL, Jodhpur, India

Parameters	*Oxygen flow rate (slpm)*	*Oxygen pressure (kg/cm²)*	*Fuel (LPG) flow rate (slpm)*	*Fuel pressure (kg/cm²)*	*Air flow rate (slpm)*	*Air pressure (kg/cm²)*	*Spray distance (cm)*	*Powder feed rate (g/min)*
Value	240–260	10–11	50–60	6.0–7.0	650–750	6.0–7.5	16	35

13.2.4 Cavitation erosion test

Coated and uncoated specimens were investigated for their cavitation erosion behavior using a high-velocity cavitation erosion test rig with configurations per ASTM G134 standards. Bansal et al. [22] provided the detailed construction and elucidation of experimental setup in one of their works. The same has been reproduced in image form in Figure 13.3. Water from the tank is taken to the nozzle through the delivery pipe with the help of a 1.5 KW power capacity phase motor. Water discharge through the nozzle was regulated by using two regulatory valves, which were mounted on the delivery pipe and are shown in Figure 13.3. Variation in water discharge through the nozzle using these regulatory valves further controlled the jet velocity as the diameter of the nozzle was fixed at 3 mm throughout the experimentation. Moreover, three pressure gauges were mounted on the nozzle to keep required surveillance on any pressure variations.

Erosion tests were performed by varying parameters, namely the velocity of the jet and stand-off distance. The velocity of the jet prodigiously affects the CE of specimens as more the velocity, more the pressure drop, and higher the bubble generation rate. Therefore, to analyze the influence of jet velocity on surface erosion, three levels of velocity were considered, that is, 20, 30, and 40 m/s. Moreover, stand-off distance affects the number of bubbles approaching the specimen surface. Therefore, it is considered one of the parameters during cavitation erosion tests with three level values of 4, 8, and 12 cm.

Further, by considering two process parameters and that too at three levels, nine different combinations of parametric values were considered using a full factorial design approach.

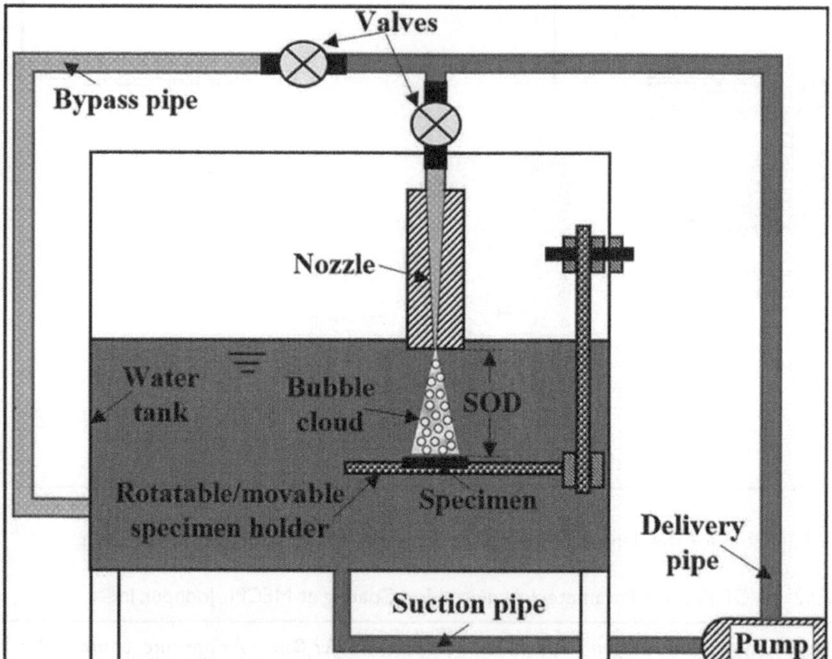

Figure 13.3 Components of construction in cavitation erosion test rig [22].

Cavitation erosion tests were performed on each specimen completely submerged in water for ten hours. Eroded specimens are then cleaned with acetone to eliminate expunged material accumulated over the surface. Then, material loss is evaluated by finding difference in weights of the specimen before and after the cavitation erosion test by using a weighing machine (Shimadzu AUW-120D) with the least count of 0.0001 g. To ensure the repeatability of experimental results, the cavitation erosion test for the same parametric combination was performed thrice, and the mass loss was observed.

13.3 Results and discussions

13.3.1 Mechanical and morphological studies of deposited coatings

Figure 13.4 shows the cross-sectional view of coatings deposited over substrate having mechanical interlocking in between them. A magnified view of the substrate-coating interface is also given adjacent to the figure. The distribution of coatings in asperities of the substrate's surface can also be seen here. Good mechanical interlocking can be observed in deposited coatings by virtue of less porosity at the interface. The reason is quite obvious that binders (CuNi-Cr) having low melting points melt before TiC particles and hence bind them together during the deposition of coatings. It further results in the complete filling of asperities with strong interlocking in between them. Good mechanical interlocking at the interface results in high bond strength of coatings. Moreover, some non-melted TiC particles can also be located in the figure. The comparatively high melting point of TiC powder particles than that of CuNi-Cr particles might be the reason for this. Tables 13.3 and 13.4 represent

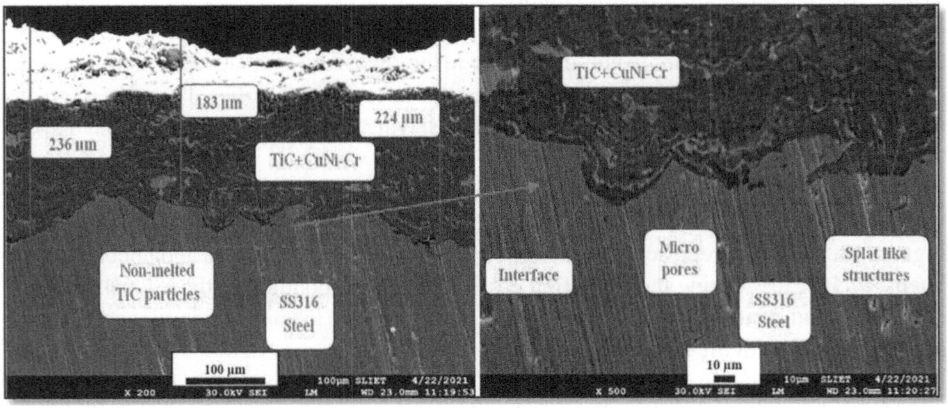

Figure 13.4 Cross-sectional SEM images of titanium carbide + 25%cupronickel-chromium coatings.

Table 13.3 Porosity Percentage and Average Thickness of Coatings Corresponding to Deposited Coatings

Sr. No.	*Composition*	*Porosity (%)*	*Thickness (μm)*
1.	(75%TiC) + (25% CuNi-Cr)	1.17	214.33

Table 13.4 Microhardness and Average Surface Roughness of Coated and Uncoated Specimens

Attributes	*Specimens*	
	SS316	*TiC + 25%CuNi-Cr (coated)*
Microhardness (HV)	155 ± 5	1015 ± 3
Average surface roughness (μm)	1.690 ± 0.17	3.792 ± 0.22

various attributes of coated and uncoated specimens, along with their recorded values. To measure the thickness of deposited coatings, ImageJ software was utilized. Coating thickness values were evaluated at ten different locations, and an average thickness of 214.33 μm was recorded.

With reference to the above-mentioned data in tabular form, it can be apprehended as TiC + 25%CuNi-Cr coatings have partially melted TiC particles in them, which are accompanied by completely melted CuNi-Cr particles. When these coatings are sprayed over the substrate, the non-melted portion of TiC particles, that is, the TiC core, is held firmly in the deposited splat due to the presence of molten binders in it. Therefore, an upcoming splat is formed over the previously deposited splat with partially melted TiC particles, which further results in the creation of pores above said coatings. These micropores at the interface can easily be located in Figure 13.4. Further, coating thickness is attributed to the number of pores in it. It may be apprehended as the more the percentage porosity in coatings, the higher the value of average coating thickness.

Moreover, porosity in deposited coatings also contributes to some value of roughness over the surface. TiC + 25%CuNi-Cr coatings are found to have an average surface roughness of 3.792 μm corresponding to a porosity percentage of 1.17 (refer to Tables 13.3 and 13.4). It can be apprehended as porosity leads to non-uniformity in the deposition of splats. This non-uniformity continues up to the top layer of coatings, resulting in surface irregularities. Therefore, a greater number of pores will end up with a coating surface having a high value of surface roughness.

13.3.2 Cavitation erosion analysis

Figures 13.5 and 13.6 represent the cavitation erosion behavior of coated and uncoated specimens. Weight loss during the cavitation erosion test corresponding to three level values of jet velocity and stand-off distance is shown in Figure 13.4 on keeping the impingement angle of the jet at 90° with the surface.

It can be clearly seen that the percentage loss in weight is excessively high in uncoated SS316 steel specimens under all considered parameters. The viable reason behind this behavior of SS316 steel is its poor mechanical properties with respect to CE. Being too soft and ductile, SS316 specimens undergo plastic deformations when battered by induced shock waves. These plastic deformations also expunge the surface material at the highest possible rates and cause maximum deterioration.

Mass loss due to cavitation is found to be very less in TiC + (25 wt.%) CuNi-Cr coated specimens. Enhanced surface properties of coated specimens are the obvious reason behind this. TiC in these coatings bestows the surface with a substantial amount of hardness which resists surface erosion due to flowing water bubbles and high-velocity jet. Further, CuNi, a

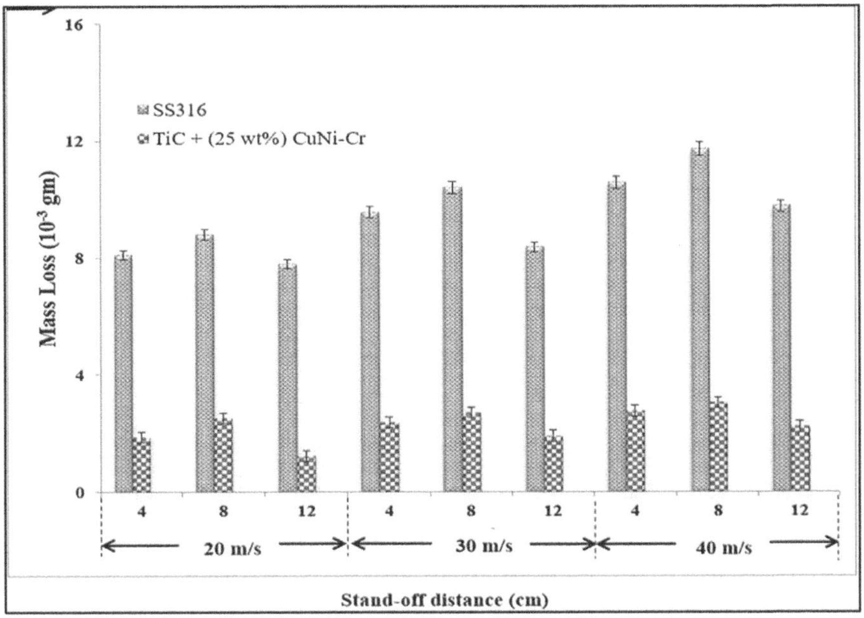

Figure 13.5 Mass loss in coated and uncoated specimens corresponding to all three levels of jet velocity and stand-off distance.

soft alloy, adds up the elastic modulus attribute in the overall coated surface, increasing its stiffness. More the stiffness, more the resistance for shock waves/cyclic stresses. Cr, in addition, enhances the corrosion resistance of the specimen's surface.

Moreover, it is quite conspicuous from the figure that an increase in jet velocity increases the mass loss of specimens for all values of stand-off distance. Maximum mass loss is observed corresponding to the highest considered jet velocity of 40 m/s. The reason behind this might be that with the increase in jet velocity, the pressure loss in flowing fluid is also increased. Higher the velocity of the jet, more the hindrance in the flow of water, and therefore, higher will be the pressure loss. It proliferates the formation of bubbles which later undermine the surface.

Further, mass loss is found to be maximum corresponding to 8 cm of stand-off distance and minimum corresponding to 12 cm of stand-off distance for all values of jet velocity. The influence of stand-off distance on mass loss can be apprehended as at low stand-off distance value (say 4 cm), generated bubbles could not persist themselves on the surface of specimens due to inertia and hence, cause less erosion due to cavitation. Moreover, at moderate stand-off distance (say 8 cm), generated bubbles find enough time to stabilize and target the surface of specimens. Surface, after getting approached, is undermined by bursting of water bubbles. Therefore, the highest mass loss is observed at 8 cm of stand-off distance.

Further increase in stand-off distance (say 12 cm), as shown in Figure 13.5, ends up with the least material removal during cavitation erosion tests. Possible reason could be that the increase in travel distance beyond 8 cm lessens the number of bubbles reaching the target surface per unit time. Most of the generated bubbles burst amid their travel far from target surface. Shock waves thus produced have less impact over the surface of specimens causing

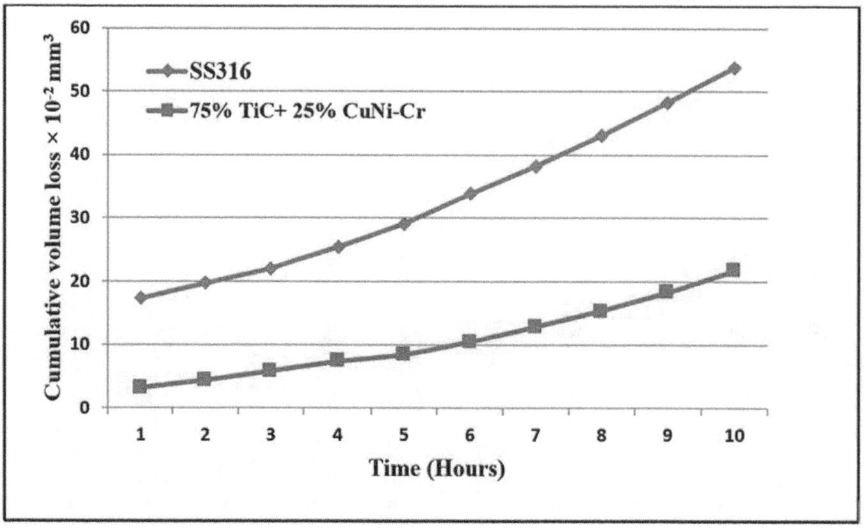

Figure 13.6 Cumulative volume loss of specimens at critical process parameters during cavitation erosion tests.

less cavitation erosion and, therefore, improves the sustainability of the component. Further, it can be deduced that the highest value of jet velocity (40 m/s in the present work) and moderate value of stand-off distance (8 cm in this case) among all three considered levels cause maximum undermining of the surfaces and is considered as a critical combination of parameters for the sustainability of component under consideration.

Furthermore, Figure 13.6 represents a cumulative volumetric loss of material in specimens during CER tests. With reference to prior discussions, it can be concluded that material loss is very high in SS316 steel specimens when compared to that in coated specimens. Moreover, the rate of material removal also increases with the passage of time during CE tests. There occurs an increase in surface area in the form of generating pits. Therefore, the upcoming bubbles have more surface to erode and hence cause an increased rate of material removal. The same is elucidated in the graph of Figure 13.6.

13.3.3 Cavitation erosion analysis of steel and TiC+25%CuNi-Cr coated surface

SEM images help decipher the pattern of surface damages and the possible reasons behind them. Therefore, an analysis of undermined surface is discussed here in order to apprehend the behavior of cavitation erosion.

13.3.3.1 SS316 steel specimen

Cavitation erosion over the surface of substrate SS316 under different values of considered factors is shown in Figure 13.7 (a) and (b). With reference to Figure 13.7 (a), micropores and pits can be seen on the surface, along with some large overlapped cavities

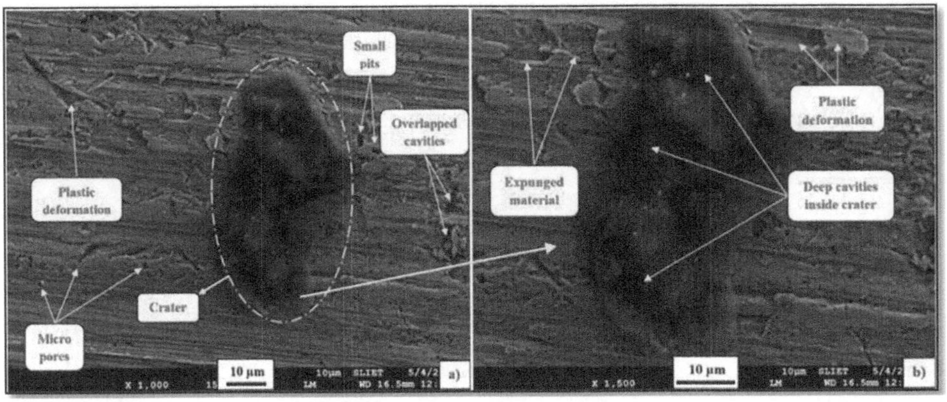

Figure 13.7 SEM images of SS316 steel specimens after cavitation erosion tests conducted for 10 hours: (a) eroded surface under critical process parameters and (b) magnified image of a crater.

corresponding to critical process parameters. The evidence of plastic deformation of soft SS316 steel specimens is also depicted in Figure 13.7 (a) and (b). A large-sized crater is found over the surface and is elucidated in Figure 13.7 (b). Being too soft, and having the least micro hardness of 155 HV, SS316 specimens prominently erode by the mechanism of plastic deformation. Bursting of bubbles impose high-intensity pressure impact over confined area of specimen, which, in turn, pushes the plastically deformed material out of the impacted zone and cause weight loss. Micro pores and small pits, as shown in Figure 13.6 (a), must be the result of collapsing of small-sized water bubbles close to the surface. The magnified SEM image of the enclosed surface is shown in Figure 13.7 (b), which explains the developed crater in detail.

13.3.3.2 TiC+25%CuNi-Cr coated surface

As discussed earlier, coated specimens are found to have very less weight loss in CE experiments when compared with that of uncoated SS316 steel specimens. Desirable mechanical properties of above-mentioned coatings are said to be the reasons for that. Figure 13.8 (a) and (c) show the eroded surfaces of two SS316 steel specimens, which are coated with TiC + 25%CuNi-Cr coatings and are tested for above said critical combination of parameters. Enclosed areas in Figure 13.8 (a) and (c) are depicted as high-magnification SEM images in Figure 13.8 (b) and (d). With reference to above-said figures, number of observations can be made. Cracks can be depicted at carbide-binders interface. Expunged material is also located on the surface (with reference to Figure 13.8 (a)), which might have accumulated as a recast during CER tests. Plastic flow of binders over the surface can also be seen in Figure 13. 8 (b) and (c). Normal impingement of bubbles (90°) and their bursting close to the surface create deep cavities on it, which are enclosed in Figure 13.8 (a) and (c) and are elucidated in details in adjacent figures.

Plastic flow of the material can easily be observed close to the crater. It verifies the existence of one of the mechanisms of material removal, which is plastic deformation here.

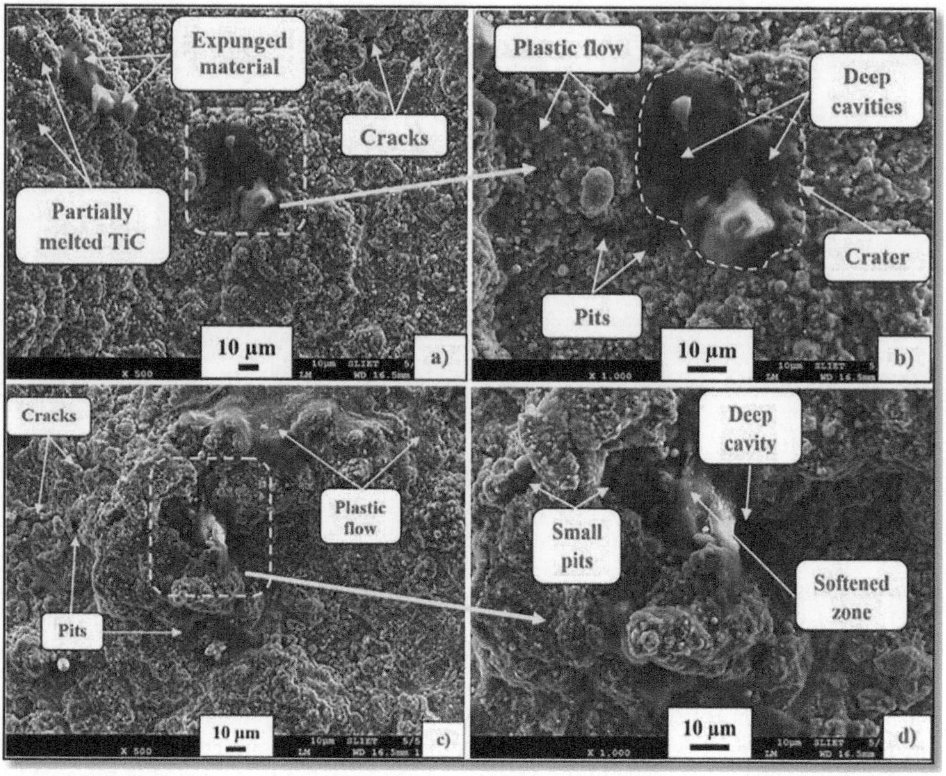

Figure 13.8 SEM images of TiC + 25%CuNi-Cr coated two SS316 steel specimens after cavitation erosion tests conducted for 10 hours of duration: (a and c) eroded surface under critical process parameters, and (b and d) magnified image of an enclosed surface.

High-frequency pressure waves generated due to bursting of bubbles soften the surface and develop a recast which can be located as softened zone in Figure 13.7 (d). Small cracks are also located on the TiC particle interface with deposited splats. Possible reason is the brittle failure of particular TiC particles due to impact caused by shock waves. Carbide-binders interface is comparatively weak in strength and therefore leads to crack initiation when battered by high-intensity shock waves. Incubated cracks then become stress-concentrated areas. These cracks are propagated with continuous bursting of bubbles. Therefore, mix mode of failure, that is, ductile and brittle can be observed in above-coated specimens. The same can be observed in above-mentioned SEM images of two specimens having the same coatings.

13.4 Conclusions

SS316 steel specimens were coated with TiC + 25%CuNi-Cr by employing the HVOF process of coating with an ambition to increase the resistance of steel specimens against cavitation erosion and to get more sustainable hydromachinery components. Cavitation erosion tests were conducted on coated and uncoated specimens for nine different combinations of parameters and for 10 hours.

- Average coating thicknesses of 214.33 μm and percentage porosity of 1.17 were obtained corresponding to the above-mentioned coatings. Maximum erosion of specimens was observed, corresponding to jet velocity of 40 m/s and stand-off distance of 8 cm. Formation of large bubbles cloud corresponding to above-stated parameters is the possible reason behind this.
- Coated specimens were reported to have less cavitation erosion. Adequate microhardness and stiffness bestowed by above-mentioned coatings might have provided the surface with maximum resistance to cavitation erosion.
- Small pits, micropores, and deep overlapped cavities were observed over the surfaces after cavitation erosion tests. Sites were also located where material has segregated from the surface.
- Mix mode of failure, that is, plastic deformation and brittle fracture was observed in case of TiC+25%CuNi-Cr coated specimens. Stress concentration at carbide-binders interface caused the formation of cracks. Further shock waves propagated the cracks along grain boundaries and caused the separation of TiC particles.

References

[1] R. G. Ugyen Dorji, "Hydro turbine failure mechanisms: An overview," *Engineering Failure Analysis*, vol. 44, pp. 136–147, September 2014.

[2] R. Pardeep Kumar, "Study of cavitation in hydro turbines—A review," *Renewable and Sustainable Energy Reviews*, vol. 14, pp. 374–383, January 2010.

[3] G. S. H. J. Amarendra, "Synergy of cavitation and slurry erosion in the slurry pot tester," *Wear*, vol. 290–291, pp. 25–31, 30 June 2012.

[4] C. A. Mahieux, "Environmental degradation of industrial composites," *Science Direct*, pp. 85–136, 2006.

[5] G. Challier, J.-L. Reboud, A. Archer, F. Patella, "Energy balance in cavitation erosion: From bubble collapse to indentation of material surface," *Journal of Fluids Engineering*, vol. 135, no. 1, p. 011303 (11 pages), 2013.

[6] P. P. Gohil, R. P. Saini, "Effect of temperature, suction head and flow velocity on cavitation in a Francis turbine of small hydro power plant," *Energy*, vol. 93, pp. 613–624, December 2015.

[7] N. Ochiai, Y. Iga, M. Nohmi, T. Ikohagi, "Study of quantitative numerical prediction of cavitation erosion in cavitating flow," *Journal of Fluids Engineering*, vol. 135, no. 1, p. 011302, 2012.

[8] I. Tzanakis, L. Bolzoni, D. G. Eskin, M. Hadfield, "Evaluation of evaluation of cavitation erosion behavior of commercial steel grades used in the design of fluid machinery," *Metall Mater Trans*, pp. 2193–2206, 2017.

[9] T. Lampke, D. Dietrich, A. Leopold, G. Alisch, B. Wielage, "Cavitation erosion of electroplated nickel composite coatings," *Surface and Coatings Technology*, vol. 202, pp. 3967–3974, 15 May 2008.

[10] K. Y. Chiua, F. T. Chenga, H. C. Man, "Cavitation erosion resistance of AISI 316L stainless steel laser surface-modified with NiTi," *Materials Science and Engineering: A*, vol. 392, pp. 348–358, 15 February 2005.

[11] A. Babu, H. S. Arora, H. Singh Grewal, "Microwave-assisted post-processing of detonation gun-sprayed coatings for better slurry and cavitation erosion resistance," *Journal of Thermal Spray Technology*, vol. 28, pp. 1565–1578, 23 September 2019.

[12] Y. Wanga, J. Liua, N. Kanga, G. Daruta, T. Poirierb, J. Stellac, H. Liaoa, M-P. Planchea, "Cavitation erosion of plasma-sprayed CoMoCrSi coatings," *Tribology International*, vol. 102, pp. 429–435, October 2016.

[13] J-b. Cheng, X-b. Liang, B-s. Xu, Y-x. Wu, "Formation and properties of Fe-based amorphous/nanocrystalline alloy coating prepared by wire arc spraying process," *Journal of Non-Crystalline Solids*, vol. 355, pp. 1673–1678, 2009.

[14] Z. Zenga, N. Sakoda, T. Tajiri, S. Kuroda, "Structure and corrosion behavior of 316L stainless steel coatings formed by HVAF spraying with and without sealing," *Surface & Coatings Technology*, vol. 203, pp. 284–290, 2008.

[15] H. J. Amarendra, M. S. Prathap, S. Karthik, B. M. Darshan, Devaraj, P. C. Girish, V. T. Runa, "Combined slurry and cavitation erosion resistance of HVOF thermal spray coated stainless steel," *Materials Today: Proceedings*, vol. 4, pp. 465–470, 2017.

[16] P. D. S. El-Eskandarany, "Utilization of ball-milled powders for surface protective coating," *Mechanical Alloying*, vol. 3, pp. 309–334, 2020.

[17] A. Krella, "Resistance of PVD coatings to erosive and wear," *Coatings*, vol. 14, pp. 80–231, 2020.

[18] U. Savitha, G. Jagan Reddy, A. Venkataramana, A. Sambasiva Rao, A. A. Gokhale, M. Sundararaman, "Chemical analysis, structure and mechanical properties of discrete and compositionally graded SS316–IN625 dual materials," *Materials Science and Engineering: A*, vol. 647, pp. 344–352, 2015.

[19] S. Zhu, L. Chen, Y. Wu, Y. Hu, T. Liu, K. Tang, Q. Wei, "Microstructure and corrosion resistance of Cr/Cr2N multilayer film deposited on the surface of depleted uranium," *Corrosion Science*, vol. 82, pp. 420–425, May 2014.

[20] M. Zmindak, Z. Pelagić, "Modeling of shock wave resistance in composite solids," *Procedia Engineering*, vol. 96, pp. 517–526, 2014.

[21] K. Cheng, F. Yang, K. Ye, Y. Zhang, X. Jiang, J. Yin, G. Wang, D. Cao, "Highly porous Fe_3O_4-Fe nanowires grown on C/TiC nanofiber arrays," *Journal of Power Sources*, pp. 260–265, 2014.

[22] S. Pandey, A. Bansal, A. Omer, A. K. Singla, D. K. Goyal, J. Singh, M. K. Gupta, "Effect of fuel pressure, feed rate, and spray distance on cavitation erosion of Rodojet sprayed Al_2O_3+50%TiO_2 coated AISI410 steel," *Surface and Coatings Technology*, vol. 410, p. 126961, 2021.

Chapter 14

Acid- and alkali-treated hierarchical silicoaluminophosphate molecular sieves for sustainable industrial applications

Rohit Prajapati, Divya Jadav, Parikshit Paredi, Rajib Bandyopadhyay, and Mahuya Bandyopadhyay

Contents

Abbreviations

SAPO-5-A1: HNO_3-treated SAPO-5
SAPO-5-A2: HCl-treated SAPO-5
SAPO-34-A1: HNO_3-treated SAPO-34
SAPO-34-A2: HCl-treated SAPO-34
SAPO-5-B1: 5 hours NaOH + TPAOH-treated SAPO-5
SAPO-5-B2: 8 hours NaOH + TPAOH-treated SAPO-5
SAPO-34-B1: 5 hours NaOH + TPAOH-treated SAPO-34
SAPO-34-B2: 8 hours NaOH + TPAOH-treated SAPO-34

DOI: 10.1201/9781003291961-16

14.1 Introduction

Microporous crystalline silicoaluminophosphate materials have been trademarked by Lok et al., which are abbreviated as SAPO-n, where n denotes the structural class [1]. To circumvent the pore size constraint of zeolites, mesoporosity generation in microporous (250 nm) and/or macropore (>50 nm) has raised the interest of researchers [2]. Comprehensive research on the functionality of SAPO materials in numerous heterogeneous catalysis has been undertaken over the past decades. These types of materials are particularly promising due to their excellent selectivity towards required outcomes and their prolonged catalytic longevity [3–5]. Various attempts have been made, including the synthesis of nano-sized zeolites [6], zeolitically-ordered mesoporous materials [7], zeolites with secondary porosity [8], desilication [9], nanozeolies created in limited space [10], and delaminated zeolites [11]. As a consequence, material adaptation to improve productivity and selectivity is still a challenging task in the modern catalytic investigation. As a matter of fact, overcoming this inherent problem is critical for improving catalytic activity. Up to this point, two key techniques have been identified as viable options. Both top-down [12, 13] and bottom-up [14, 15] methods were used to produce a hierarchical micro-porous CHA framework with extra mesoporosity. To remove framework elements while retaining crystal structure, top-down approaches often entail post-synthesis treatments such as demetallation under acid or basic conditions [16]. Zeolitic mesoporous silicoaluminophosphate materials with CHA type structure, showed exciting results as heterogeneous (acid) catalysts [17] and have been employed in the production of phenoxy propanol. Xiaoxin Chen et al. reported a top-down strategy for hierarchical CHA zeolites with secondary pores produced via etching with HF-NH_4F solutions to achieve better olefin selectivity [18]. In one recent study, microporous SAPO was used as a precursor to creating hierarchically SAPO materials, which were found to be very active for vapour phase isomerization of 1-octene [19].

In light of global industrial acceleration, coupled with decreasing global petroleum supplies and intensified global warming, the demand for alternative energy options has surged [31]. Alternative fuels for internal combustion engines have received a lot of interest, and transesterification reaction has created a lot of attention in recent years to synthesize alkyl esters with shorter chain alcohols for biodiesel fuels and value-added products [20–22]. Because their composition is similar to that of fossil fuels, triglycerides that are made using plant and animal fats can be envisaged as a potential resource for biofuel production [23, 24]. Biodiesel is a sustainable, nearly infinite alternative to fossil energy that reduces CO_2 emissions. It is a triglyceride-based fuel that could eventually replace petroleum. SAPO-34 is a great option as a catalyst for esterification reactions, producing biofuel because of its acidic properties and porous and flexible structure with a high specific surface area [25, 26]. Silicalite-1 impregnated with transition metals exhibited higher activity in organic transformations such as oxidation, epoxide conversion, oxidation of C-H bonds, and ketone ammoximation [27, 28]. Elyssa G. Fawaz et al. used hierarchical crystals with short diffusion paths, traditional microcrystals, and nanostructures of ZSM-5 zeolites to create biofuel from discarded oils [29]. Mesopores to decrease steric resistance, increased acid intensity and density, and thermal and hydrothermal robustness to reduce toxicity and leaching in the transesterification process are some of the key characteristics required for an effective heterogeneous catalyst for this purpose [30]. Typically, post-synthetic alkali modification is documented over zeolites, such as H-Y, ZSM-5, ZSM-48, and H-Beta. The enhanced catalytic behavior of these materials was observed due to hierarchical structures resulting from alkali treatment [37–39]. However, the post-synthetic modification over microporous material via acid/alkali treatments is not much explored, specifically for the catalytic application in transesterification of triacetin and epoxide ring opening reactions.

Lewis acidic materials are powerful active substances able to perform a wide range of essential activities in precise chemical synthesis, such as the opening of epoxide rings with alcohols [32]. Functionalized molecules are frequently used in ring-opening reactions of cyclic substrates. Epoxide ring opening is a key organic reaction that produces a variety of bifunctional products depending on the nucleophiles used, and numerous synthetic techniques have been documented in this area [33]. Many studies have been published on modified zeolites using various alkali treatments and subsequent catalytic applications of these materials. Desilication in the framework structure resulting from alkali treatment results in the development of mesoporosity [34–36]. In addition, employing acid and basic treatments, numerous ways to synthesize extremely efficient hierarchical zeolites have been reported [40–42]. In order to retain the structural intactness, especially for post-synthetic modification of SAPO materials, researchers have mostly focused on the removal of precursor and surface modification by incorporation of organic/inorganic functionalities. One study on hierarchical SAPO-11 was reported by Verboekend et al. in which various acid-base treatments were performed, and structure-property-activity relation was thoroughly investigated. They concluded the significant role of hierarchical SAPO material as a catalyst for toluene alkylation with benzyl alcohol [43]. Another report is available on acid-treated SAPO-5, focusing on the thermal stability after modification [44]. The conventional synthetic technique of heating epoxides with a substantial excess amine at high temperatures is less effective with weakly nucleophilic amines. Moreover, conventional strategies of amino alcohol synthesis needed activation of epoxides to boost their sensitivity to nucleophilic attack with amines due to the lack of regioselectivity, necessity for high temperatures, and excess of amine [45–46]. Aminolysis of epoxide is an efficient way to get amino alcohols that are beneficial intermediate towards synthesizing several biologically valuable substances and pesticides [47–49].

Sparked by the excellent activity of hierarchical zeolite materials via post-synthetic acid and alkali treatment, the present study focuses on the synthesis of various acid as well as alkali-treated silicoaluminophosphate materials to enhance their catalytic potential for sustainable, industrially significant reactions, such as "ring opening reaction of epoxide" and "transesterification process." To our knowledge, no detailed investigation of the catalytic usage of microporous materials that have been treated with acid and alkali in the epoxide reaction and transesterification reactions is available. Several physiochemical techniques were used to characterize the produced materials. To obtain significant conversion, both meso/micro materials have been employed in the propylene oxide ring opening reaction with nitrogen containing nucleophile and transesterification reaction of triacetin under optimal conditions by varying the amount of catalyst, process temperature, and time.

14.2 Experiments

14.2.1 Hydrothermal synthesis of SAPO-34

The inorganic precursors aluminum isopropoxide (SDFCL, 97%), H_3PO_4 acid (85 wt.% solution, Merck), and colloidal silica (Ludox 40 wt.%, Sigma-Aldrich) are employed in the synthesis of SAPO-34. TEAOH (tetraethyl ammonium hydroxide) was used as an SDA for the SAPO-34 material, which was formed through a literature-based synthesis procedure [50] with a composition of $0.6SiO_2$:$40H_2O$:$1.0Al_2O_3$:1.0SDA:$1.0P_2O_5$. The as-prepared material was calcined at 550°C.

14.2.2 Hydrothermal synthesis of SAPO-5

SAPO-5 was prepared using similar composition $0.6SiO_2$:$40H_2O$:$1.0Al_2O_3$:1.0SDA:$1.0P_2O_5$ like SAPO-34 but TEA (Triethylamine, Sigma Aldrich, 40%) was used as SDA [50].

14.2.3 Treatment with acid and alkali

14.2.3.1 Acid treatment

Post-synthetic treatment is a demetallizing procedure that often entails chemically deliberately removing a framework cation, resulting in a change in the zeolite framework compositions and acidity. SAPO-34 zeolites are difficult to tailor using typical acid and base leaching because their frameworks are often less stable than those of aluminosilicate zeolites. Acid etching was used to prepare mesoporous materials. Calcined SAPO materials were combined with a diluted nitric acid solution during the process [3]. Pure SAPO materials were stirred at 90°C, for four hours, with an appropriate amount of diluted nitric acidic. Filtration separated the SAPO materials, which were then rinsed with water with caution and dried overnight at 60°C. In another process, diluted hydrochloric acid solution was utilized to alter SAPO-34 and SAPO-5. At 100°C, the materials were stirred for 2 hours with a 0.6 mol/L diluted HCl acid solution. After that, the materials were sonicated for 10 minutes with a 0.05 mol/L diluted HCl solution. The modified material was filtered, rinsed with Milli Q water, dried at ambient temperature, and then dried overnight at 60°C.

14.2.3.2 Alkali treatment

SAPOs were treated with NaOH solution (Finar, 97%) and TPAOH (tetra propyl ammonium hydroxide, Sigma Aldrich) during alkali treatment [53]. A typical procedure involved stirring the SAPO materials with the appropriate volume of NaOH and TPAOH solution at 70°C two separate times. Finally, the product was filtered, rinsed using distilled water, and dried at 100°C. The acid and alkali treatment circumstances are summarized in Table 14.1, where acid treatment is denoted as A and alkali (base) treatment is denoted as B.

Table 14.1 Post-Synthetic Acid-Base Treatment Conditions

Sample	*Treatment*	*Reagent*	*Concentration (M)*	*Time (h)*	*Temperature (°C)*	*Sample name*
SAPO-5	A1	HNO_3	0.5 M	4	90	SAPO-5-A1
SAPO-5	A2	HCl	0.6 M	2	100	SAPO-5-A2
SAPO-34	A1	HNO_3	0.5 M	4	90	SAPO-34-A1
SAPO-34	A2	HCl	0.6 M	2	100	SAPO-34-A2
SAPO-5	B1	NaOH + TPAOH	0.1 M + 1 M	5	70	SAPO-5-B1
SAPO-5	B2	NaOH + TPAOH	0.1 M + 1 M	8	70	SAPO-5-B2
SAPO-34	B1	NaOH + TPAOH	0.1 M + 1 M	5	70	SAPO-34-B1
SAPO-34	B2	NaOH + TPAOH	0.1 M + 1 M	8	70	SAPO-34-B2

14.2.4 Catalytic activity

14.2.4.1 Epoxide ring opening reaction

All catalytic activity tests were conducted in a solvent-free reaction. Without solvent, in a round bottom flask, a propylene oxide-ring opening reaction with morpholine was performed with modified SAPO materials as a catalyst. The appropriate quantity of catalyst was used in the reaction (activated at 80–100°C) with propylene oxide and morpholine at different temperatures for different time periods. After the completion of the reaction, a syringe filter was used to separate the catalyst from the liquid product. The liquid product was analyzed using GC (Thermo Fisher Trace-1310, TG 5 column). Several reactions were performed in an attempt to improve the reaction conditions in order to maximize conversion and β-amino alcohol production.

14.2.4.2 Transesterification reaction

All these catalysts were sorely tested in the transesterification of triacetin using excess methanol. Prior to the procedure, the microporous materials were activated after being modified for 1 hour at 80–100°C. Triacetin (Finar, 99%) and excess of methanol (Rankem, 99.8%) were mixed and agitated for a while to reach the reaction temperature (methanol: triacetin molar ratio of 1:16). After that, an appropriate amount of catalyst has been added for at 68°C for 6 hours. Later following the procedure's completion, the catalyst was filtered entirely out of the reaction mixture. A Shimadzu GC-2025 as well as gas chromatography fitted with a flame ionization detector, were used to determine triacetin conversion and methyl acetate yield for each reaction.

14.3 Characterizations

Powder XRD (Ultima-IV, Rigaku) using Cu K radiation at 20 mA and 40 kV was used to analyze the crystal nature as well as the purity of the produced material. The microporous phase's patterns were produced in the 2θ range of 5–50°, with step size and speed of 0.01° and 1° per minute, respectively. Small angle X-ray diffraction patterns were also performed on modified microporous materials between 2θ range of 0.5° and 10°. For the analysis of the morphology, Zeiss Ultra 55 FE-SEM microscope was used. Sorption analysis was performed on materials using a Micromeritics-3 flex 3500 apparatus. Inductively coupled plasma optical emission spectroscopy (ICP-OES, Thermo iCAP 7000) was used to determine the elemental composition. Thermogravimetric analysis (TGA) was performed using a Hitachi machine. The material was heated at 10°C/min rate between room temperature to 800°C in a dry air flow of 30 ml/min.

14.4 Results and discussion

Figure 14.1 illustrates the XRD studies of SAPO-34 and SAPO-5 zeolites in their as-made, calcined, and acid, as well as alkali-treated forms. The X-ray patterns demonstrated all characteristic SAPOs peaks. No impurities prior or post modification processes were found. Since no impurities or amorphous phases were found, it was concluded that the phase purity of SAPO-34 and SAPO-5 crystals had been successfully maintained after post-synthetic

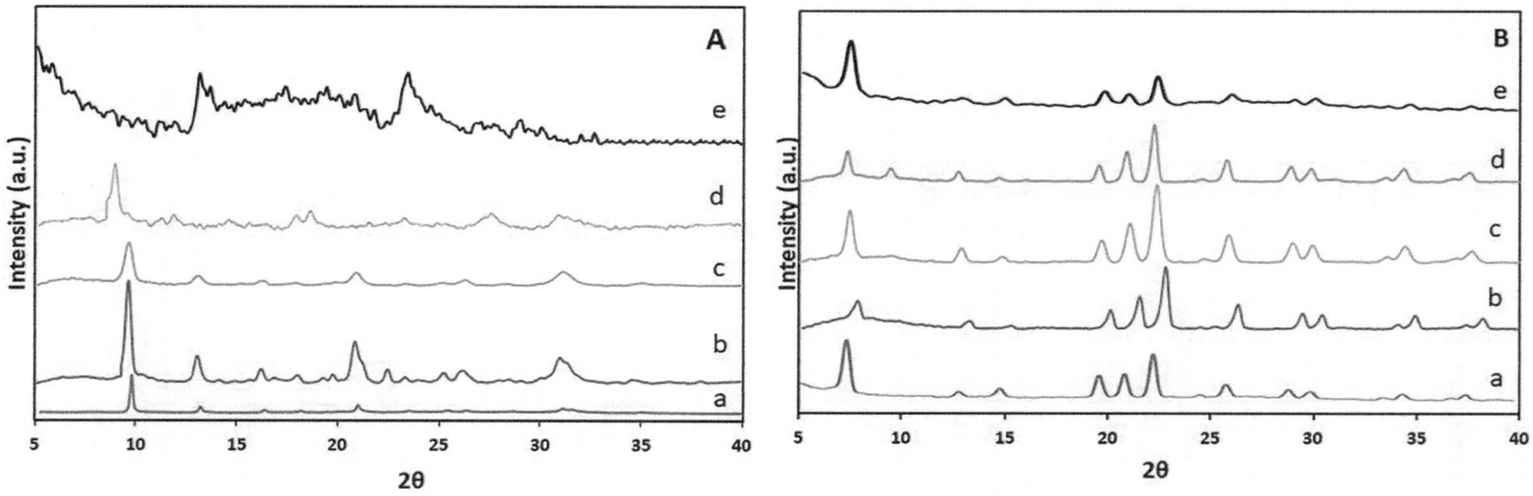

Figure 14.1 High Angle XRD of (a) as made, (b) calcined, (c) HNO_3-treated, (d) HCl-treated, and (e) alkali-treated; (A) SAPO-34 and (B) SAPO-5.

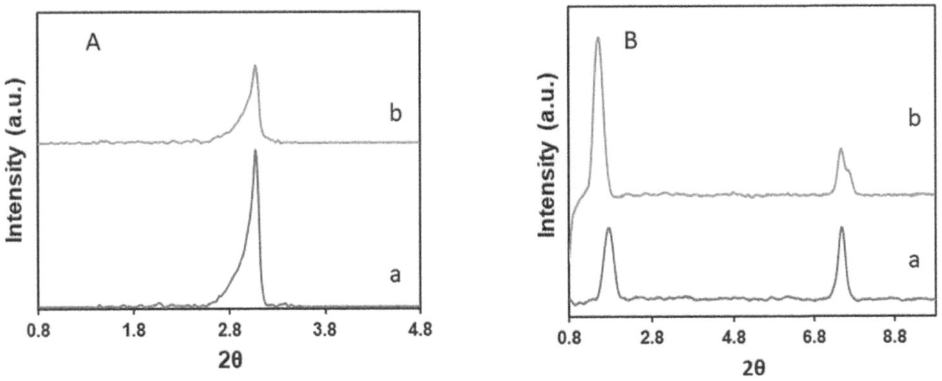

Figure 14.2 XRD (low angle) of (A) acid-treated and (B) alkali-treated: (a) SAPO-34 and (b) SAPO-5.

treatment with acid and alkali. Following the acidic procedure, the diffraction peaks' frequency reduced, indicating a lower relative crystallinity. The phase purity remained sustained even after alkali treatment with a drop in diffraction peak, as seen by XRD data of SAPO-5 material. Although substantially crystallized SAPO-34 materials were produced, after alkali processing, the crystallinity of the materials was diminished, with simply two prominent peaks detected for SAPO-34 alkali-treated material.

Low-angle XRD was used to examine microporous materials that have been acid and alkali treated, as shown in Figure 14.2. Small angle (2θ of 0.5°–10°) XRD was also conducted for HNO_3 and HCl materials. A strong peak at 3.07° 2θ is detected for acid-treated SAPO-34 and SAPO-5 materials usual of the mesoporous phase [17]. The appearance of two prominent peaks at 1.77° 2θ and 7.49° 2θ in alkali-treated materials suggests the creation of pure lamellar phase in alkali-treated material [19]. These findings explain that the material was significantly altered and that the framework structure of SAPO-5 was not affected by the additional porosity (AFI) phase; however, after treatment in SAPO-34, a single mesoporous peak was identified.

Figure 14.3 shows SEM results of pure, acid, and alkali-treated microporous materials. The result displays how both materials' morphology altered as a result of the treatment. SAPO-34 had an irregular cubic morphology when synthesized using static hydrothermal treatments. The cubic shape of SAPO-34 after acid treatment demonstrates that after modification, the SAPO-34's shape and size remained relatively unaltered. The SAPO-5 materials' SEM image shows hexagonal sheet-like plates with a length of 300 nm. Crystals may have agglomerated as an outcome of the establishment of mesoporosity within SAPO-5 with a reduction in crystallite size.

The sorption analysis result and Si/Al ratio of calcined, alkali, and acid-treated SAPO molecular sieves are shown in Table 14.2 and Figure 14.4. Pore diameter and volume rose substantially due to the development of mesoporosity in microporous structures. SAPO-5 showed a pore volume of 0.11 cm^3/g, but it rose to 0.20 cm^3/g for SAPO-5-A1 and 0.19 cm^3/g for SAPO-5-A2 after alteration. For SAPO-34, it was changed from 0.12 cm^3/g to 0.16 cm^3/g for SAPO-34-A1 and 0.62 cm^3/g for SAPO-34-A2. Following treatment with alkali, it

Figure 14.3 SEM image of (A) untreated SAPO-34, (B) untreated SAPO-5, (C) acid-treated SAPO-34, (D) acid-treated SAPO-5, (E) alkali-treated SAPO-34, and (F) alkali-treated SAPO-5.

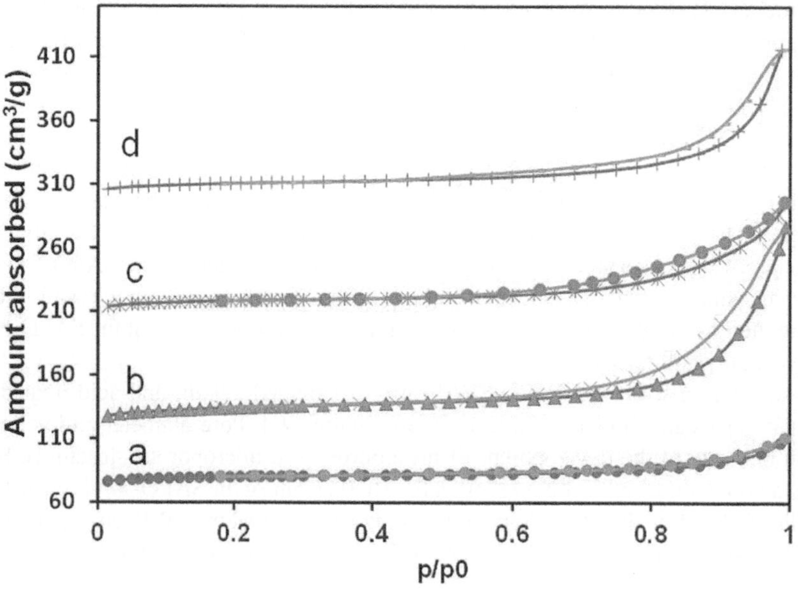

Figure 14.4 N_2 adsorption/desorption isotherm of (a) acid-treated SAPO-34, (b) acid-treated SAPO-5, (c) alkali-treated SAPO-34, and (d) alkali-treated SAPO-5.

Table 14.2 Characteristic properties derived from N_2 sorption analysis.

Catalyst	*Pore volume* (cm^3/g)[a]			*Pore diameter* (Å)[a]	*Si/Al ratio*[b]
	Total	*Micro*	*Meso*		
SAPO-34	0.12	—	—	97	0.51
SAPO-34-A1	0.16	0.11	0.05	126	0.116
SAPO-34-A2	0.62	0.03	0.59	166	0.143
SAPO-34-B1	0.47	0.002	0.47	109	0.147
SAPO-5	0.11	—	—	109	0.187
SAPO-5-A1	0.20	0.03	0.17	186	0.047
SAPO-5-A2	0.19	0.07	0.12	240	0.059
SAPO-5-B1	0.27	0.004	0.263	128	0.054

[a]BJH method, [b]ICP-OES analysis.

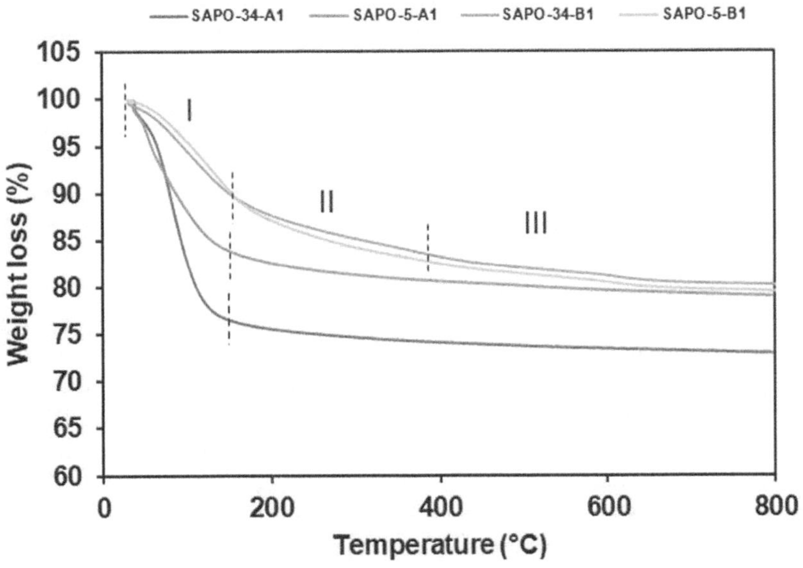

Figure 14.5 TG analysis of SAPO-34-A1, SAPO-5-A1, SAPO-34-B1, and SAPO-5-B1.

improves significantly to 0.47 cm^3/g for SAPO-34-B1 following alterations and 0.27 cm^3/g for SAPO-5-B1.

Figure 14.5 shows the results of the thermogravimetric (TG) analyses. Only a single-stage weight loss is visible in the TG analysis of modifying materials SAPO-34-A1 and SAPO-5-A2 (presented in Figure 14.5 (a) and (b)), indicating that water molecules are removed below 200°C. No further weight loss was observed because the treated material was calcined at 480°C. Since as-made material was employed for post-synthetic treatment, TG examination of SAPO-34-B1 and SAPO-5-B1 materials indicates a total weight loss of three stages. The loss of physisorbed water molecules causes the initial loss of weight below 200°C. The second weight loss may be caused by the breakdown of some parts of SDAs in SAPO materials at temperatures between 200°C and 400°C. Finally, at 400°C, the weight loss due to the exclusion of SDA and TPAOH was observed.

14.5 Catalytic activity test

14.5.1 *Performance of multiple catalysts*

The produced hierarchical materials were tested in the reaction of morpholine with propylene oxide and transesterification of triacetin with methanol. A literature survey was done with various catalysts on transesterification and ring-opening reaction of epoxide, which is given in Table 14.3. Figure 14.6 depicts the procedure system utilized in the current research to test the improvement of the activity of SAPOs after being treated with different acids. This study illustrates that with the increase of catalyst amount, significant enhancement of reactant conversion was observed in the reaction process. The importance of an epoxide ring opening process was demonstrated by analyzing the catalytic results of acid-treated materials to those of pure SAPO but also the blank process. Using 40 mg of pure SAPO-34 and SAPO-5, the procedure was conducted at 80°C for 6 hours, providing approximately 17% and 15% conversion, respectively. This reaction was also conducted under identical reaction conditions without using any catalyst, and only 7% conversion was achieved. The conversion rate increased by orders of magnitude when the reaction was carried out using modified (alkali/acid treated) microporous materials. SAPO-34-A1 and SAPO-5-A1 had morpholine conversion rates of 90% and 82%, respectively, whereas SAPO-34-A2 and SAPO-5-A2 had morpholine conversion rates of 92% and 82%, respectively. The conversion was recorded at 74% as well as 82%, respectively, when a similar procedure was carried out on materials that had been treated with alkali.

Figure 14.7 shows the results of the synthesized catalysts in the triacetin transesterification with methanol and the conversion of triacetin transesterification to methyl acetate using several catalysts. Because this process is controlled by equilibrium, a greater ratio of methanol/triacetin was employed to move the equilibrium to the right, resulting in improved yield and conversion. SAPO-5-B1 material produced 91% triacetin conversion and 69% methyl acetate yield, while for SAPO-5-B2 57% and 30% conversion and yield were obtained, respectively. SAPO-34-B1 showed 67% conversion with 38% methyl acetate yield, whereas using SAPO-34-B2, 46% conversion and 28% yield was observed. Without using any catalyst as well as using unmodified SAPOs, the same reaction was performed; no significant activity was observed, reflected in Figure 14.7, proving the importance of the designed catalyst in biodiesel production. SAPO-5 showed only 2% and SAPO-34 1% for blank reaction. From the above catalytic data, it is clearly visible that prepared materials are active in the transesterification reaction.

In light of the positive experimental results, probable mechanism pathways comprising morpholine and epoxide materials are examined. Scheme 1 [A] depicts a mechanism for S_N2 category epoxide activation through the nucleophilic nitrogen-containing attacking group in morpholine. In S_N2 approach, the reaction proceeds by involving a transition state by the formation of a carbocation. The reaction of propylene oxide with a treated SAPO catalyst typically triggers the production of 1-morpholinopropan-2-ol product through the S_N2 pathway.

The activation of alcohol as a nucleophilic reactant is required in many processes. For example, the transesterification reaction necessarily involves the deprotonation of alcohol, like methanol, which produces methoxy species and surface OH, the latter of which is assumed to be important for the nucleophilic impact on the carbonyl group on esters.

The mechanism, as shown in Scheme 1 [B], In a three-stage process, 1:3 molar ratio of triacetin and methanol is used to generate 3:1 mole of linear ester and glycerol.

Table 14.3 Literature Survey on "Epoxide Ring Opening Reaction" and "Transesterification of Triglycerides"

Reaction	*Reactants*	*Catalyst*	*Time*	*Temperature*	*Conversion*	*Selectivity*	
Transesterification	Triacetin, methanol	MCM-48-SO_3H	18 h	70°C	94%	45%	[21]
Transesterification	Triacetin, methanol	MCM-48-NH_2	4.5 h	65°C	78%	52%	[23]
Transesterification	Sunflower oil, methanol	Na/SBA-15	4 h	65°C	99%	–	[25]
Transesterification	Triacetin, methanol	KS-3	1 h	65°C	94%	98.2%	[27]
Transesterification	Waste frying oils, methanol	HZSM-5 nanosheets	4 h	180°C	–	48.29%	[29]
Transesterification	Shea butter, methanol	Sulfated FAU	3 h	200°C	–	96.89%	[30]
Transesterification	Soybean oil, ethanol	Hierarchical BCL-ZIF-8	12 h	40°C	93.4%	–	[31]
Epoxide ring opening	1,2-epoxyoctan, ethanol	(Nano-Sn-β-zeolite)	24 h	60°C	99%	57%	[33]
Epoxide ring opening	Styrene oxide, morpholine	H-Mont.-3	2 h	25°C	98%	100%	[45]
Epoxide ring opening	Styrene oxide, aniline	Meso-Zr-Beta	0.5 h	35°C	80.7%	94.7%	[46]
Epoxide ring opening	Epichlorohydrin, aniline	Pb-Beta zeolite	0.5 h	40°C	81.8%	94.3%	[48]
Epoxide ring opening	Styrene oxide, aniline	Ti-SBA-16	6 h	35°C	96%	95.1%	[49]
Epoxide ring opening	Propylene oxide, phenol	SAPO-5-N	12 h	120°C	95%	100%	[51]
Epoxide ring opening	Propylene oxide, aniline	NaMCM-22	6 h	35°C	93%	83%	[52]

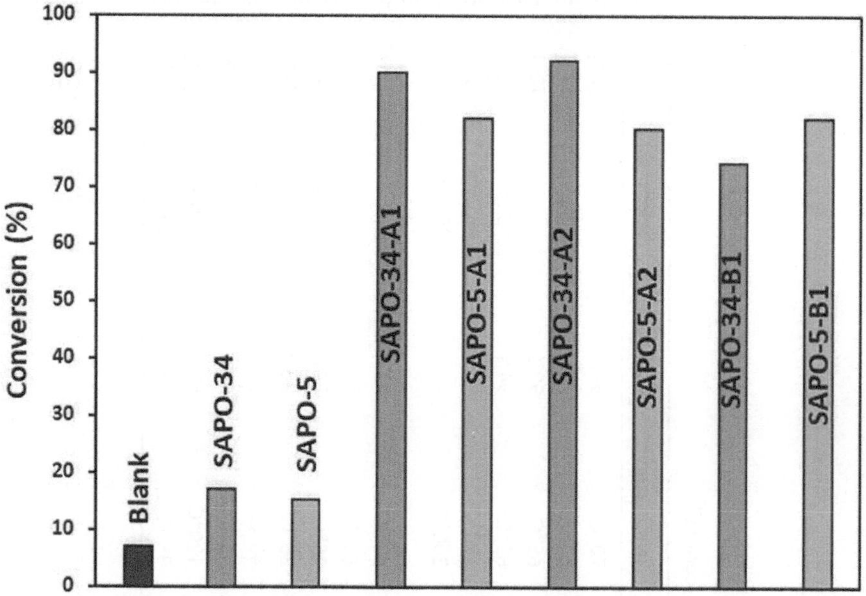

Figure 14.6 Morpholine conversion of blank, calcined SAPO-34, calcined SAPO-5, HNO_3-treated SAPO-34, HNO_3-treated SAPO-5, HCl-treated SAPO-34, HCl-treated SAPO-34, alkali-treated SAPO-34, and alkali-treated SAPO-34. Reaction conditions: 80°C, 40 mg catalyst, and 6 h.

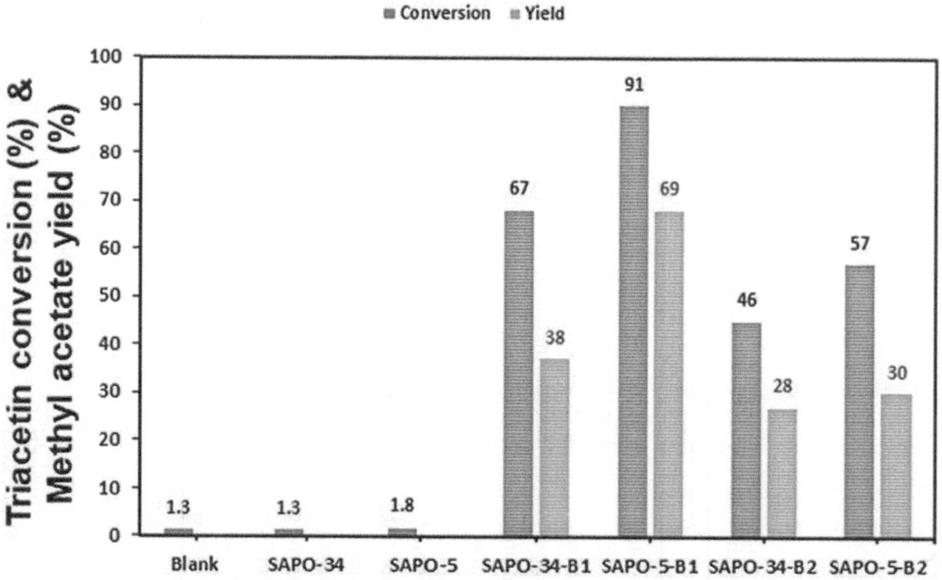

Figure 14.7 Reactant conversion and product yield of blank, as made SAPO-34, as made SAPO-5, 5 hours treated SAPO-34, 5 hours treated SAPO-5, 8 hours treated SAPO-34, 8 hours treated SAPO-5. Reaction conditions: 0.5 g catalyst amount, 6 h reaction time, 68°C temperature.

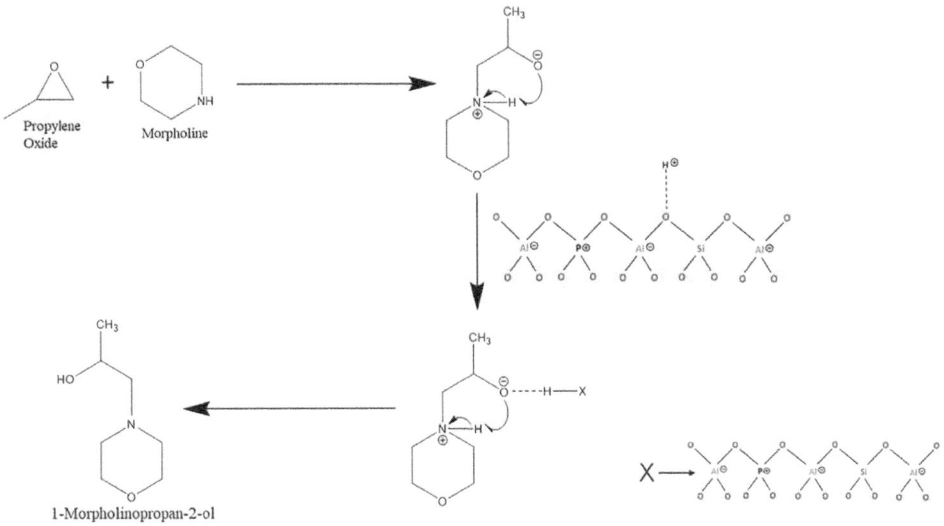

Scheme 1. [A] S_N2 mechanistic scheme of reaction of propylene oxide with morpholine using micro/meso SAPOs.

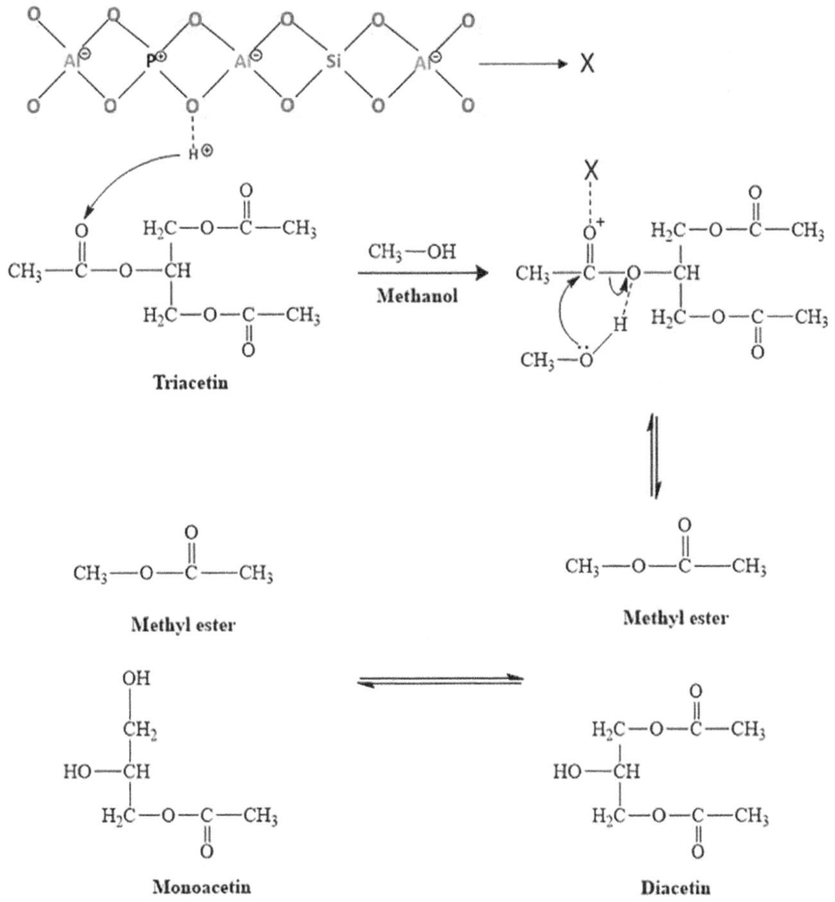

Scheme 1. [B] Transesterification of triacetin to monoacetin reaction mechanism over micro/meso SAPOs.

Using a characterized HNO_3-treated silicoaluminophosphates sample, the synthesis of 1-morpholinopropane-2-ol (P1) and 2-morpholinopropane-1-ol (P2) was studied for different reaction times at various temperatures, and the methyl acetate yield was analyzed for 6 hours using SAPOs that treated with alkali. Table 14.4 summarizes the outcomes of the reactions obtained under various experimental conditions.

Table 14.5 depicts the acid concentration of SAPO-34-A1 and SAPO-5-A1. Using the NH_3-TPD method, the acidic concentration of SAPO-34-A1 was estimated as 0.633 mmol/g

Table 14.4 Catalytic Result of Reaction Between Propylene Oxide With Morpholine and Transesterification of Triacetin

Catalyst	*Reaction*	*Reaction time (hours)*	*Catalyst amount (mg)*	*Conversion (%)*	*Selectivity (%) P1 P2*
SAPO-34-A1	Epoxide ring opening	6	20	72	98.2
SAPO-5- A1	Epoxide ring opening	6	20	77	97.3
SAPO-34-A1	Epoxide ring opening	6	30	78	99.1
SAPO-5-A1	Epoxide ring opening	6	30	78	98.2
SAPO-34-A1	Epoxide ring opening	6	40	90	98.2
SAPO-5-A1	Epoxide ring opening	6	40	82	98.2
SAPO-34-A1	Epoxide ring opening	8	40	92	98.2
SAPO-5-A1	Epoxide ring opening	8	40	84	98.2
SAPO-34-A1	Epoxide ring opening	10	40	91	98.2
SAPO-5-A1	Epoxide ring opening	10	40	84	98.2
SAPO-34-A1	Epoxide ring opening	12	40	91	98.2
SAPO-5-A1	Epoxide ring opening	12	40	87	98.2
SAPO-34-B1	Transesterification	6	200	21	–
SAPO-5-B1	Transesterification	6	200	44	–
SAPO-34-B1	Transesterification	6	300	48	–
SAPO-5-B1	Transesterification	6	300	58	–
SAPO-34-B1	Transesterification	6	500	68	–
SAPO-5-B1	Transesterification	6	500	90	–
SAPO-34-B1	Transesterification	2	500	16	–
SAPO-5-B1	Transesterification	2	500	28	–
SAPO-34-B1	Transesterification	4	500	53	–
SAPO-5-B1	Transesterification	4	500	70	–

Table 14.5 Acidic and Basic Properties of Different Catalysts

Catalyst	*Acidity (mmol/g)*[a]	*Basicity (mmol/g)*[b]
SAPO-34-A1	0.633	–
SAPO-5- A2	0.289	–
SAPO-34-B1	–	1.27
SAPO-5-B1	–	1.08

a By NH_3-TPD

b By titration

and 0.289 mmol/g for SAPO-5-A1. Using the titration method, the basicity of the treated materials was also calculated, and the basicity of SAPO-34-B1 was found to be 1.27 mmol/g and 1.08 mmol/g for SAPO-5-B1, respectively.

14.5.2 Recyclability of the catalyst

The results of a recycling test of treated silicoaluminophosphate materials under optimized reaction conditions are shown in Figure 14.8. After each cycle, the catalyst was filtered, rinsed with an ethanol system, then dried before the next cycle. SAPO-5-A1 showed 82% conversion in the recyclability test, with 79% conversion observed in the last cycle. Thus, the SAPO-5-A1 catalyst was proved to be efficient till the last round, without substantial alteration in conversion. The catalytic activity was also retained following the last recycle round. The SAPO-34-A1 catalyst, on the other hand, converted 90% during the initial stage and 88% during the last cycle. As a result, SAPO-34-A1 was shown to be highly productive through till fourth round without substantial variations in conversion as well as catalytic activity persisted during the last cycle, resulting in an 88% conversion.

Morpholine conversion did not change noticeably, and catalytic activity was retained following the last recycle round, showing 71% and 85% reactant conversion using modified SAPO-5-A2 and SAPO-34-A2, respectively. A slight reduction in conversion throughout successive runs might be due to catalyst degradation, which is thought to be caused by the product or reactant deposit within the pore size.

The first cycle of the recyclability test of triacetin transesterification with methanol over alkali-treated SAPOs shows a 90% conversion rate, which declines to 75% in the second cycle. The third cycle had a conversion rate of 65%, and the fourth cycle had a conversion rate of 40%. As a result, SAPO-5-B1 was shown to be highly productive through the 3rd run, after it caused a 40% decrease in triacetin conversion, possibly due to by-product or reactant deposit within the pore size.

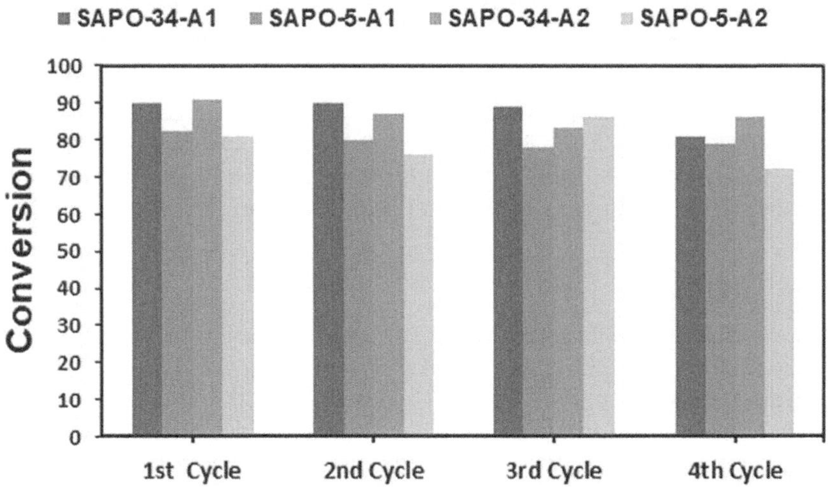

Figure 14.8 Reusability of HNO_3-/HCl-treated SAPOs.

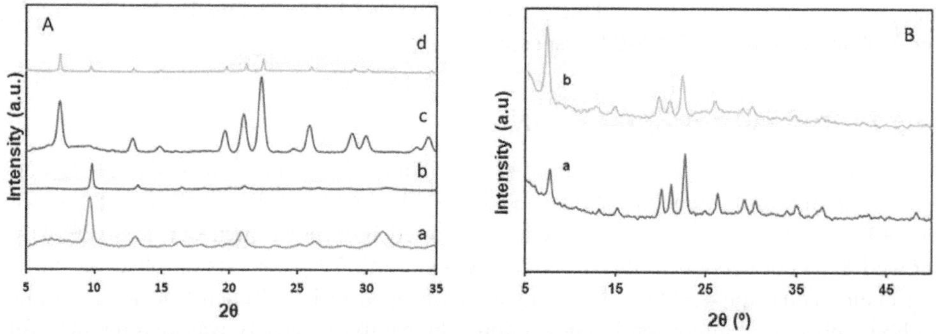

Figure 14.9 (A) Powder XRD patterns of (a) pure SAPO-34, (b) spent SAPO-34, (c) pure SAPO-5, and (d) spent treated SAPO-5 for epoxide ring opening reaction. (B) Powder XRD patterns of (a) pure SAPO-5 and (b) spent treated SAPO-5-B1 for transesterification reaction

Figure 14.9 shows the high angles of a spent catalyst's XRD pattern. The hierarchical microporous catalysts revealed negligible structure degradation in the XRD study. The sharpness of the diffraction peaks decreased in Figure 14.9 (a), and the SAPO-5 phase did not degrade structurally in the transesterification reaction (Figure 14.9 (b)). It is clear that the modified hierarchical SAPOs appeared to be a potential catalyst, for this process of chemical transformations.

14.6 Conclusion

In this work, post-synthetic treatment was used to synthesize SAPO-34 and SAPO-5 with micro/meso porous structure. To demonstrate the development of additional mesoporosity, P-XRD, scanning electron microscope, sorption analysis, and ICP-OES experiments were employed. The ring opening process of epoxide was performed using the modified silicoaluminophosphates materials. SAPO-34 treated with acid was found to be an effective catalyst with 90% conversion. Among all the catalysts that are treated with alkali solution, SAPO-5-B1 was found to be a notable and environmentally friendly catalyst, resulting in conversion and product output of 90% and 68%, respectively, and it worked very well until the fourth run. Catalytic performances in the ring opening and transesterification reactions are improved by surface modification via acid and alkali treatment.

Acknowledgments

The authors would like to thank the SRDC-PDEU, Gandhinagar, for assistance with FE-SEM measurements, and Dr. Ashish Unnatkat, Pandit Deendayal Energy University, for providing the TG analysis.

References

[1] Martens, Johan A., Piet J. Grobet, and Peter A. Jacobs. "Catalytic activity and Si, Al, P ordering in microporous silicoaluminophosphates of the SAPO-5, SAPO-11, and SAPO-37 type." *Journal of Catalysis* 126, no. 1 (1990): 299–305.

[2] Sun, Qiming, Ning Wang, Guanqi Guo, Xiaoxin Chen, and Jihong Yu. "Synthesis of tri-level hierarchical SAPO-34 zeolite with intracrystalline micro–meso–macroporosity showing superior MTO performance." *Journal of Materials Chemistry A* 3, no. 39 (2015): 19783–19789.
[3] Ren, Shu, Guojuan Liu, Xian Wu, Xinqing Chen, Minghong Wu, Gaofeng Zeng, Ziyu Liu, and Yuhan Sun. "Enhanced MTO performance over acid treated hierarchical SAPO-34." *Chinese Journal of Catalysis* 38, no. 1 (2017): 123–130.
[4] Verboekend, Danny, Maria Milina, and Javier Pérez-Ramírez. "Hierarchical silicoaluminophosphates by postsynthetic modification: Influence of topology, composition, and silicon distribution." *Chemistry of Materials* 26, no. 15 (2014): 4552–4562.
[5] Ahmed, Maqsood, and Ayyamperumal Sakthivel. "Amine functionalized AFI type microporous SAPO-5 materials: Preparation, unique method on template extraction, characterization and its catalytic application on epoxide ring opening." *Journal of Porous Materials* 26, no. 2 (2019): 319–326.
[6] Tosheva, Lubomira, and Valentin P. Valtchev. "Nanozeolites: Synthesis, crystallization mechanism, and applications." *Chemistry of Materials* 17, no. 10 (2005): 2494–2513.
[7] Laha, Subhash C., Chitravel Venkatesan, Ayyamperumal Sakthivel, Kenichi Komura, Tak Hee Kim, Sung June Cho, Shing-Jong Huang et al. "Highly stable aluminosilicates with a dual pore system: Simultaneous formation of meso-and microporosities with zeolitic BEA building units." *Microporous and Mesoporous Materials* 133, no. 1–3 (2010): 82–90.
[8] Bernasconi, Samantha, Jeroen A. van Bokhoven, Frank Krumeich, Gerhard D. Pirngruber, and Roel Prins. "Formation of mesopores in zeolite beta by steaming: A secondary pore channel system in the (001) plane." *Microporous and Mesoporous Materials* 66, no. 1 (2003): 21–26.
[9] Groen, Johan C., Louk A.A. Peffer, Jacob A. Moulijn, and Javier Pérez-Ramírez. "Mechanism of hierarchical porosity development in MFI zeolites by desilication: The role of aluminium as a pore-directing agent." *Chemistry–A European Journal* 11, no. 17 (2005): 4983–4994.
[10] Schmidt, Iver, Claus Madsen, and Claus J. H. Jacobsen. "Confined space synthesis. A novel route to nanosized zeolites." *Inorganic Chemistry* 39, no. 11 (2000): 2279–2283.
[11] Corma, A., V. Fornes, J. Martınez-Triguero, and S. B. Pergher. "Delaminated zeolites: Combining the benefits of zeolites and mesoporous materials for catalytic uses." *Journal of Catalysis* 186, no. 1 (1999): 57–63.
[12] Abdulridha, Samer, Yilai Jiao, Shaojun Xu, Rongxin Zhang, Zhongyuan Ren, Arthur A. Garforth, and Xiaolei Fan. "A comparative study on mesoporous Y zeolites prepared by hard-templating and post-synthetic treatment methods." *Applied Catalysis A: General* 612 (2021): 117986.
[13] Xi, Dongyang, Qiming Sun, Xiaoxin Chen, Ning Wang, and Jihong Yu. "The recyclable synthesis of hierarchical zeolite SAPO-34 with excellent MTO catalytic performance." *Chemical Communications* 51, no. 60 (2015): 11987–11989.
[14] Wang, Chan, Miao Yang, Peng Tian, Shutao Xu, Yue Yang, Dehua Wang, Yangyang Yuan, and Zhongmin Liu. "Dual template-directed synthesis of SAPO-34 nanosheet assemblies with improved stability in the methanol to olefins reaction." *Journal of Materials Chemistry A* 3, no. 10 (2015): 5608–5616.
[15] Newland, Stephanie H., Wharton Sinkler, Thomas Mezza, Simon R. Bare, Marina Carravetta, Ibraheem M. Haies, Alan Levy, Scott Keenan, and Robert Raja. "Expanding beyond the micropore: Active-site engineering in hierarchical architectures for Beckmann rearrangement." *ACS Catalysis* 5, no. 11 (2015): 6587–6593.
[16] Miletto, Ivana, Geo Paul, Stephanie Chapman, Giorgio Gatti, Leonardo Marchese, Robert Raja, and Enrica Gianotti. "Mesoporous silica scaffolds as Precursor to Drive the Formation of Hierarchical SAPO-34 with Tunable Acid properties." *Chemistry–A European Journal* 23, no. 41 (2017): 9952–9961.
[17] Singh, Arvind Kumar, Rekha Yadav, and Ayyamperumal Sakthivel. "Synthesis, characterization, and catalytic application of mesoporous SAPO-34 (MESO-SAPO-34) molecular sieves." *Microporous and Mesoporous Materials* 181 (2013): 166–174.
[18] Chen, Xiaoxin, Dongyang Xi, Qiming Sun, Ning Wang, Zhenyu Dai, Dong Fan, Valentin Valtchev, and Jihong Yu. "A top-down approach to hierarchical SAPO-34 zeolites with improved selectivity of olefin." *Microporous and Mesoporous Materials* 234 (2016): 401–408.

[19] Singh, Arvind Kumar, Rekha Yadav, Vasanthakumaran Sudarsan, Kondamudi Kishore, Sreedevi Upadhyayula, and Ayyamperumal Sakthivel. "Mesoporous SAPO-5 (MESO-SAPO-5): A potential catalyst for hydroisomerisation of 1-octene." *RSC Advances* 4, no. 17 (2014): 8727–8734.

[20] Ogoshi, T., and Y. Miyawaki. "Soap and related products: Palm and lauric oil." *Journal of the American Oil Chemists' Society* 62, no. 2 (1985): 331–335.

[21] Bandyopadhyay, Mahuya, Nao Tsunoji, Rajib Bandyopadhyay, and Tsuneji Sano. "Comparison of sulfonic acid loaded mesoporous silica in transesterification of triacetin." *Reaction Kinetics, Mechanisms and Catalysis* 126, no. 1 (2019): 167–179.

[22] Otera, Junzo. "Transesterification." *Chemical reviews* 93, no. 4 (1993): 1449–1470.

[23] Bandyopadhyay Mahuya, Nao Tsunoji, and Tsuneji Sano. "Mesoporous MCM-48 immobilized with aminopropyltriethoxysilane: A potential catalyst for transesterification of triacetin." *Catalysis Letters* 147, no. 4 (2017): 1040–1050.

[24] Vinh, Tran Quang, Nguyen Thi Thanh Loan, Xiao-Yu Yang, and Bao-Lian Su. "Preparation of biofuels by catalytic cracking reaction of vegetable oil sludge." *Fuel* 90, no. 3 (2011): 1069–1075.

[25] Albayati, Talib M., and Aidan M. Doyle. "Encapsulated heterogeneous base catalysts onto SBA-15 nanoporous material as highly active catalysts in the transesterification of sunflower oil to biodiesel." *Journal of Nanoparticle Research* 17, no. 2 (2015): 1–10.

[26] Ebadinezhad, Behzad, and Mohammad Haghighi. "Texture evolution of mesoporous SAPO-34 via a hard-templating sono-hydrothermal method for biodiesel production: Influence of carbon materials on nanocatalyst design." *Applied Catalysis A: General* 595 (2020): 117486.

[27] Barot, Sunita, Maaz Nawab, and Rajib Bandyopadhyay. "Alkali metal modified nano-silicalite-1: An efficient catalyst for transesterification of triacetin." *Journal of Porous Materials* 23, no. 5 (2016): 1197–1205.

[28] Iwakai, Kazuyuki, Teruoki Tago, Hiroki Konno, Yuta Nakasaka, and Takao Masuda. "Preparation of nano-crystalline MFI zeolite via hydrothermal synthesis in water/surfactant/organic solvent using fumed silica as the Si source." *Microporous and Mesoporous Materials* 141, no. 1–3 (2011): 167–174.

[29] Fawaz, Elyssa G., Darine A. Salam, Severinne S. Rigolet, and T. Jean Daou. "Hierarchical zeolites as catalysts for biodiesel production from waste frying oils to overcome mass transfer limitations." *Molecules* 26, no. 16 (2021): 4879.

[30] Alaba, Peter Adeniyi, Yahaya Muhammad Sani, Isah Yakub Mohammed, Yousif Abdalla Abakr, and Wan Mohd Ashri Wan Daud. "Synthesis and characterization of sulfated hierarchical nanoporous faujasite zeolite for efficient transesterification of shea butter." *Journal of Cleaner Production* 142 (2017): 1987–1993.

[31] Adnan, Miaad, Kai Li, Jianhua Wang, Li Xu, and Yunjun Yan. "Hierarchical ZIF-8 toward immobilizing Burkholderia cepacia lipase for application in biodiesel preparation." *International Journal of Molecular Sciences* 19, no. 5 (2018): 1424.

[32] Parulkar, Aamena, Alexander P. Spanos, Nitish Deshpande, and Nicholas A. Brunelli. "Synthesis and catalytic testing of Lewis acidic nano zeolite Beta for epoxide ring opening with alcohols." *Applied Catalysis A: General* 577 (2019): 28–34.

[33] Parulkar, Aamena, Alexander P. Spanos, Nitish Deshpande, and Nicholas A. Brunelli. "Synthesis and catalytic testing of Lewis acidic nano zeolite Beta for epoxide ring opening with alcohols." *Applied Catalysis A: General* 577 (2019): 28–34.

[34] Ogura, Masaru, Shin-ya Shinomiya, Junko Tateno, Yasuto Nara, Eiichi Kikuchi, and Masahiko Matsukata. "Formation of uniform mesopores in ZSM-5 zeolite through treatment in alkaline solution." *Chemistry Letters* 29, no. 8 (2000): 882–883.

[35] Abello, Sonia, Adriana Bonilla, and Javier Perez-Ramirez. "Mesoporous ZSM-5 zeolite catalysts prepared by desilication with organic hydroxides and comparison with NaOH leaching." *Applied Catalysis A: General* 364, no. 1–2 (2009): 191–198.

[36] Xing, Jing, Liang Song, Cong Zhang, Mingdong Zhou, Li Yue, and Xuebing Li. "Effect of acidity and porosity of alkali-treated ZSM-5 zeolite on eugenol hydrodeoxygenation." *Catalysis Today* 258 (2015): 90–95.

[37] Verboekend, Danny, Gianvito Vilé, and Javier Pérez-Ramírez. "Hierarchical Y and USY zeolites designed by post-synthetic strategies." *Advanced functional materials* 22, no. 5 (2012): 916–928.
[38] Tago, Teruoki, Hiroki Konno, Syoko Ikeda, Seiji Yamazaki, Wataru Ninomiya, Yuta Nakasaka, and Takao Masuda. "Selective production of isobutylene from acetone over alkali metal ion-exchanged BEA zeolites." *Catalysis Today* 164, no. 1 (2011): 158–162.
[39] Bjørgen, Morten, Finn Joensen, Martin Spangsberg Holm, Unni Olsbye, Karl-Petter Lillerud, and Stian Svelle. "Methanol to gasoline over zeolite H-ZSM-5: Improved catalyst performance by treatment with NaOH." *Applied Catalysis A: General* 345, no. 1 (2008): 43–50.
[40] Singh, Arvind Kumar, Rekha Yadav, and Ayyamperumal Sakthivel. "Preparation of mesoporous silicoaluminophosphate using ammonium hydroxide as the base and its catalytic application in the trans-alkylation of aromatics." *Journal of Materials Science* 51, no. 6 (2016): 3146–3154.
[41] Singh, Arvind Kumar, Rekha Yadav, Vasanthakumaran Sudarsan, Kondamudi Kishore, Sreedevi Upadhyayula, and Ayyamperumal Sakthivel. "Mesoporous SAPO-5 (MESO-SAPO-5): A potential catalyst for hydroisomerisation of 1-octene." *RSC Advances* 4, no. 17 (2014): 8727–8734.
[42] Singh, Arvind Kumar, Rekha Yadav, and Ayyamperumal Sakthivel. "Synthesis, characterization, and catalytic application of mesoporous SAPO-34 (MESO-SAPO-34) molecular sieves." *Microporous and Mesoporous Materials* 181 (2013): 166–174.
[43] Verboekend, Danny, Maria Milina, and Javier Perez-Ramirez. "Hierarchical silicoaluminophosphates by postsynthetic modification: Influence of topology, composition, and silicon distribution." *Chemistry of Materials* 26, no. 15 (2014): 4552–4562.
[44] Akolekar, Deepak B. "Thermal stability, acidity, catalytic properties, and deactivation behavior of SAPO-5 catalysts: Effect of silicon content, acid treatment, and Na exchange." *Journal of Catalysis* 149, no. 1 (1994): 1–10.
[45] Bhuyan, Diganta, Lakshi Saikia, and Dipak Kumar Dutta. "Modified Montmorillonite clay catalyzed regioselective ring opening of epoxide with amines and alcohols under solvent free conditions." *Applied Catalysis A: General* 487 (2014): 195–201.
[46] Tang, Bo, Weili Dai, Xiaoming Sun, Guangjun Wu, Naijia Guan, Michael Hunger, and Landong Li. "Mesoporous Zr-Beta zeolites prepared by a post-synthetic strategy as a robust Lewis acid catalyst for the ring-opening aminolysis of epoxides." *Green Chemistry* 17, no. 3 (2015): 1744–1755.
[47] Kore, Rajkumar, Rajendra Srivastava, and Biswarup Satpati. "Highly efficient nanocrystalline zirconosilicate catalysts for the aminolysis, alcoholysis, and hydroamination reactions." *ACS Catalysis* 3, no. 12 (2013): 2891–2904.
[48] Chai, Yuchao, Linjun Xie, Zhiyang Yu, Weili Dai, Guangjun Wu, Naijia Guan, and Landong Li. "Lead-containing Beta zeolites as versatile Lewis acid catalysts for the aminolysis of epoxides." *Microporous and Mesoporous Materials* 264 (2018): 230–239.
[49] Kumar, Anuj, and Darbha Srinivas. "Aminolysis of epoxides catalyzed by three-dimensional, mesoporous titanosilicates, Ti-SBA-12 and Ti-SBA-16." *Journal of catalysis* 293 (2012): 126–140.
[50] Bandyopadhyay, Mahuya, Rajib Bandyopadhyay, Shogo Tawada, Yoshihiro Kubota, and Yoshihiro Sugi. "Catalytic performance of silicoaluminophosphate (SAPO) molecular sieves in the isopropylation of biphenyl." *Applied Catalysis A: General* 225, no. 1–2 (2002): 51–62.
[51] Ahmed, Maqsood, and Ayyamperumal Sakthivel. "Preparation of cyclic carbonate via cycloaddition of CO_2 on epoxide using amine-functionalized SAPO-34 as catalyst." *Journal of CO_2 Utilization* 22 (2017): 392–399.
[52] Baskaran, Thangaraj, Akanksha Joshi, Gunda Kamalakar, and Ayyamperumal Sakthivel. "A solvent free method for preparation of β-amino alcohols by ring opening of epoxides with amines using MCM-22 as a catalyst." *Applied Catalysis A: General* 524 (2016): 50–55.
[53] Jadav, Divya, Rajib Bandyopadhyay, and Mahuya Bandyopadhyay. "Synthesis of Hierarchical SAPO-5 & SAPO-34 Materials by Post-Synthetic Alkali Treatment and Their Enhanced Catalytic Activity in Transesterification." *European Journal of Inorganic Chemistry* 2020, no. 10 (2020): 847–853.

Chapter 15

Reuse of waste carpet into a sustainable composite

A case study on recycling approach

Jogendra Kumar, Balram Jaiswal, Kuldeep Kumar, Kaushlendra Kumar, and Rajesh Kumar Verma

Contents

Abbreviations

VARTM	Vacuum-assisted resin transfer molding
SBR	Styrene-butadiene rubber
PP	Polypropylene
PVC	Polyvinyl chloride
PU	Polyurethane
CDF	Carpet-derived fuel

15.1 Introduction

Nowadays, carpet waste is a severe environmental and economic concern due to its harmful disposal effect and cost issues. Compared to the amount of carpet waste generated each year globally, only a small portion of its recycled (Goswami 2009; Jain et al. 2012; Atakan, Sezer, and Karakas 2018). Carpets are widely used throughout the world as floor covering in homes, offices, commercial centers, and for decorative purposes (Mohammadhosseini et al. 2018; Memon et al. 2015). Carpet waste recycling requires multiple processing steps, making it difficult and expensive (Realff, Ammons, and Newton 1999). Considering carpet waste as landfilling option becomes uneconomic due to its non-biodegradable nature and high landfilling costs (Mohammadhosseini et al. 2017). It is a type of solid waste that is

DOI: 10.1201/9781003291961-17

causing environmental issues. Thus, it becomes essential to find some solution to convert such solid waste into a functional form in society and industry interest. The burning of this fibrous waste releases high toxic fumes, resulting in danger to human health (Mishra, Das, and Vaidyanathan 2019; Ghobakhloo and Fathi 2021). An important factor contributing to global warming is waste disposal, which emits a wide range of greenhouse gases during the waste's life cycle. Landfills have traditionally been the most popular method of waste disposal in solid form. The landfills and the waste burning process generates harmful gases into the atmosphere, with methane emissions being the most dangerous of the gases released. As a result, reducing harmful gas emissions is a significant contributor to climate change.

Consequently, it tries to determine new and innovative ways of utilizing waste as a resource. It is difficult and costly to separate and reprocess carpets used as floor coverings. The multilayer composition of polymers and inorganic fillers is the primary reason. A small amount of carpet waste may not seem like much, but the low bulk density of carpet waste takes up a lot of space in the landfill. According to the carpet recycling UK report, approximately 400,000 tons of carpets are disposed of each year (Sotayo, Green, and Turvey 2015b). However, landfilling is becoming increasingly unfeasible due to rising landfill costs and environmental constraints in the UK. Few studies show that it may be possible to use carpet waste to develop composite materials as a source of raw materials. Recycling can be highly economical, inexpensive, and environmentally friendly, depending on the manufacturing processes. The development of structural composites from carpet waste is still in its early stages, with only a few studies conducted so far. These advancements will sustain carpet waste and provide new materials for the load-bearing application domain. In this way, carpet reprocessing has the potential to have a significant and positive impact on environmental sustainability.

This chapter summarizes the carpet structure, source of carpets waste, types of carpet waste, and techniques to waste management. In addition, the introduction of the carpet waste polymer composite and the method to develop the carpet waste polymer composite are discussed. This chapter also focused on properties analysis and the application of composite materials produced from carpet waste.

15.1.1 Carpet structure

Face fiber, primary backing, adhesive, and secondary backing are the four layers of a typical carpet. Natural fibers such as wool make up the top layer, known as face fiber. The face fiber yarns are attached to the primary backing via the layer into which they are looped. The face fibers are held together by an elastomeric adhesive that is applied to the underside of the primary backing and adheres to the primary backing. Styrene-butadiene rubber (SBR) is typically used as an elastomeric adhesive and can be filled with inorganic materials. Simply put, the secondary backing is the layer that is bonded to the underside of the carpet. Polypropylene (PP), polyester, polyvinyl chloride (PVC), polyurethane (PU), or jute can be used for both the primary and secondary backings. Using open loop and closed loop fiber stitching, the carpet structure is shown in Figure 15.1.

15.1.2 Carpet waste sources

Carpet waste can be collected in two forms, depending on its source of origin. The pre-consumer carpet waste is generated, such as trimmed carpet waste collected from manufacturing firms or factories. These are collected from garbage piles or dustbins. Pre-consumer carpet waste

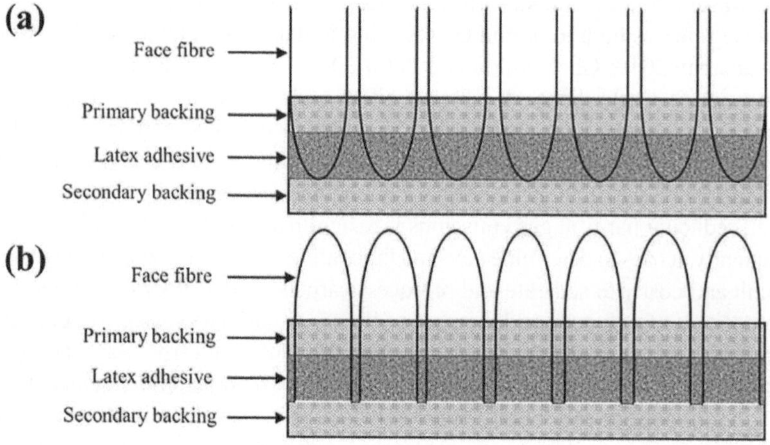

Figure 15.1 Carpet structure (Sotayo, Green, and Turvey 2015b).

includes the trimmed scrap generated at the time of manufacturing or production of carpets per customer needs and demands, such as in cinema halls, households, and educational or institutional complexes (Jaiswal et al. 2021; T. Kumar, Mishra, and Verma 2020). To make a suitable design according to the requirement, the carpet is trimmed out, and this trimmed part cannot be utilized further in the same place. This generated waste is shrunken and irregular and cannot be used further directly. About 10% of the total carpet waste is generated as scrap (Miraftab 2018; Y. Wang, Ucar, and Wang 2014). Postconsumer waste comprises carpet waste generated after it becomes useless in households and commercial buildings (Sotayo, Green, and Turvey 2015b; Dhawan et al. 2019; Cline and Friddle 1992). The users directly throw these carpets as waste. The carpet has an approximate area density of 2.3 kg/m^2. Therefore, trimmings and offcuts derived during the manufacturing process and configuration of carpets are included in pre-consumer carpet waste; however, post-consumer carpet waste includes remnants from waste streams at the end of their useful lives. In general, the normal carpet has an expected lifespan of 5–11 years, depending on the material (Jain et al. 2012). Most used carpets contain dirt, chemicals, and other materials that accumulate over time, making them 30% heavier than new carpets. Most of the carpet is made of polypropylene and SBR, with a roughly constituent proportion of 55% (Sotayo, Green, and Turvey 2015a; Mihut et al. 2001; Sotayo, Green, and Turvey 2018). The vast bulk of carpet waste is generated at home/office (after carpet use), with only 6% generated in the manufacturing sector (during fabrication/fitting). Post-consumer carpet waste is made up of 83% synthetic fibers (nylon, polypropylene, and mixed synthetics) and the rest are natural (wool blends) (Valerio, Muthuraj, and Codou 2020). Properties like fire retardants and stain-resistant chemicals can be found in used carpets from specialized sources like aircraft. Carpet waste is more difficult to reuse or recycle because of the various impurities in it. Figure 15.2 depicts a classification of the primary sources of carpet waste.

15.1.3 Technique for carpet waste management

Various carpet waste management processing options are available in manufacturing industries. Among the options are landfilling (a), incineration (b), prevention (c), carpet re-use (d), and carpet recycling. In order to recover energy from carpet waste, the carpets are shredded

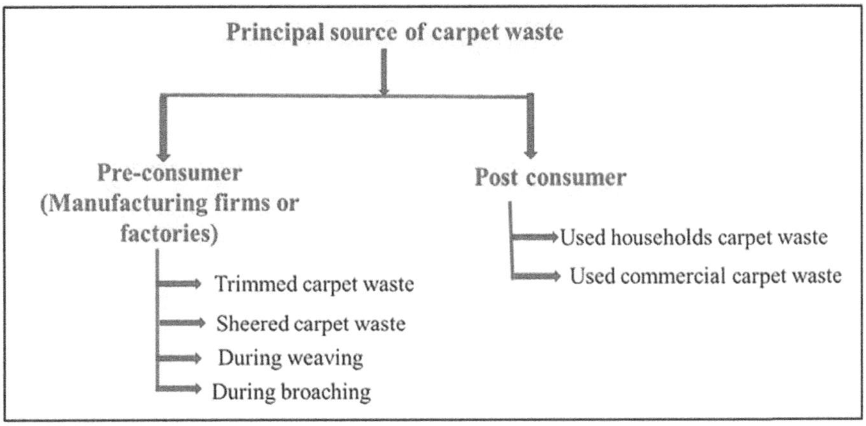

Figure 15.2 Source of carpet waste generation.

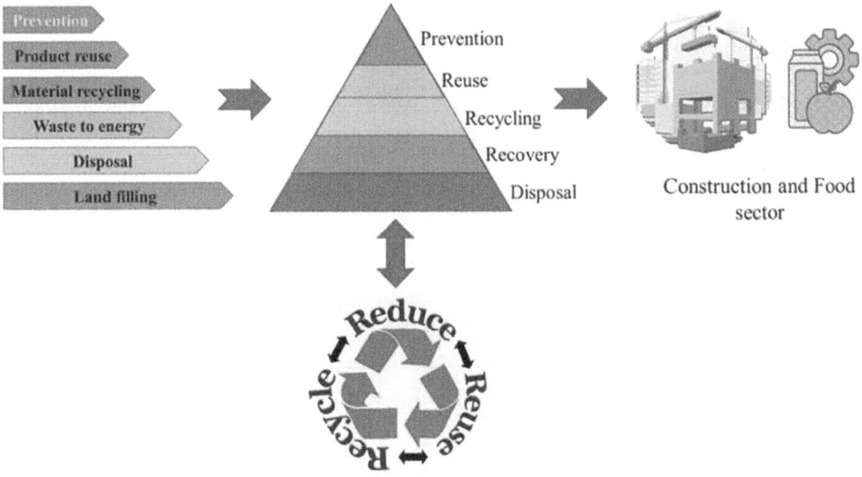

Figure 15.3 Carpet waste management hierarchy chart (Zhang et al. 2022).

and used as fuel in place of coal in cement kilns or boilers. Among the carpet waste processing options, energy recovery accounted for the highest percentage (58%) of the total (Sotayo, Green, and Turvey 2015a; Yalcin-enis, Kucukali-ozturk, and Sezgin 2019). Carpet-derived fuel (CDF) may be generated from all types of carpet waste, but there is a problem associated with the generation of toxic ash waste and ash waste. Carpets have a high calcium carbonate content (37%), which results in a large amount of ash waste (J. Wang and Mao 2012; Sotayo, Green, and Turvey 2015a). Due to the presence of heavy metals such as lead and cadmium in the ash waste, it is sent to a landfill, where it may pollute groundwater and soil, causing pollution. However, if carpet waste is used as a fuel source in kilns for cement production, the ash waste generated can be used as a raw material for cement production.

Carpet waste management is depicted in Figure 15.3 as a hierarchy leading to Carpet Recycling. The hierarchy demonstrates that the option of prevention is a preferable route. Carpets can be made more environmentally friendly by reducing the amount of raw materials used in the design and manufacturing process, and reducing the amount of waste and greenhouse gas emissions. In addition, prevention of discarded carpet landfilling, incineration, disposal damp yard, etc. can be achieved by increasing carpet waste prevention, reuse, and recycling. From Figure 15.3, the prevention, re-use, recycling, energy from waste (via incineration), and finally, landfill disposal is the most environmentally friendly carpet waste management strategies. The use of carpet waste in the development of products with an effective technique for carpet recycling.

15.2 Carpet waste polymer composite

A composite is a combination of two or more elements, one of which is rigid material reinforcement, and the other is a binder or matrix that keeps the reinforcement in place (shape/size) (J. Kumar, Verma, and Khare 2021; Malik 2007; Shin and Park 2018). When the elements (filler and matrix) are combined, the final composite material retains their individual identities (both filler and matrix), and both substantially affect the composite characteristics and performance. The term composite refers to a material system made up of two or more materials with improved material properties than the individual material. The structure of constituents is combined so that the components retain instead in the form of a chemical compound in their respective functional phases (Rajak et al. 2019). The exceptional engineering properties of polymers attracted academia and industry professionals to explore the composites research aspect in recent decades. Composites, mostly made of matrix and reinforcing materials, are well known. In this section, a significant description of matrix and reinforcement material has been described.

15.2.1 Matrix material

Essentially, the matrix is a homogenous and monolithic material into which a fiber system of a composite is incorporated. It is continuous through the composite. The matrix serves as a binding and holding medium for the reinforcements to form a solid (He et al. 2018; Choudhary and Gupta 2011). It gives finish, texture, color, durability, and usefulness while also protecting the reinforcements from environmental harm. Generally, polyester and epoxy resins have been used as matrix materials.

15.2.2 Reinforcement material

To improve the physical qualities of the finished composite material, reinforcement material has been added to the matrix material. Two forms of discarded carpet were used as reinforcement for making carpet waste composites. The first is the direct form, and the second is the random form. Waste carpet is used to develop laminated composite in "direct form". The "random" reinforcement form was created by cutting loose carpet yarn ends from carpet waste with a guillotine and placing them in the mold cavity after resin impregnation. Figure 15.4 presents the matrix material composition and reinforcement in carpet waste polymer composite.

15.3 Technique for development of carpet waste polymer composites

The epoxy phase matrix must be spread evenly over the carpet fibers if efficient composites for structural materials production are developed. The composite was made by combining epoxy (resin), hardener, and carpet waste fiber in a specified ratio. The resin improvement

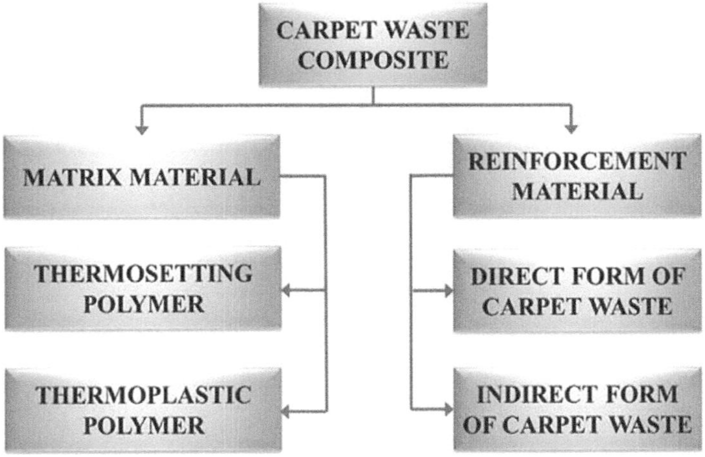

Figure 15.4 Matrix and reinforcement material in carpet waste composite.

is transferred through the carpet, which conveys the essential characteristics of bulk parts. As described in this section, vacuum-assisted resin transfer molding (VARTM) was used to transfer the epoxy (resin) into the fiber layer.

15.3.1 Vacuum-assisted resin transfer molding (VARTM) process

Structured composites may be created using a variety of processes, including pultrusion, resin transfer molding (RTM), spray layup, hand layup, and the vacuum-assisted resin transfer molding (VARTM) process. Carpet waste is utilized as a reinforcing material in developing composites from carpet waste. Figure 15.5 illustrates the matrix and reinforcement material used in carpet waste composite.

Carpet waste in its original state, as seen in Figure 15.5, is referred to as "direct form." In this form, carpet waste was employed without any extra trimming. The plies have been stacked in a non-symmetrical order. Layers of reinforcement have been stacked at angles of 0°, +45°, and −45° to the tensile test direction. The VARTM technique has been used to manufacture carpet waste composites because it produces a sample with uniform resin flow and minimum porosity. Figure 15.5 shows the VARTM setup. In order to use this method, a resin with a low viscosity was required for infusion. The thickness of the composite surfaces is consistent due to the homogeneous resin flow. Figure 15.5 illustrates the component of the VARTM setup. The carpet was trimmed to the proper size and then used as a reinforcing material to create the carpet waste composite. Releasing agent (silica gel spray) was used on the mold's bottom surface. The VARTM arrangement transferred two comparable carpet layers under the vacuum bag and over the release agent. The vacuum pump was turned on first in order to remove any remaining air from the vacuum bag. When a perfect vacuum was achieved, the resin was allowed to flow into the mold and solidify. The pressure of 1 atm provides both the driving power and the compressive force necessary for effective resin infusion into the compressed mold. The VARTM technique required a peel-ply and transfer medium to ensure that epoxy resin was evenly distributed. After the resin was fully infused, the tube was clamped shut and allowed to cure for 24–48 hours.

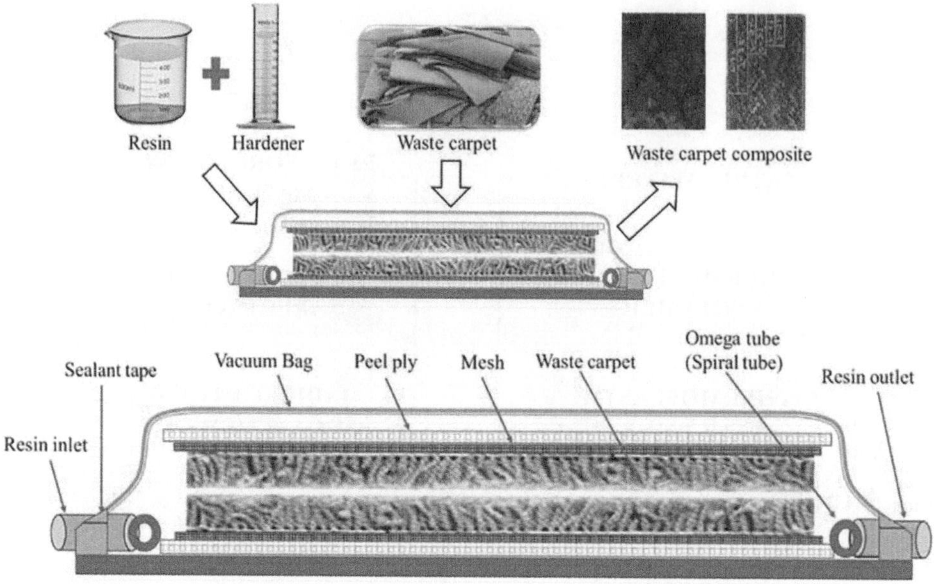

Figure 15.5 Carpet waste composite development process using VARTM setup.

15.4 Analysis of polymer composite develop from carpet waste

The polymer composite's material engineering properties significantly improved after reinforcement was incorporated with the epoxy matrix (Vaidyanathan, Singh, and Ley 2013; Sotayo, Green, and Turvey 2015a; Jaiswal, Verma, and Mishra 2022; J. Kumar et al. 2022). Modifications are influenced by processing methods, the carpet type, aspect ratio, and material. Based on the most recent reports in the literature, a particular processing method can be feasible and essential. So that the desired composite characteristics can be achieved, it is essential to optimize the various conditions involved. Several studies have been carried out in order to determine the mechanical properties of the carpet waste fiber/polymer composite under various conditions.

Carpet waste fiber polymer composite has outstanding acoustic insulation properties (Pan et al. 2016; Rahman, Siddiqua, and Cherian 2022; Sotayo, Green, and Turvey 2015a; Ghisellini and Ulgiati 2019). The acoustic test method can be used to estimate the composite sound proofing, which is controlled by the fiber design and geometry in the matrix phase. The carpet waste fiber composite was found to be more effective than the pristine epoxy composite at insulating sound (Rushforth et al. 2005; Islam and Bhat 2019). As previously stated, structural materials could be significantly strengthened by incorporating carpet waste because of their excellent sound insulation properties. The various properties of the composite material are significantly influenced by the recycling of carpet waste fiber in the polymer matrix. In the case of polymers, recycling waste carpets adds enhanced mechanical properties (Sanjeevi et al. 2021; Valerio, Muthuraj, and Codou 2020; Gowayed and El-halwagi 1995). However, the entire interpretation of reprocessing is still in its early stages. Waste carpet polymer composites need more attention from the manufacturing sector, academia, and research institutes. The summary of the carpet waste supplement in the development of polymer composites performed by eminent scholars is remarked in Table 15.1. This table described the significance of recycling of waste carpet and properties enhancement in polymer for multifunctional products.

Table 15.1 Summary of the Properties of Composite Material Developed From the Carpet Waste

Sr. No.	*Carpet (Reinforcement)*	*Matrix*	*Method*	*Remarks*	*References*
1	Nylon and olefin waste carpet	Epoxy	VARTM	Nylon face fiber-based composite exhibits higher flexural strength than olefin. And these composites can be used in light structural applications.	(Jain et al. 2012)
2	Carpet waste jute	Thermoset matrix	Compression molding	The flexural strength of the prepared sample was found to decrease by up to 40% after being immersed in water for 168 hours.	(Karaduman and Onal 2011)
3	Post-consumer nylon and olefin carpet waste	Epoxy	VARTM	The mechanical testing was performed, and it was observed that the nylon waste carpet fabricated composite exhibits higher flexural strength than the olefin waste carpet.	(Jain et al. 2012)
4	Polypropylene waste carpet	Cement mortar	Pulverization method	The addition of waste carpet face fiber to mortar reduces environmental pollution and enhances mortar's impact behavior.	(Xuan et al. 2018)
5	Different waste carpet and clay	Epoxy	VARTM	The best mechanical results were observed for clay infused in composite, and the presence of the clay in composite also increased the flame retardant behavior.	(Mishra and Vaidyanathan 2019a)
6	Carpet waste (Nylon)	Epoxy	VARTM	The fabricated carpet waste composite can be used as a low-strength structural application composite.	(T. Kumar, Mishra, and Verma 2020)
7	Carpet waste jute yarn	Epoxy	Compression molding technique	The jute waste carpet composite can serve as a potential caste more effectively than the traditional composite.	(Onal and Karaduman 2009)

(Continued)

Table 15.1 (Continued)

Sr. No.	*Carpet (Reinforcement)*	*Matrix*	*Method*	*Remarks*	*References*
8	Needlefelt carpet waste	Polyester resin	Compression molding technique	It was observed that the incorporation of the carpet waste improved the post-cracking strength of the polymer composite under the flexural load.	
9	Flax Jute	Epoxy	Compression molding techniques	The best mechanical properties were obtained by flax jute, and these types of composite materials can be used for lightweight structural applications.	(Karthi et al. 2021)
10	Waste polypropylene carpet	Portland cement	Pumping method	Carpet fibers have the potential to be used in preplaced aggregate fiber reinforced concrete (PAFRC) due to their improved transport and strength characteristics.	(Mohammadhosseini, Tahir, and Alaskar 2020)

15.5 Application of the composite materials developed from carpet waste

The main application area of polymer nanocomposite material is applied in numerous fields, including roadside barriers, home decoration, sound insulation (cinema hall), wall tiles, roof tiles, automotive, sports goods, and building materials (Singh, Ramakrishna, and Gupta 2017; Mishra and Vaidyanathan 2019b; Patti and Cicala 2021; Sotayo, Green, and Turvey 2015a; Fashandi et al. 2019; Kamble et al. 2020). Due to the high corrosion resistance, diminished weight, desired load carrying capacity, and fire retardant, specific weight, moisture, and flexibility. The series uses carpet waste polymer composites in low tooling components (Memon et al. 2015). Moreover, carpet waste is used in various appliances and reduces land filling and incineration. This has an approach towards the waste to wealth concept. A remarkable application of discarded polymer composites is shown in Table 15.2.

15.6 Conclusion remarks

The findings of this chapter describe that carpet waste can be recycled and used in structural applications. It demonstrates a high potential for altering the properties of polymer matrixes. Multiple variables, such as fiber form, carpet type, carpet material geometry, etc., impact the consistency of composites made from waste fibers and polymers. A significant amount of work has been done on carpet recycling and carpet processing techniques. A high probability of features exists for the discarded carpet additive; through these techniques, an appropriate method for improved stress transferring for fiber and matrix material is accomplished. In order to maintain the intrinsic characteristics of the carpet fiber, however, more progress must be made in this

Table 15.2 Application of Carpet Waste Composite Materials

Sr. No.	*Physical significance*	*Remark*	*References*
1	Cost	Reduced waste handling, consolidation part production time, and reduced long-term durability and efficiency	(T. Kumar, Mishra, and Verma 2020; Mishra, Das, and Vaidyanathan 2019; Mohammadhosseini et al. 2017; Dobilaite et al. 2017)
2	Weight	Light in weight, good elasticity and tensile strength, and an excellent balance of strength and weight	(Mishra and Vaidyanathan 2019a)
3	Dimension	Large sections of the body with distinct geometry	(Patel and Dave 2020)
4	Surface properties	Resistant to corrosion, adaptable to the environment, and has a custom surface finish	(Jaiswal, Verma, and Mishra 2022)
5	Thermal properties	Thermodynamically impractical and fire retardant	(Xuan et al. 2018; Mishra and Vaidyanathan 2019a; Verma et al. 2021)
6	Environment	Biodegradable, pollution-free, and controlled noise pollution	(Awal and Mohammadhosseini 2016; Mohammadhosseini et al. 2018)

area in the future. Furthermore, a specific emphasis is placed on carpet waste VARTM methods, and many recycling methods have been revealed to assist in achieving waste management goals.

For carpet waste recycling, there are currently five main methods: landfilling, energy recovery (via incineration), prevention, carpet reuse, and recycling. The previous study intended to recycle waste carpet's main issue of costly environmental pollution. It consumes lots of time, causes costly decomposition, causes environmental pollution, and has a limited exploration technique.

Recycling waste carpet in the polymer matrix was investigated in this paper during epoxy resin progression. The structural efficiency of a manufacturing process is typically associated with the performance of the production process. It has been observed that the mechanical properties of the polymer matrix have improved when using VARTM, which produces a sustainable polymer composite. Additional significant improvements to carpet waste fiber/epoxy composites include acoustic, fire, moisture, and thermal insulation.

Besides, the proposed methods could adequately resolve carpet waste management such as land filling and incineration.

A higher potential for application in manufacturing sectors was demonstrated by the improved characteristics of the carpet fiber when compared to the pristine polymer composite. A more significant analysis of the properties of composites is expected to improve their performance at a low cost.

Recycling to the management of waste materials is another emerging area for practitioners and academia, as it can be used in various parts production. Hence, the developed material can be explored for the zero-waste generation process. The current study may be summarized by claiming that the proposed technique can be recommended for critical recycling problems in manufacturing and industrial engineering case studies. The proposed technique can be recommended for other recycling materials for the developed composites.

References

Atakan, Raziye, Serdar Sezer, and Hale Karakas. 2018. "Development of Nonwoven Automotive Carpets Made of Recycled PET Fibers with Improved Abrasion Resistance." *Journal of Industrial Textiles* 49: 835–857. doi:10.1177/1528083718798637.

Awal, A. S. M. Abdul, and Hossein Mohammadhosseini. 2016. "Green Concrete Production Incorporating Waste Carpet Fiber and Palm Oil Fuel Ash." *Journal of Cleaner Production* 137. Elsevier Ltd: 157–166. doi:10.1016/j.jclepro.2016.06.162.

Choudhary, Veena, and Anju Gupta. 2011. "Polymer/Carbon Nanotube Nanocomposites." In *Carbon Nanotubes—Polymer Nanocomposites*, edited by Siva Yellampalli, 65–90. London: Intech Open. doi:10.5772/18423.

Cline, C. D., and J. D. Friddle. 1992. "Specialty Textile Coatings—a Formulation Overview." *Journal of Industrial Textiles* 22 (1): 32–41. doi:10.1177/152808379202200104.

Dhawan, Ridham, Brij Mohan Singh Bisht, Rajeev Kumar, Saroj Kumari, and S. K. Dhawan. 2019. "Recycling of Plastic Waste into Tiles with Reduced Flammability and Improved Tensile Strength." *Process Safety and Environmental Protection* 124. Institution of Chemical Engineers: 299–307. doi:10.1016/j.psep.2019.02.018.

Dobilaite, Vaida, Gene Mileriene, Milda Juciene, and Virginija Saceviciene. 2017. "Investigation of Current State of Pre-Consumer Textile Waste Generated at Lithuanian Enterprises." *International Journal of Clothing Science and Technology* 29 (4): 491–503. doi:10.1108/IJCST-08-2016-0097.

Fashandi, Hossein, Hamid Reza, Masoud Latifi, Hamid Reza Pakravan, and Masoud Latifi. 2019. "Application of Modified Carpet Waste Cuttings for Production of Eco-Efficient Lightweight Concrete." *Construction and Building Materials* 198. Elsevier Ltd: 629–37. doi:10.1016/j.conbuildmat.2018.11.163.

Ghisellini, Patrizia, and Sergio Ulgiati. 2019. "Circular Economy Transition in Italy. Achievements, Perspectives and Constraints." *Journal of Cleaner Production*, no. September. Elsevier Ltd: 118360. doi:10.1016/j.jclepro.2019.118360.

Ghobakhloo, Morteza, and Masood Fathi. 2021. "Industry 4.0 and Opportunities for Energy Sustainability." *Journal of Cleaner Production* 295. Elsevier Ltd: 126427. doi:10.1016/j.jclepro.2021.126427.

Goswami, K. K. 2009. *Advances in Carpet Manufacture*. Edited by K. K. Goswami. Advances in Carpet Manufacture. 2nd ed. Kidlington: Elsevier B.V. doi:10.1533/9781845695859.

Gowayed, Yasser, and Mahmoud M. El-halwagi. 1995. "Synthesis of Composite Materials from Waste Fabrics and Plastics." *Journal of Elastomers & Plastics* 27: 79–90. doi:10.1177/009524439502700106.

He, Runqin, Qiuxiang Chang, Xinjun Huang, and Jin Bo. 2018. "Improved Mechanical Properties of Carbon FIber Reinforced PTFE Composites by Growing Graphene Oxide on Carbon FIber Surface." *Composite Interfaces* 25 (11). Taylor & Francis: 995–1004.

Islam, Shafiqul, and Gajanan Bhat. 2019. "Environmentally-Friendly Thermal and Acoustic Insulation Materials from Recycled Textiles." *Journal of Environmental Management* 251. Elsevier: 109536. doi:10.1016/j.jenvman.2019.109536.

Jain, Abhishek, Gajendra Pandey, Abhishek K. Singh, Vasudevan Rajagopalan, Ranji Vaidyanathan, and Raman P. Singh. 2012. "Fabrication of Structural Composites from Waste Carpet." *Advances in Polymer Technology* 31 (4): 380–389. doi:10.1002/adv.20261.

Jaiswal, Balram, Vijay Kumar Singh, Sanjay Mishra, and Rajesh Kumar Verma. 2021. "Study on Polymer (Epoxy) Composite Using Carpet Waste for Lightweight Structural Applications: A New Approach for Waste Management." *Materials Today: Proceedings* 44 (February). Elsevier Ltd: 2678–2684. doi:10.1016/j.matpr.2020.12.681.

Jaiswal, Balram, Rajesh Kumar Verma, and Sanjay Mishra. 2022. "Use of Discarded Carpet Material in the Development of Polymer (Epoxy) Composites for Structural Functions." *The Journal of The Textile Institute*. Taylor & Francis: 1–11. doi:10.1080/00405000.2021.2025302.

Kamble, Zunjarrao, Bijoya Kumar Behera, Teruo Kimura, and Ino Haruhiro. 2020. "Development and Characterization of Thermoset Nanocomposites Reinforced with Cotton Fibres Recovered from Textile Waste." *Journal of Industrial Textiles*: 1–27. doi:10.1177/1528083720913535.

Karaduman, Y., and L. Onal. 2011. "Water Absorption Behavior of Carpet Waste Jute-Reinforced Polymer Composites." *Journal of Composite Materials* 45 (15): 1559–1571. doi:10.1177/0021998310385021.

Karthi, N., K. Kumaresan, S. Sathish, L. Prabhu, S. Gokulkumar, D. Balaji, N. Vigneshkumar, et al. 2021. "Effect of Weight Fraction on the Mechanical Properties of Flax and Jute Fibers Reinforced Epoxy Hybrid Composites." *Materials Today: Proceedings* 45 (9). Elsevier Ltd.: 8006–8010. doi:10.1016/j.matpr.2020.12.1060.

Kumar, Jogendra, Kuldeep Kumar, Balram Jaiswal, Kaushlendra Kumar, and Rajesh Kumar Verma. 2022. "Investigation on the Physio-Mechanical Properties of Carpet Waste Polymer Composites Incorporated with Multi-Wall Carbon Nanotube (MWCNT)." *The Journal of The Textile Institute*. Taylor & Francis: 1–10. doi:10.1080/00405000.2022.2062860.

Kumar, Jogendra, Rajesh Kumar Verma, and Prateek Khare. 2021. "Graphene-Functionalized Carbon/Glass Fiber Reinforced Polymer Nanocomposites: Fabrication and Characterization for Manufacturing Applications." In *Handbook of Functionalized Nanomaterials*, edited by Chaudhery Mustansar Hussain and Vineet Kumar, 57–78. Amsterdam: Elsevier. doi:10.1016/B978-0-12-822415-1.00011-1.

Kumar, Tejendra, Sanjay Mishra, and Rajesh Kumar Verma. 2020. "Fabrication and Tensile Behavior of Post-Consumer Carpet Waste Structural Composite." *Materials Today: Proceedings* 26 (2). Elsevier Ltd.: 2216–2220. doi:10.1016/j.matpr.2020.02.481.

Malik, P. K. 2[00]7. *Fiber Composites Reinforced Materials, Manufacturing, and Design*. Edited by P. K. Malick. 3rd ed. New York: Taylor & Francis.

Memon, Hafeezullah, Zamir Ahmed Abro, Arsalan Ahmed, and Nazakat Ali Khoso. 2015. "Considerations While Designing Acoustic Home Textiles: A Review." *JTATM* 9 (3): 1–29.

Mihut, Corina, Dinyar K. Captain, Francis Gadala-Maria, and Michael D. Amiridis. 2001. "Review: Recycling of Nylon from Carpet Waste." *Polymer Engineering and Science* 41 (9): 1457–1470. doi:10.1002/pen.10845.

Miraftab, M. 2018. "Rec[yc]ling Carpet Materials." In *Advances in Carpet Manufacture*, edited by K. K. Goswami. 2nd ed., 65–77. Kidlington: Elsevier Ltd. doi:10.1016/B978-0-08-101131-7.00005-8.

Mishra, Kunal, Sarat Das, and Ranji Vaidyanathan. 2019. "The Use of Recycled Carpet in Low-Cost Composite Tooling Materials." *Recycling* 4: 12 (1–8). doi:10.3390/recycling4010012.

Mishra, Kunal, and Ranji Vaidyanathan. 2019a. "The Influence of Nanoclay on the Flame Retardancy and Mechanical Performance of Recycled Carpet Composites." *Recycling* 4 (2): 22 (1–10). doi:10.3390/recycling4020022.

Mishra, Kunal, and Ranji K. Vaidyanathan. 2019b. "Application of Recycled Carpet Composite as a Potential Noise Barrier in Infrastructure Applications." *Recycling* 4: 9 (1–11). doi:10.3390/recycling4010009.

Mohammadhosseini, Hossein, Jamaludin Mohamad, Abdul Rahman, Mohd Sam, A. S. M. Abdul Awal, Jamaludin Mohamad Yatim, Abdul Rahman Mohd, et al. 2017. "Durability Performance of Green Concrete Composites Containing Waste Carpet Fibers and Palm Oil Fuel Ash." *Journal of Cleaner Production* 144. Elsevier B.V.: 448–458. doi:10.1016/j.jclepro.2016.12.151.

Mohammadhosseini, Hossein, Mahmood Tahir, and Abdulaziz Alaskar. 2020. "Enhancement of Strength and Transport Properties of a Novel Preplaced Aggregate Fiber Reinforced Concrete by Adding Waste Polypropylene Carpet Fibers." *Journal of Building Engineering* 27: 101003. doi:10.1016/j.jobe.2019.101003.

Mohammadhosseini, Hossein, Mahmood Tahir, Abdul Rahman, Mohd Sam, Nor Hasanah, Abdul Shukor, and Mostafa Samadi. 2018. "Enhanced Performance for Aggressive Environments of Green Concrete Composites Reinforced with Waste Carpet Fibers and Palm Oil Fuel Ash." *Journal of Cleaner Production* 185 (1). Elsevier B.V.: 252–265. doi:10.1016/j.jclepro.2018.03.051.

Onal, L., and Y. Karaduman. 2009. "Mechanical Characterization of Carpet Waste Natural Fiber-Reinforced Polymer Composites." *Journal of Composite Materials* 43 (16): 1751–1768. doi:10.1177/0021998309339635.

Pan, Gangwei, Yi Zhao, Helan Xu, Bomou Ma, and Yiqi Yang. 2016. "Acoustical and Mechanical Properties of Thermoplastic Composites from Discarded Carpets." *Composites Part B* 99. Elsevier Ltd: 98–105. doi:10.1016/j.compositesb.2016.06.018.

Patel, Himanshu V., and Harshit K. Dave. 2020. "Study of Permeability and Resin Flow Front during the Fabrication of Thin Composite Using VARTM Process." *International Journal of Modern Manufacturing Technologies* 7 (1): 125–130.

Patti, Antonella, and Gianluca Cicala. 2021. "Eco-Sustainability of the Textile Production : Waste Recovery and Current Recycling in the Composites World." *Polymers* 13: 134 (1–25).

Rahman, Saadman Sakib, Sumi Siddiqua, and Chinchu Cherian. 2022. "Sustainable Applications of Textile Waste Fiber in the Construction and Geotechnical Industries: A Retrospect." *Cleaner Engineering and Technology* 6 (January). Elsevier Ltd: 100420. doi:10.1016/j.clet.2022.100420.

Rajak, Dipen Kumar, Durgesh D. Pagar, Pradeep L. Menezes, and Emanoil Linul. 2019. "Fiber-Reinforced Polymer Composites: Manufacturing, Properties, and Applications." *Polymers* 11 (10): 1667 (1–37). doi:10.3390/polym11101667.

Realff, Matthew J., Jane C. Ammons, and David Newton. 1999. "Carpet Recycling: Determining the Reverse Production System Design." *Polymer-Plastics Technology and Engineering* 38 (3): 547–567. doi:10.1080/03602559909351599.

Rushforth, I. M., K. V. Horoshenkov, M. Miraftab, and M. J. Swift. 2005. "Impact Sound Insulation and Viscoelastic Properties of Underlay Manufactured from Recycled Carpet Waste." *Applied Acoustics* 66: 731–749. doi:10.1016/j.apacoust.2004.10.005.

Sanjeevi, Sekar, Vigneshwaran Shanmugam, Suresh Kumar, Velmurugan Ganesan, Gabriel Sas, Deepak Joel Johnson, Manojkumar Shanmugam, et al. 2021. "Effects of Water Absorption on the Mechanical Properties of Hybrid Natural Fibre/Phenol Formaldehyde Composites." *Scientific Reports* 11 (1). Nature Publishing Group UK: 1–11. doi:10.1038/s41598-021-92457-9.

Shin, Yeaheun, and Youngmi Park. 2018. "Preparation and Application of Polymer-Composited Yarn and Knit Containing CNT/Ceramic." *Clothing and Textiles Research Journal* 36 (1): 3–16. doi:10.1177/0887302X17737839.

Singh, Sunpreet, Seeram Ramakrishna, and Munish Kumar Gupta. 2017. "Towards Zero Waste Manufacturing: A Multidisciplinary Review." *Journal of Cleaner Production* 168. Elsevier B.V.: 1230–1243. doi:10.1016/j.jclepro.2017.09.108.

Sotayo, Adeayo, Sarah Green, and Geoffrey Turvey. 2015a. "Carpet Recycling : A Review of Recycled Carpets for Structural Composites." *Environmental Technology & Innovation* 3. Elsevier B.V.: 97–107. doi:10.1016/j.eti.2015.02.004.

———. 2015b. "Carpet Recycling: A Review of Recycled Carpets for Structural Composites." *Environmental Technology and Innovation* 3. Elsevier B.V.: 97–107. doi:10.1016/j.eti.2015.02.004.

———. 2018. "Development, Characterisation and Finite Element Modelling of Novel Waste Carpet Composites for Structural Applications." *Journal of Cleaner Production* 183. Elsevier Ltd: 686–697. doi:10.1016/j.jclepro.2018.02.095.

Vaidyanathan, Ranji, Raman P. Singh, and Tyler Ley. 2013. "Recycled Carpet Materials for Infrastructure Application." https://rosap.ntl.bts.gov/view/dot/27055/dot_27055_DS1.pdf

Valerio, Oscar, Rajendran Muthuraj, and Amandine Codou. 2020. "Strategies for Polymer to Polymer Recycling from Waste: Current Trends and Opportunities for Improving the Circular Economy of Polymers in South America." *Current Opinion in Green and Sustainable Chemistry* 25. Elsevier B.V.: 100381. doi:10.1016/j.cogsc.2020.100381.

Verma, Rajesh Kumar, Balram Jaiswal, Rahul Vishwakarma, and Kuldeep Kumar. 2021. "Polymer Composite Developed from Discarded Carpet for Light Weight Structural Applications: Development and Mechanical Analysis." *E3S Web of Conferences* 309: 01154 (1–5). doi:10.1051/e3sconf/202130901154.

Wang, Jian, and Qianchao Mao. 2012. "Methodology Based on The PVT Behavior of Polymer for Injection Molding." *Advances in Polymer Technology* 32: 474–485. doi:10.1002/adv.

Wang, Youjiang, Mehmet Ucar, and Youjiang Wang. 2014. "Utilization of Recycled Post Consumer Carpet Waste Fibers as Reinforcement in Lightweight Cementitious Composites." *International Journal of Clothing Science and Technology* 23 (4): 24–248. doi:10.1108/09556221111136502.

Xuan, Weihong, Xudong Chen, Guo Yang, Feng Dai, and Yuzhi Chen. 2018. "Impact Behavior and Microstructure of Cement Mortar Incorporating Waste Carpet Fibers after Exposure to High." *Journal of Cleaner Production* 187. Elsevier B.V.: 222–236. doi:10.1016/j.jclepro.2018.03.183.

Yalcin-enis, Ipek, Merve Kucukali-ozturk, and Hande Sezgin. 2019. "Risks and Management of Textile Waste." In *Nanoscience and Biotechnology for Environmental Applications*, edited by K. M. Gothandam, Shivendu Ranjan, Nandita Dasgupta, and Eric Lichtfouse, 29–53. Switzerland: Springer Nature Switzerland. doi:10.1007/978-3-319-97922-9.

Zhang, Chunbo, Mingming Hu, Francesco Di Maio, Benjamin Sprecher, Xining Yang, and Arnold Tukker. 2022. "An Overview of the Waste Hierarchy Framework for Analyzing the Circularity in Construction and Demolition Waste Management in Europe." *Science of the Total Environment* 803. The Authors: 149892. doi:10.1016/j.scitotenv.2021.149892.

Chapter 16

Additive manufacturing of waste plastic

Current status and future directions

Pankaj Sawdatkar and Dilpreet Singh

Contents

DOI: 10.1201/9781003291961-18

16.1 Introduction

Climate change and global warming are due to the unchangeable habits of humans in consuming the natural resources of the earth. The race to simplify human life has led to many inventions in the last three decades. One such invention is plastics. Plastic materials have very significantly contributed to the development of many technologies. Plastics have very good properties, such as light weight, durable, easy to handle, cost-effective, higher chemical resistance, and good strength. Due to these properties, plastics are used in many industries, such as automotive, packaging, construction, and medicine. Over the past years, plastics have become a very useful commodity due to their excellent properties (low weight, electrical and thermal insulation, etc.) and vast application scope at affordable prices. Plastic waste is generally defined as the end of life of a usable plastic product. It has been cited by the United Nations Environment Programme (UNEP) that over 400 million tonnes of plastic waste is generated globally (UNEP 2022). According to a survey, it was estimated that 6.3 billion tonnes of plastic waste were generated, and only 9% of the waste was recycled, while the remaining was disposed in landfills or other natural ecosystems (Brooks, Wang, and Jambeck 2018). Plastic waste has a complex structure. It comprises various composite materials, harmful emissions, and residual ash (Eriksen et al. 2018). The management of plastic waste is complicated, as, most often, the mixing of other contaminants limits its recyclability (A. Kumar et al. 2018). In countries like India, the collection and separation operations are conducted by informal sectors such as multi-tier players. This creates further difficulty in identifying the flow of waste and its traceability in different steams of waste (Balaji and Liu 2021). The American Society of Plastics Industry has categorized plastic into seven groups, namely high-density polyethylene, low-density polyethylene, vinyl/polyvinyl chloride, polypropylene, polystyrene, polyethylene terephthalate, and other types of plastic materials (such as nylon, polycarbonate, and ABS) (Bhattacharya et al. 2018). Plastic waste possesses very dire consequences for human health and the ecological environment. Since the natural degradation of plastic is a very time-consuming process, it remains there in the environment for a very long time. Therefore, it causes damage to the ecosystem and biodiversity and causes long-term social and ecological problems (Friot and Boucher 2017). Therefore, the management of plastic waste is very much essential for achieving a pollution-free environment.

16.2 Importance of plastic waste management and circular economy

Recycling plastic waste includes five processes. These processes are based on the quality of the product manufactured after recycling plastic waste: upgrading, recycling (open/closed loop), downgrading, waste to energy, and dumpsites or landfilling (Chidepatil et al. 2020). Figure 16.1 depicts the processing route of plastic waste.

The linear economy model dictates the "take-make-use-dispose" approach to the product life cycle. On the other hand, the circular economy uses the perception of integrating economy and environment in a sustainable path that focuses on "make-use-recycle-remake" (Sikdar 2019). The prime objective of this model is to (a) design products to minimize plastic waste, (b) recollect used materials and components, and (c) natural system regeneration. There are six "Rs" of circular economy: reduce, reuse, recycle, recover, redesign, and

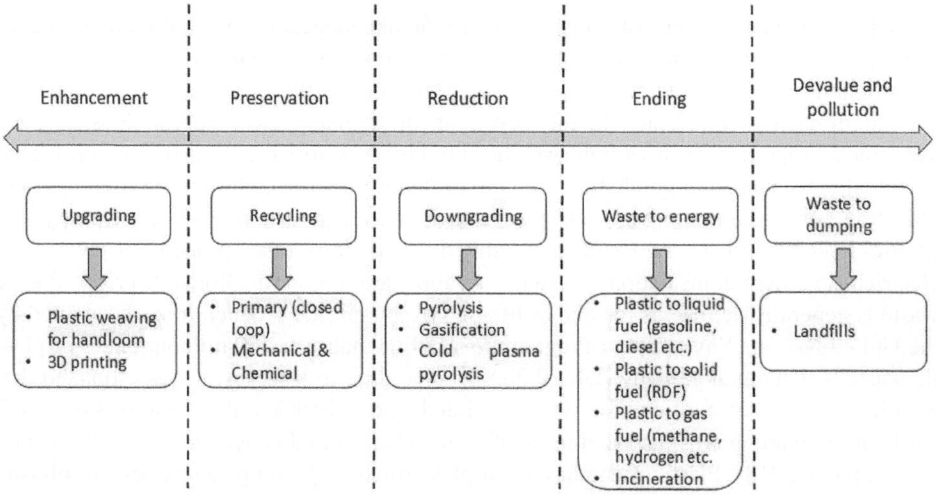

Figure 16.1 Processing routes for plastic waste (adapted from Chidepatil et al. 2020).

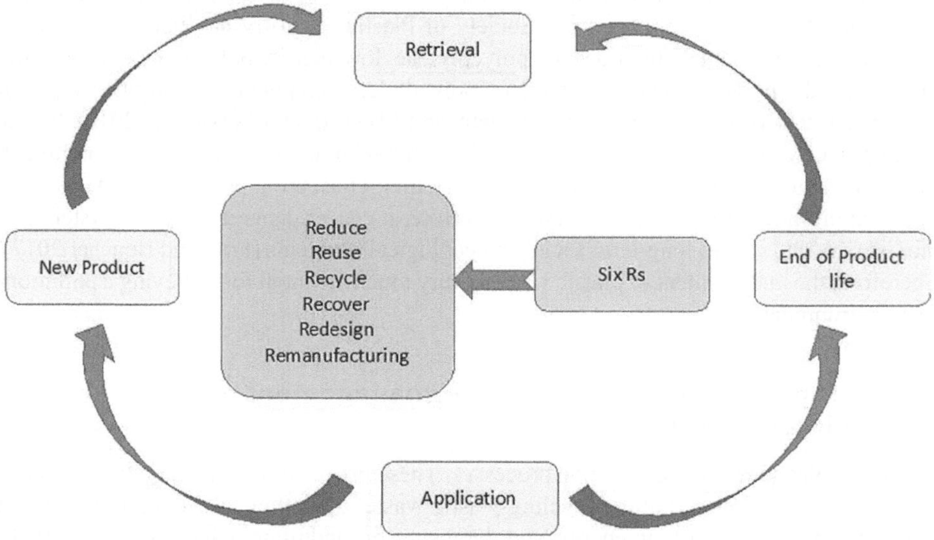

Figure 16.2 Model for the circular economy (adapted from Mikula et al. 2021).

remanufacturing, as depicted in Figure 16.2. The circularity is affected by various factors: (a) the raw material used to make plastic, (b) the type of plastic material, and (c) the socio-political framework wherein the plastic waste is generated, used, and managed. Therefore, it is very important to know the various facets of the circular economy.

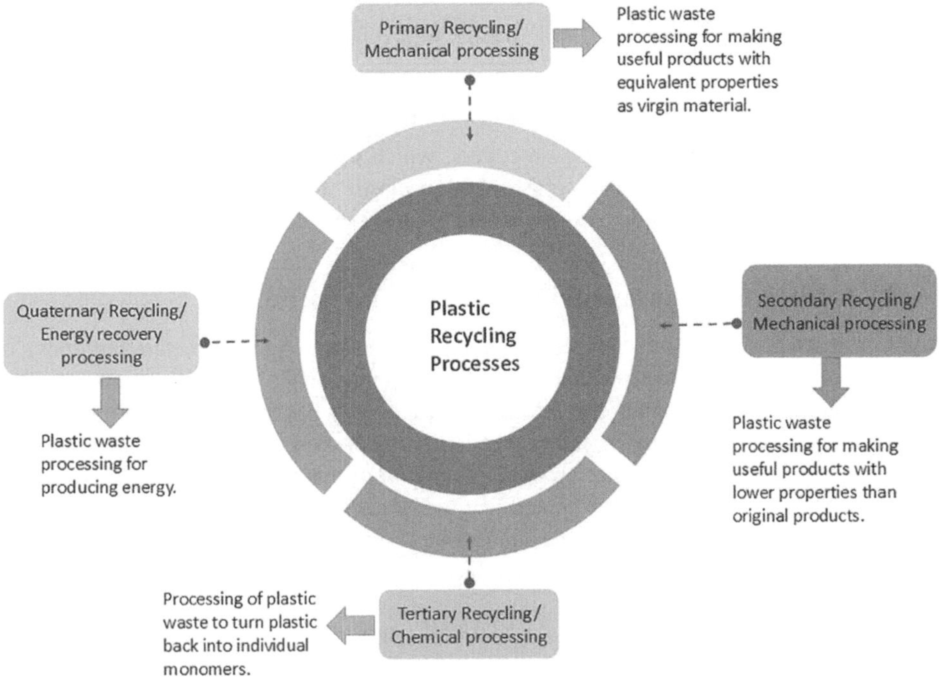

Figure 16.3 Various types of plastic recycling processes.

16.3 Methods of recycling plastic waste

There are different methods of recycling plastic waste. According to the ASTM D7209 and ISO 15270:2008 standards, plastic waste recycling is divided into four basic components. The same is depicted in Figure 16.3.

16.3.1 Primary recycling

It is a very fundamental way of processing plastics. In this, the received product is washed and cleaned using a standardized cleaning process. These cleaned products are then reused for walling new products, such as PET and bottles. In this, flow plastics are sorted and cleaned, shredded, etc.

16.3.2 Secondary recycling

This is the most widely used recycling process for the waste plastics. In this process, the thermoplastics are recycled by using the mechanical recycling method. There are various steps followed for this type of recycled plastic.

16.3.3 Tertiary recycling

In this type of recycling, recycling is performed using the chemical processing of plastics. In this type of process, plastics are used as feedstock material for basic chemicals and fuels.

This process involves various processes, such as solvolysis, thermolysis, microwave irradiation, pyrolysis gasification, thermal cracking, hydrolysis methanolysis and chemolysis (R. Singh and Kumar 2022).

16.3.4 Quaternary recycling

It is a technique that involves energy recovery, which involves incineration to obtain the energy component of plastic waste. Combustion of plastic with energy recovery is the most effective way of reducing waste, severely contaminated, and low-quality mixed plastics.

16.4 Additive manufacturing for circular economy

Additive manufacturing is the process wherein the material (plastic/metals/ceramics) is deposited layer by layer. In this technology, a three-dimensional (3D) part is created using a computer-aided design model as the input. This technology has the potential to disrupt and create value for manufacturing processes by eliminating the waste generated in the processes (R. Singh and Kumar 2022). Since an optimal amount of material is only deposited to create the 3D shape of the object to be manufactured. In this chapter, the terms "3D printing" and "additive manufacturing" are used interchangeably. Both names imply the same in this chapter. The various additive manufacturing technologies are illustrated in Figure 16.4.

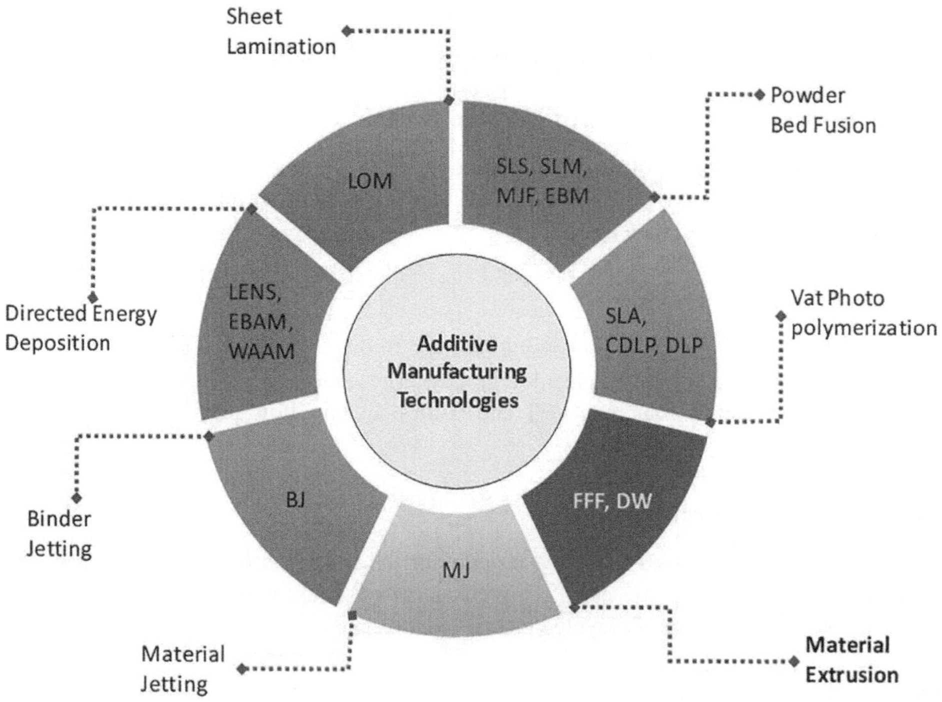

Figure 16.4 Various types of additive manufacturing technologies (adapted from Rafiee, Farahani, and Therriault 2020).

16.4.1 Powder bed fusion (PBF)

Powder bed fusion is an additive manufacturing technology in which some light source is used to fuse the powder material to form a 3D part. The various technologies that come under this domain are selective laser sintering (SLS), selective laser melting (SLM), multi-jet fusion (MJF), and electron beam melting (EBM).

16.4.2 Vat photopolymerization

Generally, liquid resins are used in this method, and the model is printed layer by layer using some light sources. The methods of vat photopolymerization include (a) stereolithography (SL), (b) digital light processing (DLP), and (c) digital light synthesis (DLS).

16.4.3 Material extrusion (ME)

By using this method, paste or semi-liquid material can be extruded through a nozzle to form a layer-by-layer 3D structure or product. Fused filament fabrication (FFF), fused granular fabrication (FGF) or fused granular deposition (FGD), and digital writing (DW) are some of the proven material extrusion technologies.

16.4.4 Material jetting (MJ)

It is one of the additive manufacturing methods in which different/multiple materials are sprayed through multi-jets of very fine size. The material is then cured using UV light to form a solid object. This method is similar to a 2D inkjet printer (D. Singh, Pandey, and Kalyanasundaram 2018).

16.4.5 Binder jetting (BJ)

Binder jetting is a method of 3D printing wherein a liquid-like bonding agent is put on the surface of the powder so that the powder will adhere to it and a layer is formed. In this way, layer by layer object/product can be created.

16.4.6 Directed energy deposition (DED)

In this method, focused thermal energy is used to bind the material by melting and deposition processes.

16.4.7 Laminated object manufacturing (LOM)

It is the process in which sheets of adhesive materials are glued together to form layers and then cut into the desired shape with the help of a cutter or by using laser cutting. This technology is now rarely used due to the complexity of the process.

The extrusion-based additive manufacturing method, such as fused filament fabrication, is the most accessible among other types of additive manufacturing methods. This is due to its versatility of part design (D. Singh et al. 2022), low complexity (Rafiee, Farahani, and Therriault 2020), lower investment cost (R. Singh, Singh, and Singh 2016), multi-material printing (Lopes, Silva, and Carneiro 2018), and product customization properties.

Some researchers have used the waste powder from the SLS to make filament for the fused filament fabrication process. Their research has brought the scope of consuming the waste of one additive manufacturing process to be used in other processes for sustainability (Mägi, Krumme, and Pohlak 2016).

16.5 Methods for using recycled plastics FFF and FGD

Alva et al. used waste plastics as disposable plastic bottles and shred them in a shredder. The shredded plastic was then used in an extruder to produce a filament. The model proposed by the authors was technically viable since the defective products can be reused to produce a filament and then reprinted (Romero-Alva, Alvarado-Diaz, and Roman-Gonzalez 2018). Likewise, Lee et al. studied a recycling model to be used for producing filament made from ABS and PLA waste plastic. The model proposed different setups for shredding the materials and extruding them in the form of a filament of the desired diameter. The researchers observed that the dimensional appearance of parts produced by recycled plastic was stable; however, they observed discoloration of the parts after some time. The strength was reduced by ~11% with an increase in its stiffness and fragility (Lee et al. 2019). Chong et al. studied the suitability of very commonly used recyclable plastic high-density polyethylene (HDPE) and compared the results with the most commonly 3D printed material ABS. The author showed in their research that recycled HDPE could be used for 3D-printed filament material (Chong et al. 2017). Lanzotti et al. compared the mechanical properties of the 3D-printed virgin PLA components with the 3D-printed recycled PLA. The research group maintained that recycling a filament twice does not decrease the mechanical properties and durability of the components produced using recycled 3D-printed PLA. However, they stated that recycling it further decreased the mechanical properties (Lanzotti et al. 2019).

Singh et al. developed a 3D printing filament produced using waste plastic using SiC/ Al_2O_3 as reinforcement in the polymer. The strength testing depicted that the addition of the filler materials improves the mechanical properties of the recycled plastic (N. Singh, Singh, and Ahuja 2018a). Anderson compared the mechanical properties such as tensile yield strength, elastic modulus, yield strength in shear, and hardness. The author observed that the mechanical properties of recycled PLA were at par with the virgin PLA. These encouraging results shall help the adoption of recycled 3D printed materials for reducing plastic waste using additive manufacturing technology (Anderson 2017). Paciorek-Sadowska et al. successfully performed experiments to form new poly foams based on bio-polylactide waste for thermal insulation applications (Paciorek-Sadowska, Borowicz, and Isbrandt 2019). Mosaddek et al. successfully developed and tested recycled 3D-printed drones for remote sensing applications. The filament was the composition of 90% waste materials and 10% virgin material (Mosaddek, Kommula, and Gonzalez 2018).

Sanchez et al. invented a methodology for using recycled plastic waste as a feedstock for 3D printers. The applied strategy was then implemented to produce the recycled PLA waste feedstock and used for 3D printing using fused filament fabrication technology. The authors observed that recycling slightly degrades the mechanical properties (Cruz Sanchez et al. 2017). Exconde et al. studied various types of virgin and recycled plastics, such as LDPE, HDPE, PET, and PP, for its feasibility in producing 3D printing filament through multi-criteria decision-making. The author maintained that recycled plastics demonstrated better properties than virgin material. These findings demonstrate that adopting recycled filament is a viable solution to single-use waste plastics (Exconde et al. 2019).

It is depicted from the literature above cited that the general procedure to recycle waste plastics for 3D printing is to shred the parts into small pieces and then extrude them into the form of a filament. The other waste plastic technique is mixing it with virgin plastic and even metal reinforcements. This method has proven to enhance the mechanical properties of recycled plastic for its application in the 3D printing technique. It is a well-known fact that recycled plastic's mechanical properties deteriorate every time the material goes through shearing and thermal cycles. The degradation of the mechanical properties is also due to the natural aging of the plastics due to repeated exposure to UV light during their service life and the addition of additives. It has been proven that in normal situations, plastic can undergo up to five recycling cycles without the necessity of adding virgin materials or other additives to enhance its mechanical properties (A. L. Woern et al. 2018).

16.6 FFF and FGD method

Fused deposition modeling (FDM), or Fused Filament Fabrication (FFF) is a material extrusion method of additive manufacturing in which material is extruded through the nozzle and joined together to create a 3D object. In a typical FDM process, the polymer-based filament is extruded through a heated nozzle, which melts the material and deposits it in a layer-by-layer manner on the build platform, and ultimately these layers fuse with each other and create a 3D part. FDM is one of the simplest technology to 3D print because FDM is accessible, reasonably efficient, and widely popular. The basic setup used for FDM is depicted in Figure 16.5 (a). Figure 16.5 (a) shows that a filament is extruded through a heated extruder and nozzle. The extruded material is deposited layer by layer.

Generally, the FDM technology follows the FFF technique and uses filaments for 3D printing, but the making of filaments from every material is not possible to pass through the extruder due to complex geometries (Valkenaers et al. 2013). Due to this, FFF is restricted to

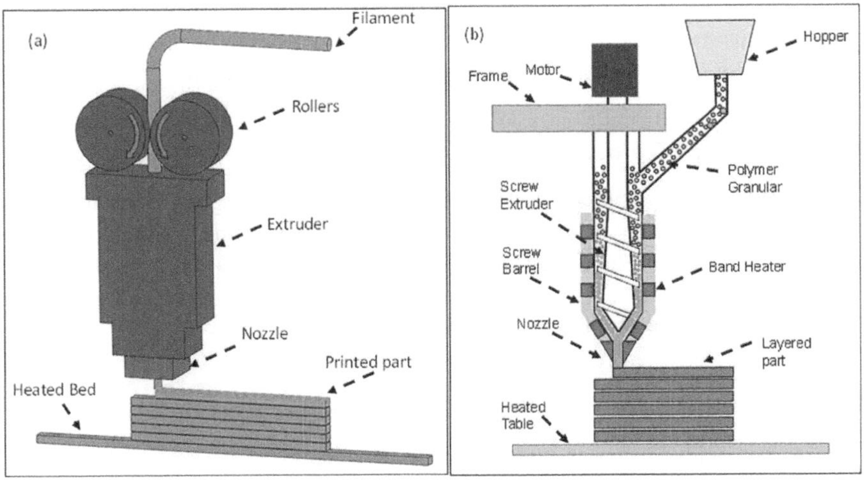

Figure 16.5 (a) Setup for fused filament fabrication (FFF) technology. (b) Setup for fused granular fabrication (FGF) technology.

use in manufacturing functional materials. To solve this issue, direct FDM or Fused Granular Fabrication (FGF) or Fused Granular Deposition (FGD) is one of the prominent alternatives to the FFF system, in which plastic powder, plastic pellets, plastic flakes, and shredded plastics are directly used for additive manufacturing (Justino Netto and Silveira 2018). This system uses screw-based print heads to transport the molten material by auger screw (Valkenaers et al. 2013).

During filament extrusion, when the filament material is extruded through liquefaction, any variations in the diameter of the filament material can cause congestion. A larger diameter feedstock can block the nozzle, while a small diameter feedstock may not provide sufficient materials to reach the walls of the extruder and can cause materials to short in the printed product (Valkenaers et al. 2013). Buckling of the filament may also cause hindrances during the printing process. The availability of polymer feedstock is also one of the limitations of the FFF 3D printing process (Bellini, Güçeri, and Bertoldi 2004). Contrarily, materials for direct deposition processes are widely used because the materials are no longer limited by the filament's mechanical properties (Whyman, Arif, and Potgieter 2018). A typical system for FGF technology-based 3D printer is depicted in Figure 16.5 (b).

As the name implies, screw assist systems include different types of screw extruders, like single-screw extruders and twin-screw extruders. A plastic extruder generally has three zones: a solid conveying zone, a melting zone, and a metering zone (Agassant et al. 2017). Solid pellets are transported from the hopper to the screw flight in the solids haul zone. These pellets are then compressed and move down the channel. The process of compressing the material is only possible if the friction on the cylinder surface is greater than the friction on the screw surface (L. Pan et al. 2012). The drum's friction contributes to the feedstock movement along the nozzle direction. In the absence of running friction, the pellet speed will be slower than the screw speed. This means that the pellet cannot achieve axial thrust. Finally, the pellet is melted and processed into a homogeneous mixture. This mixture is pushed from the nozzle.

16.7 Steps involved in recycling of plastics

There are various steps involved in the recycling of the plastics to be processed through the route of additive manufacturing. The detailed steps are enlisted below.

16.7.1 Recovery of plastic

It is the initial stage where plastic waste is mainly collected and logistic. Here it is also considered that sorting different material types is necessary. The sorting is either manual or can be automated. The second important thing is that the supply chain must be shorter so that the transportation cost and social and environmental impact can be reduced. For that reason, now a day's design is done on the basis of closed-loop approach.

16.7.2 Preparation for recycling

This phase deals with sorting, in which useable recycled plastic material taken a side and then process of identification, washing, sorting and size reduction takes place. Identification is very important step and is done as per the recycled code framework. Although different types of plastic having different properties so while creating complex 3D printing plastic code (1–7) must be known. This numbering during the 3D printing may be helpful for environmental awareness about plastic recycling. Then sorting is also an important step after identification, in sorting, specific waste material is identified. Proper methodology must be implemented for sorting and cleaning. Size reduction is the process where different methods and design are used for shredding recycled material. In the cleaning process of waste plastics, detergent and water is used to wash dirt particles from waste plastic surfaces followed by drying.

16.7.3 Compounding of plastics

It is the process for the development of single and composite material for decentralized recycling via additive manufacturing. To check feasibility of mono-plastic recyclability different studies have been conducted. The main conclusion from these studies is that it is technical not feasible to mix plastic which have different processing window. Therefore, recycled mono plastics were used for 3D printing through various methods, however, all the thermoplastics are not sharing the same properties for 3D printing. As an example, the high melting temperature, crystalline, moisture absorption, and weak interfacial bonding between the layers of recycled PET made the printing process difficult. The recycled PET structure may degrade more quickly if there are low levels of pollutants and chemicals present. However, recycled plastic materials such as ABS, PLA, and HDPE proven to be additive manufacturing compatible materials. Due to bed adhesion, distortion, and weak interfacial welding between printed layers, further research may improve the 3D printing capability of PP and PET. Defects in the 3D printing methods frequently occur at the interfaces because the solidification process and the diffusion of polymeric chains in the layers determine its printability.

In the case of composites, 3D printing is favorable due to its affordability, environmental friendliness, and ability to accommodate flexible filament materials, special emphasis has been dedicated to the manufacturing of polymer and composite filament. Mixing virgin/recycled materials in different ratios by extruding can be an economical way to enhance the reuse of recycled plastic materials.

16.7.4 Feedstock materials preparation

The main purpose of this phase is to get proper recycling material used for printing. Open-source machine setup of this design (free, self-replicating, modular) is sustainable for the wide adoption of the technology. It is worth noting that this approach will reduce costs and simplify the setup's production and assembly by ensuring process reproducibility. For example, Woern et al. developed a waste-design plastic extruder that can produce commercial quality (Recyclebot) 3D-printed filament. Specified filament manufacturing conditions were 0.4 kg/h, consuming 0.24 kWh/kg and providing a diameter accuracy of ± 4.6% (A. Woern and Pearce 2018). In this regard, Zhonge et al. studied the Recyclebot extruder for its energy returns time using the materialized energy of PLA and ABS plastic materials. Also, using a recycler to manufacture printed filaments from recycled plastics has proven itself an efficient method to save energy (Zhong, Rakhe, and Pearce 2017). Similarly, Woern and Pearce studied a pelletizer 3-D printable chopper system. This open-source system provides the compounding of filament and feedstock for fused granular printing purposes (A. Woern and Pearce 2018).

16.7.5 3D printing

Fused filament fabrication has a strong track record of printing plastics. However, the technical advancement of fused granular fabrication (FGF) may represent a significant step toward demonstrating the recyclable nature of plastic trash. Since FGF can print from pellets, filament production is no longer necessary. These FGF technologies have effectively recycled virgin polymers as well as plastic waste in a "green fabrication laboratory" context (Byard et al. 2019). A piston-drive head and screw-based extrusion techniques are two other forms of direct extrusion. The establishment of this technology for its widespread adoption will require further research.

16.7.6 Part quality

Quality is one of the important aspects of any product. Similarly, for a 3D printed part from waste plastic, there are three areas where quality is a major issue, which are, (1) raw material, (2) 3DP feedstock, and (3) quality of the printed part.

The raw material properties allow researchers to estimate the initial quality of the recycled material. Important properties such as flowability and thermal characteristics, Kumar and Czekanski studied the melt flow index (MFI) of three different amounts of recycled HDPE reinforced with SiC/Al_2O_3 with the purpose of having equivalent rheological properties as a market-available filament (S. Kumar and Czekanski 2017).

Concerning about 3DP feedstock, properties of filament, like diameter of filament and liner density of filament, are important characteristics while considering the quality. Mechanical properties, such as tensile strength and elastic modulus, are important properties for 3D printing. In the case of composite filament, a morphological test is also necessary to evaluate the distribution of matrix and filler in a filament.

While considering the quality of the 3D-printed object, printability is the most important parameter. It means the ability of waste material to extrude and retain the dimensional shape after the extrusion and having sufficient strength and mechanical properties. In 3D printing of waste plastic, various issues may occur, like warping, deformation, or buckling. The various past literature works are compiled in Section 16.9, where researchers have improved the strength of the recycled filaments by adding additives.

16.8 Commercially available 3D printing filament in the market

The plastics that are currently proven to be useful as the feedstock material for the FFF method are listed in Table 16.1.

16.9 Improving the strength of the recycled plastic by adding additives

Environmental pollution is a major problem due to plastic waste. Many methods have been developed to develop new materials for 3D printing. Different additives were added to enhance the molecular weight and improve waste recycled plastics' mechanical strength. There are two major issues, which are (1) extending polymer chains and (2) permitting the formation of free radicals by adding additives in the form of peroxides.

Table 16.1 List of Commercially Available Recycled 3D Printing Filaments (Adapted From Mikula et al. 2021)

Manufacturer (Country)	*Recycled plastic*	*Raw material*	*Content of recycled plastic (%)*
B-PET (Spain)	PET	Bottles made from PET	100
Filamentive (United Kingdom)	PLA	Waste stream of factory	55
	ASA	NA	50
	PETg	PET bottles	99.5
	PET	PET bottles	100
	ABS	NA	64
Fila-cycle (United Kingdom)	PLA	Yogurt pots	100
	ABS	Automotive waste	NA
	PET	PET bottles	
	HIPS	Automotive waste	
Refil (Cyprus, Europe)	HIPS	Fridge	100
	ABS	Dash board of automotive	
	PLA	Packaging of food	
	PET	Water storage bottles	
Innofil3D (Netherlands)	PET	Recycled PET materials	100
Fishy filaments, Porthcurno (United Kingdom)	Nylon	Fish nets	100
Tridea (Belgium)	PET	Food containers	100
	PLA	NA	
CREAMELT (Switzerland)	TPU	Ski boots	100

The impact of adding lignin was investigated in the morphological, mechanical, and thermal properties of recycled PLA. Lignin is added to ground PLA before being extruded at 180–190°C. Compared to samples composed of pure PLA, the inclusion of biopolymer enhances melting characteristics, reduces tensile strength (18%), and reduces Elastic modulus by ~6% (Gkartzou, Koumoulos, and Charitidis 2017). Additionally, the material was strengthened using carbon fibers. In comparison to the original, the recycled material's bending strength was 25% higher. The carbon fiber-reinforced material recovery rate was 100%, and the PLA material recovery rate was 71% (Tian et al. 2017). Dopamine, which is readily absorbed by most surfaces, was employed. This characteristic also permits polymer coatings. For 4 hours, ground PLA is agitated in a solution of dopamine in water. The PLA is then dried and extruded after this compared to pure PLA, the mass distribution of PLA containing dopamine began at a temperature of 200°C. The material's coating also contributed to a 20% boost in tensile strength (P. Zhao et al. 2018; X. G. Zhao et al. 2018). Stabilizers that oxidize recyclable materials are employed to enhance their qualities. Tropolone and hydroquinone both have this capability.

The introduction of a biocarbon additive led to a notable increase in the mechanical characteristics of the recycled material. Recycled PET bottles were combined with biocarbon, then heated. The addition of a second component resulted in a 32% improvement in the material's tensile strength. It was discovered that the material has stronger resistance to heat and oxidizing conditions and an increased modulus of elasticity (by about 60%) (Idrees, Jeelani, and Rangari 2018). The natural origin of PLA is one of the polymers that use biocarbon to reinforce it. When combined with the additives of natural origin, it was discovered that the additions increased the samples' stiffness by 8% (Notta-Cuvier et al. 2014). In the process of developing new kinds of floor mats, filaments based on hemp, harakeke, or recycled gypsum (0–50 wt.%) were independently added to raise the values of recycled polypropylene (PP). The best outcomes were obtained for filaments constructed of harakeke fibers (30 wt.%; tensile strength: 39 MPa; Young modulus: 2.8 GPa). However, those materials tend to lose some qualities when printed (Stoof and Pickering 2018).

In literature, Fe, Si, Cr, and Al nano-crystalline powders were added to PP and HDPE filament extrusion to improve the quality of recycled polymers. The researchers studied that adding 1% of a powder mixture (Fe-Si-Cr or Fe-Si-Al) improved the yield strength (37%) and young's modulus (17%) compared to the values of the base materials. Metals also lessen the possibility of crack formation (G. T. Pan et al. 2018). The addition of SiC/Al_2O_3 was also observed to increase the strength of recycled waste HDPE material significantly. Waste HDPE was reinforced with SiC/Al_2O_3, and paraffin wax was used as a binding agent. In a screw extruder, the prepared material was introduced into the additive. The additive has a minor effect on the thermal properties of the waste material; however, this increases the strength significantly (N. Singh, Singh, and Ahuja 2018b). Using zirconium oxide as an additive to recycle HDPE, mixed and ground using a ball mill. The filament was extruded at 190°C. It was reported that the coefficient of friction was 40% higher as compared with the polymer with zirconium oxide. The developed plastic material can be used to build low-temperature bearings (N. Singh, Singh, and Ahuja 2018b).

16.10 Role of additive manufacturing in tackling the waste problem

Conventionally, recycling involves the collection and segregation of plastic materials. Plastic waste recycling can be performed mechanically, chemically, and through energy recovery. However, mechanical recycling is the most widely used method, which uses extrusion processes to make pallets (R. Singh, Singh, and Singh 2016). There are some limitations to conventional recycling, such as fumes emissions and disposal of waste. In the present situation, closed loop recycling by 3D printing shows a prominent route of recycling (P. Zhao et al. 2018; X. G. Zhao et al. 2018). Additive manufacturing/3D printing in thermoplastics recycling can be achieved with the enhancement of materials' thermal, mechanical, and tribological characteristics by using additives and short fibers as reinforcements, ceramics, metals, or glass particles (Boparai et al. 2016). The additive manufacturing can produce value-added products with customized shapes and sizes and of very complex geometry.

16.10.1 Decentralized recycling

This approach depends on recycling consumer waste by upcycling the plastic waste into making filament for 3D printing applications. This recycling can be achieved using small units of recycling extruders. In the market, now, there are extruders with the facility to make filament from the waste generated by discarded plastic components in 3D printing. These are basically called benchtop-based filament-making extruders. Using this approach,

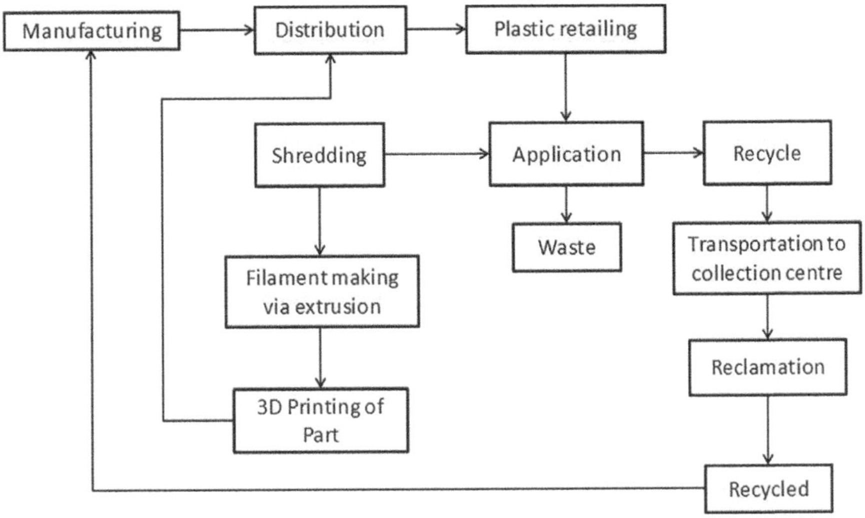

Figure 16.6 Process flow diagram of recycling standard plastics and 3D printing.

the lab-scale waste plastics produced using 3D printing can be recycled. However, the steps involved in decentralized recycling are explained in Section 16.7 (recycling of plastic). Figure 16.6 illustrates the process flow diagram for recycling using collection centers (conventional approach) and decentralized recycling.

16.11 Future directions and applications

Plastics have gained immense popularity in the last fifty years. Plastics have good mechanical properties, are lighter in weight and durable, and have a lower price. However, these are long-persisting (non-biodegradable) materials in the environment after their service life, thereby making them harmful to the environment (R. Singh, Singh, and Singh 2016). The Annual production of plastic is over 320 million Metric tonnes globally, and it is expected to surpass even 850 million Metric tonnes by 2050 (Zhong, Rakhe, and Pearce 2017; Mikula et al. 2021). The increase in the production of plastic is also going to increase plastic waste production. This alarms us about actively developing methods and strategies to tackle plastic waste to save the environment for future generations.

Previous studies have directed us to explore future research directions. One such area would be the multi-material mixer of second-use and virgin materials. In this aspect, very few studies have been conducted. The existing studies are on printing the multi-materials through multiple nozzles and polymer blending. In future, it would be interesting to study the mixing of "x" times recycled materials with "y" times recycled materials of the same polymer or different polymer. In addition, the extent of virgin material into the x and y times recycled polymer or same class or different class. There is a lot of scope for studying and standardizing these approaches to obtain the best properties of recycled materials.

Another area for future growth is direct printing of the shredded recycled material without making a filament. A few emerging studies and designs discussed in this chapter exist in the literature, such as fused granular fabrication (FGF) systems. However, many more studies in progress will show the efficient system and designs of direct printing systems. There is a scope for developing direct printing machines that can print different materials in parallel and blend the materials directly into them. This will open up a new regime of material properties. This can even take up the mixing of thermosets reinforcements into the thermoplastics. Thermosets can enhance the thermal properties of thermoplastics.

Implementing artificial intelligence into recycling polymers for additive manufacturing applications can bring out the new area for enhancing mechanical properties. With regards to the mechanical properties enhancement of different materials, big data can be produced. The data set will come from multivariable situations and multi-processing cycles for different material combinations. These data can be utilized to train the algorithms and models of machine learning. Depending upon the accuracy of the model, the predictions can be made for the optimal design and mechanical properties.

16.12 Conclusions

Plastics are used in our day-to-day life, and the products made using plastics can assist in simplifying our life. However, if not used wisely, plastic can severely threaten the environment. Therefore, the central idea behind this chapter was to discuss the technological intervention using additive manufacturing to solve the problem of plastic waste management. The decentralized recycling concept discussed in this chapter is key to realizing the solution to waste plastics using additive manufacturing. In this chapter, the importance of plastic waste management was highlighted. Recycling plastic waste is the key idea for plastic waste management. Various plastics recycling methods were discussed. The importance of additive manufacturing for the circular economy was explained. Various additive manufacturing processes were discussed. The additive manufacturing process used for solving the plastic waste problem was discussed. Previous literature work using additive manufacturing for plastic waste management was reported. Various steps involved in decentralized recycling were presented. The current market products using recycled plastics were enlisted. The role of additive manufacturing in tackling the plastic waste problem was also discussed. Lastly, the future directions in plastic waste management using additive manufacturing were discussed.

References

Agassant, Jean-François, Pierre Avenas, Pierre J. Carreau, Bruno Vergnes, and Michel Vincent. 2017. "Single-Screw Extrusion and Die Flows." *Polymer Processing* 301–432. doi:10.3139/9781569906064.005.

Anderson, Isabelle. 2017. "Mechanical Properties of Specimens 3D Printed with Virgin and Recycled Polylactic Acid." *3D Printing and Additive Manufacturing* 4 (2): 110–115. doi:10.1089/3dp.2016.0054.

Balaji, Anand Bellam, and Xiaoling Liu. 2021. "Plastics in Circular Economy: A Sustainable Progression." In *An Introduction to Circular Economy*, 159–178. Singapore: Springer. doi:10.1007/978-981-15-8510-4_9.

Bellini, Anna, Selçuk Güçeri, and Maurizio Bertoldi. 2004. "Liquefier Dynamics in Fused Deposition." *Journal of Manufacturing Science and Engineering* 126 (2). doi:10.1115/1.1688377.

Bhattacharya, R. R. N. Sailaja, Kaushik Chandrasekhar, M. V. Deepthi, Pratik Roy, and Ameen Khan. 2018. "Challenges and Opportunities: Plastic Waste Management in India." *The Energy and Resources Institute* 1–18, New Delhi.

Boparai, K.S., R. Singh, F. Fabbrocino, and F. Fraternali. 2016 "Thermal characterization of recycled polymer for additive manufacturing applications." *Composites Part B: Engineering*, 106: 42–47. doi:10.1016/j.compositesb.2016.09.009.

Brooks, Amy L., Shunli Wang, and Jenna R. Jambeck. 2018. "The Chinese Import Ban and Its Impact on Global Plastic Waste Trade." *Science Advances* 4 (6). doi:10.1126/sciadv.aat0131.

Byard, Dennis J., Aubrey L. Woern, Robert B. Oakley, Matthew J. Fiedler, Samantha L. Snabes, and Joshua M. Pearce. 2019. "Green Fab Lab Applications of Large-Area Waste Polymer-Based Additive Manufacturing." *Additive Manufacturing* 27. doi:10.1016/j.addma.2019.03.006.

Chidepatil, Aditya, Prabhleen Bindra, Devyani Kulkarni, Mustafa Qazi, Meghana Kshirsagar, and Krishnaswamy Sankaran. 2020. "From Trash to Cash: How Blockchain and Multi-Sensor-Driven Artificial Intelligence Can Transform Circular Economy of Plastic Waste?" *Administrative Sciences* 10 (2). doi:10.3390/admsci10020023.

Chong, Siewhui, Guan Ting Pan, Mohammad Khalid, Thomas C. K. Yang, Shuo Ting Hung, and Chao Ming Huang. 2017. "Physical Characterization and Pre-Assessment of Recycled High-Density Polyethylene as 3D Printing Material." *Journal of Polymers and the Environment* 25 (2): 136–145. doi:10.1007/s10924-016-0793-4.

Cruz Sanchez, Fabio A., Hakim Boudaoud, Sandrine Hoppe, and Mauricio Camargo. 2017. "Polymer Recycling in an Open-Source Additive Manufacturing Context: Mechanical Issues." *Additive Manufacturing* 17: 87–105. doi:10.1016/j.addma.2017.05.013.

Eriksen, M. K., K. Pivnenko, M. E. Olsson, and T. F. Astrup. 2018. "Contamination in Plastic Recycling: Influence of Metals on the Quality of Reprocessed Plastic." *Waste Management* 79: 595–606. doi:10.1016/j.wasman.2018.08.007.

Exconde, Mark Keanu James E., Julie Anne A. Co, Jill Z. Manapat, and Eduardo R. Magdaluyo. 2019. "Materials Selection of 3D Printing Filament and Utilization of Recycled Polyethylene Terephthalate (PET) in a Redesigned Breadboard." *Procedia CIRP* 84: 28–32. doi:10.1016/j.procir.2019.04.337.

Friot, Damien, and Julien Boucher. 2017. "Primary Microplastics in the Oceans." *IUCN Library System*. https://portals.iucn.org/library/node/46622.

Gkartzou, Eleni, Elias P. Koumoulos, and Costas A. Charitidis. 2017. "Production and 3D Printing Processing of Bio-Based Thermoplastic Filament." *Manufacturing Review* 4: 1. doi:10.1051/mfreview/2016020.

Idrees, Mohanad, Shaik Jeelani, and Vijaya Rangari. 2018. "Three-Dimensional-Printed Sustainable Biochar-Recycled PET Composites." *ACS Sustainable Chemistry and Engineering* 6 (11). doi:10.1021/acssuschemeng.8b02283.

Justino Netto, Joaquim M, and Zilda de C Silveira. 2018. "Design of an Innovative Three-Dimensional Print Head Based on Twin-Screw Extrusion." *Journal of Mechanical Design* 140 (12). doi:10.1115/1.4041175.

Kumar, Atul, S. R. Samadder, Nitin Kumar, and Chandrakant Singh. 2018. "Estimation of the Generation Rate of Different Types of Plastic Wastes and Possible Revenue Recovery from Informal Recycling." *Waste Management* 79: 781–790. doi:10.1016/j.wasman.2018.08.045.

Kumar, Sanjay, and Aleksander Czekanski. 2017. "Development of Filaments Using Selective Laser Sintering Waste Powder." *Journal of Cleaner Production* 165. doi:10.1016/j.jclepro.2017.07.202.

Lanzotti, Antonio, Massimo Martorelli, Saverio Maietta, Salvatore Gerbino, Francesco Penta, and Antonio Gloria. 2019. "A Comparison between Mechanical Properties of Specimens 3D Printed with Virgin and Recycled PLA." *Procedia CIRP* 79: 143–146. doi:10.1016/j.procir.2019.02.030.

Lee, Dongoh, Younghun Lee, Kyunghyun Lee, Youngsu Ko, and Namsu Kim. 2019. "Development and Evaluation of a Distributed Recycling System for Making Filaments Reused in Three-Dimensional Printers." *Journal of Manufacturing Science and Engineering, Transactions of the ASME* 141 (2). doi:10.1115/1.4041747.

Lopes, L. R., A. F. Silva, and O. S. Carneiro. 2018. "Multi-Material 3D Printing: The Relevance of Materials Affinity on the Boundary Interface Performance." *Additive Manufacturing* 23: 45–52. doi:10.1016/j.addma.2018.06.027.

Mägi, Piret, Andres Krumme, and Meelis Pohlak. 2016. "Recycling of PA-12 in Additive Manufacturing and the Improvement of Its Mechanical Properties." In *Engineering Materials and Tribology XXIV* 674: 9–14. Key Engineering Materials. Trans Tech Publications Ltd. doi:10.4028/www.scientific.net/KEM.674.9.

Mikula, Katarzyna, Dawid Skrzypczak, Grzegorz Izydorczyk, Jolanta Warchoł, Konstantinos Moustakas, Katarzyna Chojnacka, and Anna Witek-Krowiak. 2021. "3D Printing Filament as a Second Life of Waste Plastics—a Review." *Environmental Science and Pollution Research* 28 (10): 12321–12333. doi:10.1007/s11356-020-10657-8.

Mosaddek, Akif, Hrushi K. R. Kommula, and Felipe Gonzalez. 2018. "Design and Testing of a Recycled 3D Printed and Foldable Unmanned Aerial Vehicle for Remote Sensing." In *2018 International Conference on Unmanned Aircraft Systems, ICUAS 2018,* 1207–1216. doi:10.1109/ICUAS.2018.8453284.

Notta-Cuvier, D., J. Odent, R. Delille, M. Murariu, F. Lauro, J. M. Raquez, B. Bennani, and P. Dubois. 2014. "Tailoring Polylactide (PLA) Properties for Automotive Applications: Effect of Addition of Designed Additives on Main Mechanical Properties." *Polymer Testing* 36. doi:10.1016/j.polymertesting.2014.03.007.

Paciorek-Sadowska, Joanna, Marcin Borowicz, and Marek Isbrandt. 2019. "New Poly(Lactide-Urethane-Isocyanurate) Foams Based on Bio-Polylactide Waste." *Polymers* 11 (3): 481. doi:10.3390/polym11030481.

Pan, Guan Ting, Siewhui Chong, Hsuan Ju Tsai, Wei Hua Lu, and Thomas C. K. Yang. 2018. "The Effects of Iron, Silicon, Chromium, and Aluminum Additions on the Physical and Mechanical Properties of Recycled 3D Printing Filaments." *Advances in Polymer Technology* 37 (4). doi:10.1002/adv.21777.

Pan, L., M. Y. Jia, P. Xue, K. J. Wang, and Z. M. Jin. 2012. "Studies on Positive Conveying in Helically Channeled Single Screw Extruders." *Express Polymer Letters* 6 (7). doi:10.3144/expresspolymlett.2012.58.

Rafiee, Mohammad, Rouhollah D. Farahani, and Daniel Therriault. 2020. "Multi-Material 3D and 4D Printing: A Survey." *Advanced Science* 7 (12): 1–26. doi:10.1002/advs.201902307.

Romero-Alva, Victor, Witman Alvarado-Diaz, and Avid Roman-Gonzalez. 2018. "Design of a 3D Printer and Integrated Supply System." *Proceedings of the 2018 IEEE 25th International Conference on Electronics, Electrical Engineering and Computing, INTERCON 2018* 1–4. doi:10.1109/INTERCON.2018.8526458.

Sikdar, Subhas. 2019. "Circular Economy: Is There Anything New in This Concept?" *Clean Technologies and Environmental Policy* 21 (6): 1173–1175. doi:10.1007/s10098-019-01722-z.

Singh, Dilpreet, Bhavuk Garg, Pulak Mohan Pandey, and Dinesh Kalyanasundaram. 2022. "Design and Development of 3D Printing Assisted Microwave Sintering of Elbow Implant with Biomechanical Properties Similar Tohuman Elbow." *Rapid Prototyping Journal* 28 (2): 390–403. doi:10.1108/RPJ-05-2021-0116.

Singh, Dilpreet, Pulak Mohan Pandey, and Dinesh Kalyanasundaram. 2018. "Optimization of Pressure-Less Microwave Sintering of Ti6Al4V by Response Surface Methodology." *Materials and Manufacturing Processes* 33 (16): 1835–1844. Taylor & Francis. doi:10.1080/10426914.2018.1476765.

Singh, Narinder, Rupinder Singh, and I. P. S. Ahuja. 2018a. "Recycling of Polymer Waste with SiC/Al_2O_3 Reinforcement for Rapid Tooling Applications." *Materials Today Communications* 15: 124–127. doi:10.1016/j.mtcomm.2018.02.008.

Singh, Narinder, Rupinder Singh, and I. P. S. Ahuja. 2018b. "Recycling of Polymer Waste with SiC/Al_2O_3 Reinforcement for Rapid Tooling Applications." *Materials Today Communications* 15. doi:10.1016/j.mtcomm.2018.02.008.

Singh, Rupinder, and Ranvijay Kumar. 2022. *Additive Manufacturing for Plastic Recycling*, Boca Raton: CRC Press. doi:10.1201/9781003184164.

Singh, Rupinder, Jagdeep Singh, and Sunpreet Singh. 2016. "Investigation for Dimensional Accuracy of AMC Prepared by FDM Assisted Investment Casting Using Nylon-6 Waste Based Reinforced Filament." *Measurement: Journal of the International Measurement Confederation* 78. doi:10.1016/j.measurement.2015.10.016.

Stoof, David, and Kim Pickering. 2018. "Sustainable Composite Fused Deposition Modelling Filament Using Recycled Pre-Consumer Polypropylene." *Composites Part B: Engineering* 135. doi:10.1016/j.compositesb.2017.10.005.

Tian, Xiaoyong, Tengfei Liu, Qingrui Wang, Abliz Dilmurat, Dichen Li, and Gerhard Ziegmann. 2017. "Recycling and Remanufacturing of 3D Printed Continuous Carbon Fiber Reinforced PLA Composites." *Journal of Cleaner Production* 142. doi:10.1016/j.jclepro.2016.11.139.

UNEP. 2022. "Single-Use Plastics: A Roadmap for Sustainability." Accessed July 6. www.unep.org/resources/report/single-use-plastics-roadmap-sustainability.

Valkenaers, Hans, Frederik Vogeler, Eleonora Ferraris, André Voet, and J. P. Kruth. 2013. "A Novel Approach to Additive Manufacturing: Screw Extrusion 3D-Printing." *10th International Conference on Multi-Material Micro Manufacture* Research publishing, Singapore, 235–238.

Whyman, Sean, Khalid Mahmood Arif, and Johan Potgieter. 2018. "Design and Development of an Extrusion System for 3D Printing Biopolymer Pellets." *International Journal of Advanced Manufacturing Technology* 96: 9–12. doi:10.1007/s00170-018-1843-y.

Woern, Aubrey L., Dennis J. Byard, Robert B. Oakley, Matthew J. Fiedler, Samantha L. Snabes, and Joshua M. Pearce. 2018. "Fused Particle Fabrication 3-D Printing: Recycled Materials' Optimization and Mechanical Properties." *Materials* 11(8): 1413. doi:10.3390/ma11081413.

Woern, Aubrey L., and Joshua Pearce. 2018. "3-D Printable Polymer Pelletizer Chopper for Fused Granular Fabrication-Based Additive Manufacturing." *Inventions* 3 (4): 78. doi:10.3390/inventions3040078.

Zhao, Peng, Chengchen Rao, Fu Gu, Nusrat Sharmin, and Jianzhong Fu. 2018. "Close-Looped Recycling of Polylactic Acid Used in 3D Printing: An Experimental Investigation and Life Cycle Assessment." *Journal of Cleaner Production* 197. doi:10.1016/j.jclepro.2018.06.275.

Zhao, Xing Guan, Kyung Jun Hwang, Dongoh Lee, Taemin Kim, and Namsu Kim. 2018. "Enhanced Mechanical Properties of Self-Polymerized Polydopamine-Coated Recycled PLA Filament Used in 3D Printing." *Applied Surface Science* 441. doi:10.1016/j.apsusc.2018.01.257.

Zhong, Shan, Pratiksha Rakhe, and Joshua Pearce. 2017. "Energy Payback Time of a Solar Photovoltaic Powered Waste Plastic Recyclebot System." *Recycling* 2 (2): 10. doi:10.3390/recycling2020010.

Index

Note: Page numbers in **bold** indicate a table on the corresponding page.

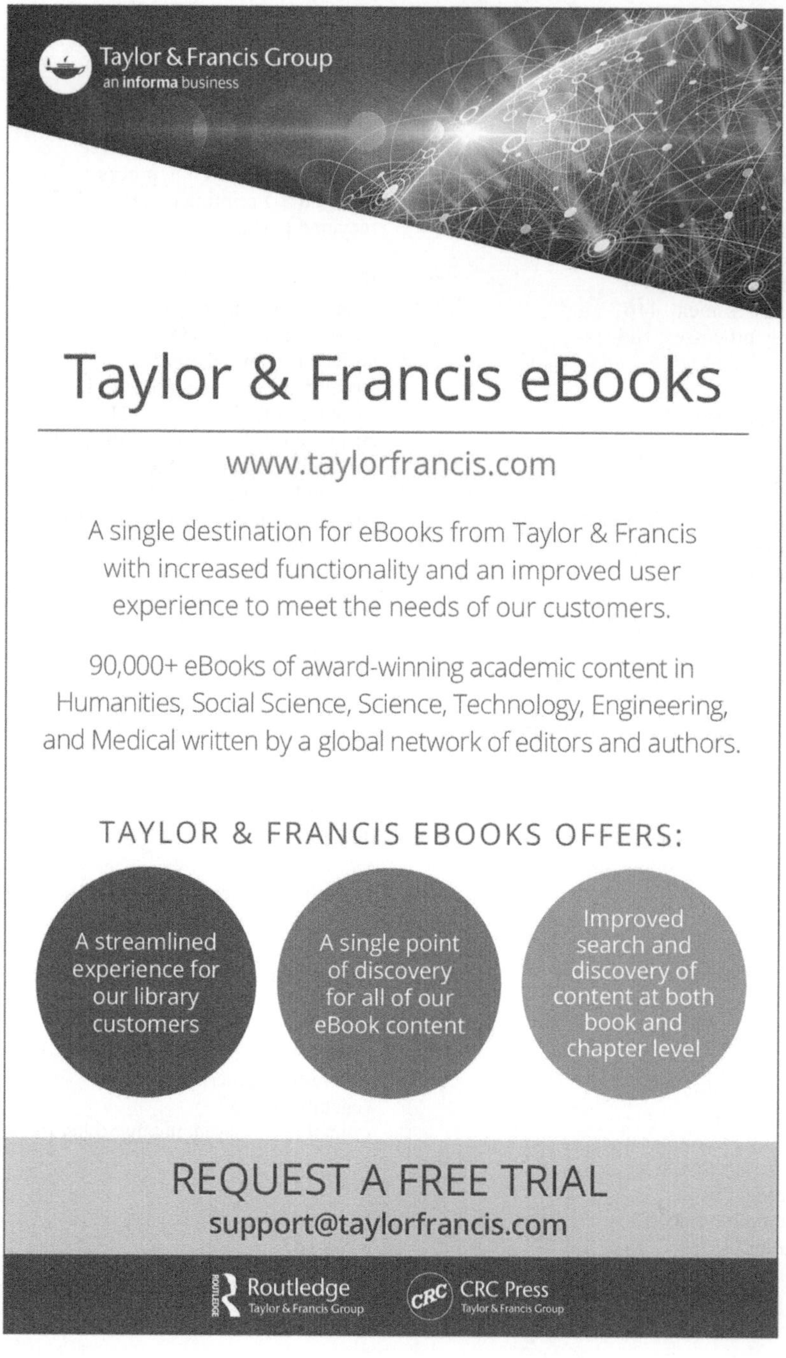
Taylor & Francis Group
an informa business
Taylor & Francis eBooks
www.taylorfrancis.com
A single destination for eBooks from Taylor & Francis with increased functionality and an improved user experience to meet the needs of our customers.
90,000+ eBooks of award-winning academic content in Humanities, Social Science, Science, Technology, Engineering, and Medical written by a global network of editors and authors.
TAYLOR & FRANCIS EBOOKS OFFERS:
A streamlined experience for our library customers
A single point of discovery for all of our eBook content
Improved search and discovery of content at both book and chapter level
REQUEST A FREE TRIAL
support@taylorfrancis.com
Routledge
Taylor & Francis Group
CRC Press
Taylor & Francis Group